# Volta

# Volta

SCIENCE AND CULTURE

IN THE

AGE OF ENLIGHTENMENT

*Giuliano Pancaldi*

PRINCETON UNIVERSITY PRESS   PRINCETON AND OXFORD

*Library of Congress Cataloging-in-Publication Data*

Pancaldi, Giuliano.
Volta : science and culture in the age of
Enlightenment / Giuliano Pancaldi.
p.   cm.
Includes bibliographical references and index.
ISBN: 0-691-09685-6 (acid-free paper)
1. Volta, Alessandro, 1745–1827. 2. Physicists—Italy—Biography.
3. Electricity—History. I. Title.
OC515.V8 P36 2003
537′.092—dc21
[B]      2002074874

British Library Cataloging-in-Publication Data available.

This book has been composed in Palatino

Printed on acid-free paper.∞

www.pupress.princeton.edu

Printed in the United States of America

1   3   5   7   9   10   8   6   4   2

*To Paola, Vera, and Luca*

■

# C O N T E N T S

## Chapter 4

### VOLTA'S SCIENCE OF ELECTRICITY
Conception, Laboratory Work, and Public Recognition   110

## Chapter 5

### THE COSMOPOLITAN NETWORK
Volta and Communication among Experts
in Late Enlightenment Europe   146

## Chapter 6

### THE BATTERY
Invention, Instrumentalism, and Competitive Imitation   178

# ACKNOWLEDGMENTS

I am deeply grateful: To John L. Heilbron, for having shown the way and shared liberally with me on many occasions his invaluable knowledge of Volta and modern science. To Robert Fox, for insightful, gentle guidance and generous encouragement when an earlier version of this book was the doctoral dissertation of an already mature student at the University of Oxford. To Laurence B. Brockliss and Geoffrey N. Cantor, for having made valuable criticism of the said dissertation. To Roger Hahn, for having helped convince me that biography adds to our understanding of science and technology, and for his comments on chapter 5. To Frederic L. Holmes, on several occasions, for having shared with me his fascination for fine-grained analyses of scientific creativity. To Tore Frängsmyr, for much valuable advice on Enlightenment historiography. To Roderick W. Home, for having read and commented on early drafts of chapter 3, offering guidance through the tricky notions of eighteenth-century electricity.

Special thanks are due to I. Bernard Cohen and Charles Coulston Gillispie, who read the manuscript for Princeton University Press, offering generous and helpful advice.

Plans for this book began during an inspiring sojourn at the Office for the History of Science and Technology of the University of California at Berkeley, made possible by a Fulbright Fellowship. The book started to materialize in the friendly atmosphere of Wolfson College, Oxford. Subsequent research was sustained by grants accorded by the Italian Ministry for Education, University and Research, and the University of Bologna. The final touches to the book were given while I was fellow of the Dibner Institute for the History of Science and Technology, on the campus of MIT. The Dibner, its Burndy Library, and MIT provided the best environment and facilities a scholar in the field could dream of.

The present work has benefited from the generous advice and criticism of many colleagues during seminars and lectures I have given over the years in Bari, Berkeley, Berlin, Boston, Como, Edinburgh, Oxford, Paris, Pavia, Rome, San Diego, Stockholm, Uppsala, as well as at Stanford and Yale universities. My gratitude extends to all those who shared their knowledge with me on those occasions. If I have been unable to cope with all their advice, or avoid the mistakes they pointed out, this is my responsibility.

I owe similar gratitude (and the same disclaimer) to my colleagues at the University of Bologna, especially Anna Guagnini, Carlo Poni, Pietro Redondi, and Niccolò Guicciardini (now at the University of Siena), as well as Gabriele Baroncini, Marta Cavazza, Antonio Santucci, Raffaella Simili, Walter Tega, Luigi Turco, and a host of doctoral students with whom I have interacted fruitfully in recent years: Stefano Belli, Paola Bertucci, Marco Bresadola, Luca Ciancio, Lucia De Frenza, Michelangelo Ferraro, and Raffaella Seligardi.

For having helped polish my English, and reminding me of the rights of the reader, I am grateful to David Allen, Ruey Brodine, Charles Hindley, Stephen Jewkes, and Ross McLean.

I am deeply indebted to Joseph Wisnovsky, executive editor of Princeton University Press, for much valuable advice throughout the preparation of this book.

The librarians and archivists of the Istituto Lombardo, Accademia di Scienze e Lettere, Milan; the Archivi di Stato of Como, Milan, Pavia, and Turin; the Archivio Storico della Società di Gesù, Rome; the Biblioteca del Politecnico, Milan; the University Libraries of Bologna, Pavia, Uppsala, and Leiden; the Royal Society Archives, London; the Bodleian Library and the Museum of the History of Science Library, Oxford; the British Library; the Bibliothèque Nationale; the Staatsarchiv, Vienna; the Down House Archives, Downe (Kent); the Smithsonian Institution Libraries, Washington; the Bancroft Library, Berkeley; the Yale University Library; the Burndy Library of the Dibner Institute, Cambridge, Mass.; the Dartmouth College Library, Hanover (New Hampshire); and above all the librarians and staff of the Department of Philosophy and the International Center for the History of Universities and Science (CIS) at the University of Bologna, deserve my thanks for the generosity they displayed in meeting my requests. Permission to quote from material in the possession of libraries and archives is acknowledged in the appropriate place.

This book is dedicated to Paola, Vera, and Luca: I owe it mostly to them if, in the pages that follow, I was able to recapture some of the pleasures enjoyed by enlightened natural philosophers.

An earlier version of chapter 5 appeared as an article, with the title "Electricity and Life: Volta's Path to the Battery," in Historical Studies in the Physical and Biological Sciences, 21 (1990): 123–160. Chapter 7 is an expanded, more elaborate version of a paper bearing a similar title in Instruments, Travel, and Science, edited by Marie-Noelle Bourguet, Christian Licoppe, and H. Otto Sibum, for Routledge (2002). Chapter 8 includes materials published in G. Pancaldi, "The Social

Uses of Past Science: Celebrating Volta in Fascist Italy," in *Natural Sciences and Human Thought*, ed. R. Zwilling (Berlin: Springer-Verlag, 1995), pp. 218–224. Chapter 9 contains excerpts from a commentary originally published in *Solomon's House Revisited: The Organization and Institutionalization of Science*, ed. Tore Frängsmyr, Nobel Symposium 75 (Science History Publications and the Nobel Foundation, 1990), pp. 65–71. I am grateful to the University of California Press, Routledge, Springer-Verlag, Science History Publications, and the Nobel Foundation for permission to reprint.

For the titles of current history of science journals, we use the abbreviations of the *Isis Current Bibliography*, published by the University of Chicago Press for the History of Science Society.

ASC    Archivio di Stato, Como.

ASM    Archivio di Stato, Milan.

ASSI    Archivio Storico della Società di Gesù, Rome.

BCC    Biblioteca Comunale, Como.

BL    British Library, London.

*DBI*    *Dizionario Biografico degli Italiani* (Rome: Istituto della Enciclopedia Italiana, 1960–).

*DNB*    *Dictionary of National Biography*, 28 vols. (London: Oxford University Press, 1949–50).

*DSB*    *Dictionary of Scientific Biography*, 18 vols. (New York: Scribners, 1970–90).

*PT*    *Philosophical Transactions of the Royal Society of London.*

RSA    Royal Society Archives, London.

SAV    Staatsarchiv, Vienna.

*VE*    Volta, Alessandro, *Epistolario*, ed. F. Massardi, 5 vols. (Bologna: Zanichelli, 1949–55). When specified *VE* includes: Volta, Alessandro, *Indici delle opere e dell'epistolario di Alessandro Volta*, ed. A. Ferretti Torricelli, 2 vols. (Milan: Rusconi, 1974–76); Volta, Alessandro, *Aggiunte alle opere e all'epistolario di Alessandro Volta*, ed F. Massardi and A. Ferretti Torricelli (Bologna: Zanichelli, 1966).

VMS    Volta Papers at the Istituto Lombardo, Accademia di Scienze e Lettere, Milan

*VO*    Volta, Alessandro, *Le opere*, 7 vols. [(Milan: Hoepli, 1918; reprint: *The Sources of Science*, no. 70 (New York and London: Johnson, 1968)].

*VE* and *VO* are now also available on CD-ROM: Istituto Lombardo Accademia di Scienze e Lettere, Alessandro Volta, *Edizione Nazionale delle Opere e dell'Epistolario in 15 volumi (1918–1976)*, ed. Fabio Bevilacqua, Gianni Bonera, and Lidia Falomo, Dipartimento di Fisica "A. Volta," Università di Pavia (Milan: Hoepli, 2002).

# Volta

How did it happen that an age which proclaimed itself "enlightened," but had developed no electrical industry, ended up with an invention that would make of electric lighting an everyday marvel? The age in question historians call the Enlightenment,[1] and the invention was the electric or voltaic battery, the earliest examples of which were built toward the end of 1799. The present book revolves around wide-ranging questions like the one just raised, and it seeks the answers in a detailed study of the cultures of science and technology in late-eighteenth- and early-nineteenth-century Europe.

Many of the threads linking the Enlightenment tradition to the science of electricity are already well known. "The Enlightenment liked to play with electricity," it has been noted by a leading historian in the field,[2] and a figure like Benjamin Franklin was represented already by his contemporaries as embodying an alliance between the science of electricity and "enlightened" political endeavors.[3] The present book focuses on the specific threads linking the battery to Enlightenment culture on one hand, and to later industrial societies on the other. The book argues that a key factor in the process that led to the battery was the cultural, technological, and social ferment that, around 1800, inspired a mixed population of natural philosophers, physicists, physicians, instrument makers, and amateurs interested in electricity, belonging to different cultural traditions and scattered over several European countries. The book further argues that the role of the inventor in that process was, to an important extent, one of interpreting and mediating between the different agendas pursued within that mixed population.

By emphasizing the diversity of the cultures involved in the process that led to the battery—and the unintended consequences attached to diversity—the book outlines a picture of late Enlightenment science and technology that goes beyond the interpretive framework provided by the traditional conflicting views held by the supporters or critics of the Enlightenment and their present-day descendants.[4] The book argues that the diversity of the cultures and goals involved in the pursuit of science and technology around 1800 was just as important as the discipline propounded by some of the followers of the Enlightenment as a key factor in the scientific enterprise. The book also argues that the unintended consequences of diversity (and thus the difficulty of orienting and predicting scientific and technological change) characterized the earlier, no less than the later, stages of the age of electricity. As in other fields of comparatively free human endeavor, too in scientific and technological pursuits both diversity and unintended consequences seem to be at work, in spite of the efforts of the advocates of a normative view of science, rationality, and social order, or the understandable denunciations leveled by those critics of the Enlightenment tradition who fear the power of elites won over to the notion of useful knowledge.

The book's narrative follows three main lines. The first is biographical: it is the story of Alessandro Volta, a leading figure in the history of the science of electricity during the late Enlightenment, and the inventor of the battery. His personal, scientific, and social endeavors are dealt with, offering a glimpse of what it meant to be a "natural philosopher" (the old phrase for scientist) and an inventor of electrical instruments (mostly useless, by our early-twenty-first-century standards) in late-eighteenth-century Europe.

Insofar as it is a biography, this is a biography in context. Volta's work and instruments are appraised both as products of specialist endeavor and in connection with the reforms of the educational system and the public administration then under way in several European countries. Volta, whom historiography has singled out as a key figure in the transition of the science of electricity from amateur to professional enterprise, was also a well-known figure in the Republic of Letters, to which enlightened "philosophers" felt they belonged. We thus explore Volta's network of acquaintances and his frequent travels, during which he felt as much at home in London, Paris, Geneva, Berlin, Göttingen, or Vienna, where he would perform his electrical experiments, as in Milan, Pavia, or Como, in Lombardy (then part of the Austrian empire), where he was born.

The second line the book follows is an inquiry into some long-term features of the culture of science and technology as they developed in the early age of electricity. This entails a reassessment of the legacies of the Enlightenment based on a field of historical evidence—placed at the intersection between cultures, natural philosophies, and machines—that has received comparatively little attention by the interpreters of the Enlightenment tradition so far. Seen from a long-term perspective, the case of Volta and the battery can be regarded as a paradigmatic episode in the economy of invention enforced by Enlightenment ideas and practices: a case study of some of the consequences brought about by Enlightenment values and notions, which resulted in the introduction of the nineteenth-century figure of the "scientist" and the partial eclipse of the "natural philosopher."

The third line of inquiry to be pursued is epistemological and an-thropological. It is an investigation into the material culture of science and technology exemplified in Volta's machines. It describes their early, "private" development, as well as their public career, using the rich documentation offered by Volta's laboratory notes and by the ne-gotiations he undertook to obtain recognition via a varied network of expert and amateur electricians, patrons, and reward-dispensing col-leagues scattered in several countries. As an epistemological and an-thropological study, an approach has been adopted that goes beyond the opposition between realism and constructivism that has nurtured protracted scholarly controversies over the past decade.[5] Throughout the book it will be assumed that scientific and technological objects are both real *and* historical: real in the sense of pragmatic realism;[6] histori-cal in the sense that scientific and technological knowledge and prac-tices are in any case embedded in human culture, as constructivists emphasize.[7]

Moving to a review of the detailed contents of the book, chapter 1 examines what was distinctive in Volta's personal, social, and cultural endeavors. His career from amateur to expert, to professor and public servant, is outlined, as are some of the features of his emotional life, his attitude toward religion, and his scientific work. Chapter 2 is a study of the Italian scientific community from 1770–1795, based on a survey of seventy-four natural philosophers active south of the Alps. The aims of the institutional reforms being carried out in the Italian peninsula in the second half of the eighteenth century, and their constraints, are assessed through the testimony of insiders and foreigners. Chapters 3 and 4 review Volta's contributions to the science of electricity in the 1770s and 1780s. His day-to-day work is reconstructed from a wealth of surviving manuscript notes. His qualities as an investigator are pre-sented as stemming from an adjustment between his ambition to be regarded as a natural philosopher and the more modest role of inven-tor that some of his colleagues were inclined to assign him. In chapter 4 a detailed discussion of Volta's techniques, and his reactions to Cou-lomb's celebrated memoirs on electricity, is used to outline both the peculiar traits of the "quantifying spirit" Volta subscribed to and the plurality of the strategies that late-eighteenth-century physicists adopted in their pursuit of a quantitative science of electricity.

Chapter 5 describes the cosmopolitan network of contacts that Volta developed in his dual role of expert in search of recognition and public servant involved in the imitation-competition game that several gov-ernments of Enlightenment Europe played in their well-advertised support of fashionable science. The chapter shows how closely Volta's assessment of scientific merit, and his choice of foreign experts and scientific institutions from which to expect reward, linked up with his perception of their broad cultural and political leanings.

Chapter 6 is a fine-grained study of Volta's path to the battery. It begins with a detailed discussion of Volta's reactions to Galvani's experimental results and interpretations of "animal electricity" in the 1790s. It continues with a description of the sophisticated measuring techniques that Volta developed while engaged in the hunt for "weak" electricity generated by the controversy over galvanism. The chapter concludes with a circumstantial reconstruction, based on Volta's laboratory notes, of the cognitive and manipulative steps, and expert negotiations, that led Volta to conceive and build the battery in 1799, and announce it to the Royal Society of London in the spring of 1800. This took only a few months once Volta had read a paper by William Nicholson suggesting a method for "imitating the electric fish" (the torpedo) by means of an electrical and mechanical apparatus based on one of Volta's own earlier machines. The circumstance that the battery (later depicted as a momentous turning point in the history of physics) was the outcome of a program that included the goal of "imitating the electric fish" is used as a cautionary tale against the sort of history of science that prevailed when the disciplinary boundaries imposed by twentieth-century developments oriented the work of historians.

A comparative study of the early reception of the voltaic battery in five European countries is developed in chapter 7. It is an essay on how scientific instruments and their interpretations travel across cultural frontiers. It shows how easily simple machines like the voltaic battery could be replicated, and how—easy replication notwithstanding—they could be adopted within widely different research programs, occasionally far removed from the interests and goals pursued by the inventor. The easy replication of the voltaic battery, and its appropriation within different cultural contexts, are discussed as evidence that a revision is needed in the way historians and sociologists have dealt with the issue of replication in scientific practice, and with the role of local cultures and expert knowledge in the production and diffusion of scientific instruments. The revision recommended combines the realist notion that the voltaic batteries built throughout Europe in the early nineteenth century were basically the same apparatus, with the constructivist notion that indeed any expert electrician in Europe could bend the battery to his or her own particular intellectual and social needs.

The early reception of the voltaic battery shows that consensus on how the battery worked, and on its implications for natural philosophy, was conspicuously lacking in both expert and amateur circles around Europe in the first few years after the introduction of the new instrument. Despite this, experts, amateurs, and even heads of state like general Bonaparte—often informed of Volta's achievement merely through the daily press—showed themselves able and willing to celebrate the merits of Volta's contribution to expert knowledge before Europe's cultivated elites.

The mechanisms that regulate the assessment and celebration of scientific and technological achievement among lay audiences are the subject of chapter 8. It explores the ways in which, in nineteenth- and twentieth-century Italy, Volta became a sort of national hero. The chapter shows that some long-established rituals, used to celebrate achievement in general, were easily—and superficially—adapted during the "age of progress" to celebrate scientists and the new technologies. The chapter also shows how, within these celebrating mechanisms, the cultural, social, and political needs of the celebrating people were often imposed on the celebrated hero.

The conclusions (chapter 9) offer an overview of the economy of invention emerging from the present study, and its implications for an assessment of the legacies of the Enlightenment. The chapter provides an analysis of the system of values and utilitarian concerns that prepared the shift from the classical and Enlightenment figure of the natural philosopher to the nineteenth-century (and present-day) figure of the scientist. It also shows that, within the broad framework provided by Enlightenment notions like "useful knowledge" and "the quantifying spirit," several different cultural and research traditions shaped the endeavors of physicists like Volta.

Accordingly, the conclusions advocate an interpretation of the history of science and technology that reintroduces those elements of diversity and contingency that the critics of the Enlightenment, and some of its supporters, have removed from their narratives in response to their respective agendas. The supporters of the Enlightenment tradition (to be found more often among scientists, engineers, and philosophers of science) have often been led to remove diversity, and contingency because they wanted to impose a normative conception of science and technology, and their proclaimed rationality. The critics of the Enlightenment (to be found more often among historians and sociologists), on the other hand, have claimed that that same normative view was typical of the Enlightenment in order to expose more easily what they regard as the cynical use, by interested elites, of the prestige of science and technology to enforce an authoritarian view of science, rationality, and social order.

The present book argues that neither party has done proper justice to the wealth, diversity, and unpredictability of the intellectual, technological, and social ferment that, under the banner of the vague but compelling Enlightenment notions of useful knowledge and the quantifying spirit, led to the age of electricity, and to our industrial societies.

# The Making of a Natural Philosopher

## FROM AMATEUR, TO EXPERT, TO PUBLIC SERVANT

The present chapter is devoted to a biographical sketch of Alessandro Volta, a leading figure in the history of the science of electricity during the late Enlightenment and the inventor of the electric battery, focusing on his formative years. Our goal is to pursue the different threads that combined in making up Volta's personality as an experimental natural philosopher, committed to the ideals of the Enlightenment and involved—as a professor and a public servant—in the reform of the public administration carried out by the enlightened governments of Austrian Lombardy from the 1760s to the 1780s. More information of biographical and contextual interest will be found in the following chapters, especially in chapters 4, 5, and 7. Here we attempt what is both the major challenge and reward of a biographical approach—the search for what was distinctive in the personal, social, and cultural traits that made up Volta's personality.[1]

Volta indeed shared several traits in common with the men historian Franco Venturi has depicted in his fascinating, collective portrait of "The Reformers" in Austrian Lombardy in the age of the Enlightenment.[2] A cadet from the lesser nobility having no direct control over the mediocre rents he relied on, Volta had been educated for a while by the Jesuits and was soon to become a self-trained adept of Enlightenment philosophy and ideology. With this background and the family connections that offered him aristocratic patronage in Milan and Vienna, he easily qualified to join the expanding ranks of educational experts created by the reform initiatives of Maria Theresa and Joseph II. The first in his family to live on state appointment rather than on private or Church rents, he was throughout his life a committed, motivated professor and public servant. He clearly perceived the fulfillment of his duties as going some way to promote the ideals of the "more enlightened age" to which he felt he belonged. Consistent with

this self-ascribed role, he managed to ensure that his latent, unorthodox views on religion, philosophy, and society—views linked to the
radical side of Enlightenment culture—did not come to the surface and
disrupt his career. Respect for traditional religious, political, and social
values thus characterized Volta's public figure throughout his life, pur-
portedly making of him a typical representative of the moderates
among the intellectuals of the age of the Enlightenment. This public
image of Volta prevailed among his contemporaries as well as biographers,[3] and seemed perfectly in keeping with his chosen field of expertise: the apparently innocuous science of electricity, typically indifferent to the hot religious and political issues debated by the radical
*philosophes*.

Though these traits of Volta's public image were and remain real,
the new material on Volta's education, private life, and career that I
have dug out of the archives in his home-town paints a new picture of
him, shedding light on other, no less important aspects of his personality and work. The new picture invites a better understanding of what
was distinctive of Volta, both as a person *and* a natural philosopher.
These distinctive features, seldom touched upon by previous biographers and discussed in this chapter, can be summed up under three
headings.

One is the extent of the interior conflict going on beneath the surface
of Volta's smooth, carefully administered public career. As we shall see,
the conflict derived from a tension between Volta's willingness to conform to the social and religious traditions of his group and age, and
his inclination to try—in his own private life and experience—the new
values and habits proclaimed by the secularized philosophies of the
Enlightenment. The conflict between tradition and innovation present
in Volta's own private life invites reflection on the amount of strain
the otherwise optimistic goals of moderate Enlightenment carried with
them in the people who joined in their realization. The tension between
tradition and innovation in Volta's private life also offers new insights
into his scientific personality. It will help us understand what we will
describe in the following chapters as his attitude of a "reluctant theorist," a none-too-convinced adept of the instrumentalist bent of late-
eighteenth-century physics.

A second, distinctive trait of Volta's personality, the material discussed in this chapter will help bring out, is the close relationship between his role as an ambitious public servant and his vocation for natural philosophy. This relationship seems to have been fostered, above
all, by the entrepreneurial spirit with which Volta pursued his career,
both as a scientist and a public servant. For this entrepreneurial
spirit—which brought him to develop his contacts and initiatives on a

European scale, as we shall see in chapter 5—he was indebted to the most dynamic among his acquaintances in the aristocratic, merchant, and banking circles of Lombardy.

By offering new insights into Volta's emotional life, finally, this chapter provides grounds for understanding what was perhaps the most distinctive of Volta's features as an investigator: the exceptional, conceptual, and manipulative skills that he displayed as an inventor of new machines, a brilliant performer of electrical experiments, and a successful lecturer in natural philosophy. As will be pointed out in the following pages, only by uncovering Volta's rich emotional life can we fully appreciate the traits that made him an extraordinary "virtuoso" in the theory and practice of electricity; an investigator who engaged in his pursuits body and soul, and combined the thrust of the natural philosopher with qualities that remind us of the musicians and skilled artisans of the late eighteenth century with whom he liked to mix in his intense social life.

## The Town

Alessandro Volta was twenty-four and had just published his first, ambitious essay "on the attractive force of the electric fire," when Joseph II, associated with his mother Maria Theresa in the ruling of the Austrian empire including Lombardy, visited the Italian domains.[4] It was in June 1769, and the future emperor was shown around Como and its lake, Volta's birthplace, by the owners of the largest wool factory in town. A report prepared for the occasion[5] took pride in stating that the small town of Como (fig. 1.1), situated a few miles from the Swiss and Piedmontese borders, two dozen miles from Milan (the state capital), and counting about 14,000 inhabitants, had 155 silk looms, 25 looms producing wool cloth and serge, 11 producing cotton cloth, 2 producing stockings. The list also mentioned 30 silk mills, 4 dye works, 3 factories producing hats, 4 producing soap, one producing glass, another producing wax, 3 leather tanneries, and 2 printing works. The people employed in factories amounted to 2,570, plus 56 masters. According to a contemporary observer, a survey extended to the entire district of Como—to include the many valleys around the town in easy contact with the rich Lombard plain to the south and with central Europe to the north—reinforced the image of a dynamic region, participating in the modernizing efforts of Maria Theresa and Joseph II.[6] The silk factories in particular, traditionally the strong engine of Como's economy, continued to prosper in the second half of the eighteenth century despite a few ups and downs. The number of silk looms active in and

1.1. Como, ca. 1730. Etching in BCC, Collezione Stampe, Vedute generali 74, Acq. 220723 (*Biblioteca Comunale, Como*).

around the town passed from 275 in 1772, to 1,035 in 1787, 553 in 1788, 646 in 1789, 905 in 1790, and 1,333 in 1795.[7] Commerce, another strength of Como—through which a prominent part of the goods traded between Lombardy and Germany, Switzerland and the Grigioni region passed—remained lively throughout the second half of the century. Though relatively weak in the mountains near Como, agriculture, by far the major source of income in eighteenth-century Lombardy, had been given an impulse by the introduction of a new register of landed property, decreed by Maria Theresa and completed in 1760.[8]

Reports of the kind prepared in 1769 were clearly intended to please the imperial visitor—who was indeed impressed by Como's industrious inhabitants—and to kindle municipal pride. Certainly, they depicted only one side to Como's reality. For example, the family who had the honor of showing the emperor around, the Guaitas, had not yet been given the privilege by their fellow citizens of sitting in the local assembly, still firmly controlled by the nobility to which the Volta family belonged instead. However, a Marta Guaita, married to a local nobleman and public servant, would later become Volta's mother-in-law. Members of the Guaita family were merchants in Amsterdam and Brussels, and, as we shall see, were occasionally instrumental in transmitting Volta's publications to Priestley in England and van Marum in the Netherlands. Because of the aristocratic ties, Volta's own family connections abroad were more likely to be found around the court of Vienna.

The town council of Como, where the Voltas traditionally held a seat, was made up of forty members drawn from the nobility. Since some families had more than one representative on the council (occasionally even five), about two dozen families ran the public affairs of Como. They exercised their power as they had for centuries, effectively resisting the efforts made by successive foreign rulers (the Spanish in the seventeenth century, the Austrians in the next) to curtail their influence.[9]

Despite the habit of occasionally admitting to the nobility people who had distinguished themselves in the accumulation of wealth (often acquired through commerce) over several generations, the main source of income of the Lombard nobility in the eighteenth century was still land and property revenues. Lifestyle was molded by the *rentier* habits together with a few other factors. One of these was the involvement of members of the nobility in the system of magistracies, relatively autonomous from the central power, through which the local administration was run. The system occasionally engendered within the nobility a sense of commitment to public service that we shall find well exemplified in Alessandro Volta. Another salient trait of Lombard

nobility was the way the nobles frequently entered the church or its religious orders. The church was a major social, economic, and cultural factor in Como's life, the influence of which was only slowly being eroded during Volta's lifetime by the secularizing measures introduced by Maria Theresa and, above all, by Joseph II from 1780 to 1790.

A catalogue (drawn up in 1767) of the clergy living under special rules (*clero regolare*) listed 25 monasteries and male religious institutions within the district of Como, involving 330 people.[10] Gian Domenico Volta, a Dominican and one of Alessandro Volta's four siblings who entered the church, figured in the catalogue as the person in charge of the church's inquisition in the region. The 1767 list did not include the common (*secolare*) clergy, to which two other brothers of Alessandro's, Luigi and Giovanni, belonged, nor did it include nunneries and other female religious institutions, to which one of the three sisters of Volta's belonged. Including the common clergy and female institutions, the figures of people serving the church in Como more than doubled the number mentioned in the 1767 catalogue.

The chief characteristics of Lombard nobility—its *rentier* habits, its occasional commitment to public administration through the old system of magistracies, and its frequent service in the church—were all present in Alessandro Volta's family and helped shape his early life and education.

## The Family

Little is known of Volta's father, Filippo, other than two facts. First, before marrying in 1733, he had joined the Society of Jesus for eleven years,[11] and second, after marrying, he was prodigal and, when he died, left his wife and seven surviving children not particularly well off.[12] Evidence of Filippo's prodigality, and a glimpse into the habits of Volta's family, are offered by the menus which have survived of two sumptuous dinners Filippo organized for his acquaintances in Como on two subsequent weeks during Carnival 1750.[13] Alessandro was at that time not yet five. As biographers concur, he did not inherit his father's lavishness, but he certainly inherited a liking for feasts, especially Carnival celebrations, which was to last well into his adult life.

Volta's father died when Alessandro was still a child, certainly before June 1756.[14] After that Alessandro—together with his mother and two sisters, Marianna and Chiara—lived with an uncle (also named Alessandro Volta), who was archdeacon of the Como cathedral. Not much is known of this uncle whom Alessandro, sixteen, depicted as a stepfather.[15] The uncle was still alive in 1770, when his canonry was

recorded as having a rent of 962.6 lire, about 230 lire less than what his nephew would earn four years later as newly appointed state supervisor to the schools of Como.[16]

Neither the name nor the figure of the father ever emerge in Volta's extant recollections and correspondence, except on one occasion, when he mentions a small pension left to him by Filippo.[17] Volta's mother, Maddalena Inzaghi, apparently had a similar, indecisive presence in Alessandro's life. She was from a noble family, members of which lived in Graz.[18] Testimonials of past traveling habits of members of the Inzaghi family were passed on to young Volta. As he fondly remembered later in his life—when he himself embarked on frequent European tours—he had spent many evenings as a child by the fireplace listening to an old "father Inzaghi" telling stories of his traveling adventures.[19]

Though of noble origin, Volta's mother was not herself rich. After her husband died, she brought up the children acting as a mere mediator in the administration of riches that her sons had inherited from a wealthy relative in 1756. Her seemingly subordinate role is conveyed by a few surviving letters in which she applied to the will executor, in the name of her sons, for more money (and cheese) for approaching festivities.[20] Two other episodes suggest that she did not play a major role in Volta's emotional life. One was the attempt (a failure as it turned out) to enroll Volta, sixteen, in the Jesuit order—an episode (of which more below) she does not seem to have exercised a decisive influence. The other episode reflects, rather, a degree of indifference on the part of her son. Alessandro was just back in Milan after his big European tour of 1781–82, when he learnt that his mother was ill and her life in danger. Instead of rushing the few miles from Milan to Como, he stayed in Milan looking after his affairs there and waiting for further news from home. His mother died a few days later, without seeing her son again.[21]

The weaker the role played on young Volta by the parents, the stronger were the affective and interest ties linking the "Volta brothers," as several documents preserved in the notarial archives of Como call them. The phrase referred especially to Luigi and Alessandro who, as beneficiaries (since they were fifteen and eleven, respectively) of the deed of trust from which most of the family's revenues derived, were the pivots around which the family revolved. Indeed, for a while, till the summer of 1756, Alessandro had enjoyed the privilege of being the appointed, universal heir of the richest of the relatives, uncle Nicolò Stampa, who had probably chosen Alessandro because he was his youngest male great-nephew.[22] On 13 July 1756, however, for reasons that are far from clear, the uncle changed his will and appointed Luigi instead, leaving Alessandro and his male descendants as substitutes in case Luigi died without issue.[23] From the mid-1760s, when Luigi was ordained a priest, the role of Alessandro as the only remaining propagator of the Voltas and, in the end, sole heir to their possessions must have been apparent to young Alessandro himself and relatives. Yet, until he married at forty-nine and secured from his elder brother Luigi—on marrying—half of the rents stemming from the Stampa heritage,[24] Alessandro occupied a subordinate position in family affairs, a fact which often affected his decisions as documented by several episodes reported below.

## Lifestyle

The inheritance of the Volta brothers allowed them to enjoy the rents and leisure offered by some nine different houses and small estates scattered across the provinces of Como and Milan,[25] plus the old family house in Como and, very likely, a number of minor properties in town. While these properties taken together did not place the Voltas beyond the rank of the lesser nobility, they did allow them to enjoy or, at any rate, imitate as closely as possible the lifestyle of Lombard nobility. The Voltas spent a long part of the year, from springtime to November, touring the family properties, each of which had its own special attraction depending on the season. The hottest months were spent in Gravedona, a spot by the lake north of Como. Campora was preferred with the coming of autumn when walking and hunting in the woods and mountains nearby was rewarding. As winter approached, Lazzate, a populous village on the plain near Milan, offered better opportunities for leisure. Despite the many estates and the touring habits, daily life in the family was unpretentious. *Otium* was the keynote and habits were affected by the large number of priests in and around the family. Volta often recalled later how family life was spent amid masses, card games, and glasses of wine. In the only surviving letter to his mother, written from Aix-la-Chappelle in November 1781, Alessandro depicted the family life at home in the following terms:

> By the time you receive this letter, my brothers will be perhaps in Gravedona, the holidays in Campora being over. You've probably had the usual company of priests; so I am sure that, to pass the rainy days, a good deal of time must have been spent at mass, tarots, breviaries and glasses of wine.[26]

A degree of impatience with the closeted family routine emerges in Volta's correspondence from time to time, and from childhood he displayed a determination to break with that routine. His early literary ambitions, as we shall see, expressed his desire to imitate the habits of the upper nobility and to adopt the cultural models of the cosmopolitan Republic of Letters. Similar impatience with the narrowness of the family milieu was revealed in his early scientific efforts, which led him to approach the leading authorities of the day in the field of electricity when he was not yet eighteen, and still imperfectly trained. The eagerness he displayed as an adult in cultivating the rich and powerful probably had its roots in that same drive.

As we shall see, in Volta ambition mingled with restlessness and a tendency to lean toward unorthodox views and behavior. His formal adhesion to Catholicism throughout his life cohabited with a partial, deep-seated predilection for unorthodox views. Also, his determination—in the end repressed—to make what was regarded as an unconventional marriage revealed his potential inclination to break with conventions. However, in keeping with one of his personality traits that we shall become acquainted with, he never allowed his impatience with sheltered family life and orthodox views to spill over into contempt, much less revolt. Apparently, the urge to imitate different cultural models and the determination not to break with his own roots were both embedded in the education he received.

## Education

Apart from a few recorded episodes, direct evidence of Volta's education is scanty. From surviving correspondence we know for certain that, at fifteen, he attended as a boarder the Jesuit College of Como.[27] It is also known that in the following year he moved, still as a boarder, to one of the three seminaries in town.[28] Apparently, no records of the students at these institutions during the 1750s and 1760s have been preserved: the information on this part of Volta's education is therefore indirect.

Volta's early education must have been imparted at home. This was common at the time. Volta himself, in his later role as superintendent to the public schools in Como, suggested that this habit should be tolerated if not encouraged by the state. He himself, as a father, would devote considerable energy in personally seeing to the education of his children. Young Volta, living as he did in the house of an uncle who is said to have studied either law or medicine, and who enjoyed the none-too-busy life of a canon in Como's cathedral, may have received some basic education from that uncle, who was fifty-nine when he was ten. Where needed, additional instruction was probably provided by some of the many priests with whom the family was in touch.

Indeed, before the reforms carried out by Maria Theresa and Joseph
II after 1765 paved the way for the introduction of lay teachers appointed by the state, elementary education in Lombardy was overwhelmingly the preserve of priests. Priests made up the great majority
of teachers running the small elementary schools scattered in and
around the towns, supported by funds coming from the wills of donors
who had left part of their wealth to the church for just such a purpose.[29]
The quality of the teaching in these schools was decidedly uneven.
Learning to read and write first in Latin and then in Italian was their
basic goal, a goal that was not always achieved. Teaching to count was
often the business of different schools, which taught the arithmetic
needed by merchants, and which, according to a later survey commissioned by the state, were found to be of good quality.[30] In the light of
the comparatively good level of instruction available to Volta at home,
and the uneven quality of elementary instruction offered by the
schools in Como, it was probably a good thing for Volta to have had his
early education in the family until he entered the local Jesuit College.

A biographer who had access to family documents now apparently
lost, writes that Volta entered Como's Jesuit College in 1758 to attend
the "rhetoric" classes, and subsequently, in 1760, the "philosophy"
course which other extant evidence shows he actually attended.[31]
Como's Jesuit College, established in 1561, was a relatively small institution. When Volta completed his first "philosophy" year in July 1761,
the staff running the college comprised ten people. The institution was
devoted to education only, which meant teaching "in genere Pueros, et
in specie Logicam seu Philosophiam, quod prestat addita etiam titulo
charitatis Theologia Morali, et Speculativa."[32]

In the Jesuit curriculum typical of the eighteenth century, "rhetoric"
was a third-year course, following those devoted to "grammar" and
"humanity," and based on the teaching of (mainly Latin) *oratoria* and
poetry. In the same context, "philosophy" meant a two- to three-year
course for students between sixteen and nineteen, followed by a theology course. The philosophy course included, in addition to classical
Aristotelian philosophy, a substantial amount of natural philosophy
and physics in the modern sense. The larger Jesuit colleges had a special professor in physics.[33] When no such professor existed, the philosophy professor took charge, as was the case at the college in Como
where the philosophy professor was a certain Girolamo Bonesi.[34] For
reasons connected to what appeared for a time to be Volta's inclination
to join the Society of Jesus, a special relationship developed between
the philosophy professor and young Alessandro.

No detailed evidence survives of how Volta was taught philosophy
in college. We know for certain only that during the summer holidays

of 1761 Bonesi passed on to Volta some manuscript lectures on "animastics," that is, psychology, to be copied. Volta must have taken a deep interest in the subject if a friend of his was right in recalling, several years later, that Volta at sixteen had his own special view on a controversial issue: he believed that animals too had a soul, conceived of as a true spiritual substance.[35] The issue was much debated at the time. Volta's stance was unorthodox enough to scandalize his more conventional friend. This same friend many years later (when Volta had gained celebrity) attributed to young Volta the authorship of eleven (lost) manuscript notebooks on the same topic, which he had donated to his friend.[36] The notebooks might well have been a copy with comments of the lecture notes Bonesi had given Volta; or they might indeed have been more than that if we accept the friend's story that Volta, years later—during the French revolutionary era—searched for his friend in vain to get them back.

Animastics was usually taught during the second year of the philosophy course. In the first year—the one attended by Volta in 1760–61—the typical curriculum provided for the teaching of "Logic and Introduction to Physics." This meant "the first and second books of Aristotle's *De interpretatione*, mostly in summary; first notions of science: its divisions, abstractions, theories, practices; the diverse methods of proceeding in physics and mathematics."[37]

In the absence of more direct evidence of the kind of teaching the College of Como provided in these fields, additional evidence can be evinced from the books kept in the college library. The library shows that, despite the provincial setting, the Jesuit College in Como had a respectable tradition in the teaching of physics. A catalogue drawn up in 1774 (at the request of a committee that Volta was a member of, after the Society of Jesus had been suppressed) shows that the college had accumulated, since its foundation, a number of interesting classics of science for its library.[38] These included Lucretius's *De rerum natura* and seventeenth-century authors like Galileo, Descartes, Borelli, Cavalieri, Riccioli, and Cabeo. Eighteenth-century textbooks included Musschenbroek, Nollet, and Gravesande. Recent works by Boscovich, Frisi, and Jacopo Riccati were also in their possession. A treatise on electricity by Giambatista Beccaria, whom young Volta chose as mentor in his early scientific correspondence, was held, in two copies, in the college library. This latter also held a set of the proceedings of the Turin Academy of Sciences, which included recent, first-rate papers on mathematics and physics. There is no way of ascertaining how much or how little of the books kept in the college library were available to the average philosophy student. Their presence, however, suggests that some of the college professors took a real interest in the teaching of science as part of the philosophy course.

By the summer of 1761 Volta, sixteen, had acquired a nonephemeral interest in philosophy. When his family mooted the possibility of taking him from the Jesuit College to enroll him in the law courses provided by the local Collegio de' Dottori—a college intended for members of the aristocracy aspiring to the magistracies—he put up resistance, inclined as he was to continue philosophy instead.[39]

In the summer of 1761 Volta's predilection for simple toy tools is also documented. When relatives found in his pockets a penitence chain the Jesuit philosophy professor had secretly given him to test his religious calling, the chain was easily taken for one of the iron tools young Volta was often seen playing with.[40] Since the chain was a common sort of device for electrical experiments of the time, it is tempting to speculate that Volta was experimenting on electricity already in 1761. However, there is no need to speculate too much on the date of his early interest in the subject; in a couple of years he startled contemporaries by directly addressing two of the leading authorities on electricity, Nollet and Beccaria.

In the autumn of 1761 Volta was taken from the Jesuit College to avoid what relatives viewed as a conspiracy by the philosophy professor to make him join the Society of Jesus. As mentioned, young Volta had a hand in deciding his own future: he was not sent to study law, as relatives and especially uncle Alessandro wished, but to an institution offering a curriculum not too dissimilar from the one taught by the Jesuits. This was a seminary—we do not know for certain whether that of Santa Caterina or another called Seminary Benzi, run by the common clergy—hosting also boarders who did not intend to pursue a career in the church.

As in the case of the Jesuit College, no record of Volta's presence as a student in one of the two seminaries has survived. We know that the seminary of Santa Caterina in the 1760s had teachers of grammar, rhetoric, philosophy, and dogmatics.[41] As regards the teaching of philosophy and science, the books kept in the library of the seminary of Santa Caterina suggest that teaching levels in the seminaries were inferior to those of the Jesuits. Philosophy was represented almost exclusively by compendia, and science was hardly represented at all, apart from ten books on practical medicine.[42] Volta, however, may have benefited in those years from resources of another library to which he very likely had access, and which—as a superintendent to the schools in Como—he commended a few years later: the library of the Collegio de' Dottori. Though devoted mainly to the teaching of civil and canon law, this college also kept in its library books of scientific interest, including the collection of the *Commentarii* of the Bologna Academy of Sciences.[43]

In sustaining Volta's interest in natural philosophy during these years, an informal network of acquaintances and friends must have been no less influential than the formal curriculum Volta was pursuing as a student. Among Volta's acquaintances, Giulio Cesare Gattoni played a particularly important role. Gattoni—the friend who did not share Volta's belief that animals too had a soul—had also been trained by the Jesuits and stayed on at the college when Volta left, eventually becoming a canon. Slightly older than Volta, a generous, eclectic and somewhat extravagant character, Gattoni took an interest in every curious topic he came across, including electricity, all of which he shared with his friend. The major achievements of Gattoni were the setting up of the first lightning rod in Como in 1768, and a collection of natural and antiquarian objects that earned him some renown.[44]

Gattoni was not the only one of Volta's early acquaintances having a penchant for natural philosophy. Throughout Volta's biography one comes across several amateur philosophers among the array of aristocrats, canons, and intellectuals he knew in and around Como. Among such amateurs an interest for natural philosophy merged with an often stronger interest in literature, especially poetry, an interest nurtured by the persistent tradition of Arcadian groups of poets. For a long while Volta too paid homage to that tradition: it is no coincidence that the first document we have of Volta's views on natural philosophy is a poetical composition, written in Latin. The tone of the composition, however, places it at a considerable distance from the lyric mood of Arcadia, and close to the scientific and technological spirit of the Enlightenment.

## "A More Enlightened Age"

Volta's first known literary composition, part in prose and the rest in 492 verses, bears no title or date.[45] It is usually ascribed to the early 1760s, perhaps between 1762 and 1764, when Volta—about eighteen—also started up his scientific correspondence. The composition is inspired by classical models like Virgil, and especially by Lucretius's *De rerum natura*, from which several verses are taken almost verbatim. The genre is that of the didactic poem—the earliest form of science popularization—conceived in the light of a *motto* from Torquato Tasso: "truth dispensed in soft verses has persuaded through allurement even the most reluctant."[46] Another likely inducement that led Volta to tackle scientific issues in verse was a statement by Francesco Algarotti, in his well-known dialogues on Newton's *Opticks*, to the effect that the exact sciences could hardly be treated in verse.[47] The challenge to prove the successful Algarotti wrong on this point must have stimulated the ambition of young Volta.

The topics treated in Volta's composition were "the newest discoveries of recent philosophers." To young Volta, contemporary natural philosophers appeared as the heralds of a new, enlightened age. The notion of a "clarior," "sapientior aetas" recurs time and again in the poem, coupled with the notion that the new age is exploding "blind superstition" and the "people's deliria" of old times.[48] The intellectual *and* moral ingredients of the new philosophy exalted by Volta were in fact a combination of classical Lucretian and Enlightenment ideals. The most revered of human gifts were "acumen of reason" (*rationis acumen*) and "understanding" (*ingenium*), but also "virtue" (*virtus*) and "fortune" (*fortuna*). The secular theme of "fortune" was linked to Volta's other remark, that creativity itself, as displayed in the achievements of the new philosophy, was *not* a divine gift.[49]

The main goal pursued throughout the poem was to offer rational explanations for phenomena usually regarded as wonderful and frightening, such as gunpowder, fulminating gold, fireworks, and will-o'-the-wisp. The conceptual framework adopted was that of corpuscularism, to which the notion of an "attractive force" was added in phrases suggesting that the author's commitment to the notion was lukewarm. All the phenomena dealt with were explained basically in terms of atoms and void, by arguing that bodies contained different amounts of "elementa ignis" (elements of fire) depending on their different textures. The particles of fire contained in bodies were deemed to account for the explosions and other reactions typical of the phenomena described.[50]

Rational explanations and theories, in any case, were not the only themes pursued by Volta in his juvenile poem. He clearly liked the actual procedures required to carry out the experiments. The vividness of his descriptions leaves no doubt that he tried some of the experiments himself, especially those of a chemical nature, and those carried out with the air pump. He must also have observed the uses of gunpowder in mines, probably in the quarries on the mountains near Como. The art of blowing glass, in the glassworks in town or in and around Milan, must also have been known to him.[51] Indeed, the useful arts had pride of place in the poem, another homage young Volta paid to the enlightened age he was proud to belong to.

The "philosophers" Volta cited explicitly had all dealt at some time with electricity—von Guericke, Gray, Dufay, Musschenbroek, Nollet, Franklin, Delor, Dalibard, and Beccaria. However, in the poem he touched upon electricity only to remark that the topic was so important that he planned to devote another, special poem to it.[52] Of this other composition nothing else is known. In the extant poem, however, Volta ascribed to the science of electricity a key role in his portrait of the age of reason. He maintained that, by showing that lightning was an electrical phenomenon, not caused by sulphur or by element of fire as Lucretius and many others had believed since antiquity, modern philosophers had demonstrated their superior understanding. The science of electricity, Franklin's special achievements, and the recurrent theme of "a more enlightened age" were all knit together in the rhetoric of the following verses:

Fire often ignites itself in fat matters
even without any spark being brought to them,
and the luminous globes of the flames develop spontaneously:
not otherwise the ancients and many among recent authors deemed
when teaching that burning meteors derive from a sulphurous disposition.
Lightning too was believed to have such an origin;
yet wrongly so, because since a more enlightened age has shone,
and new arts have been disclosed to the geniuses of our times,
how transformed the face of Sophia has turned out from then![53]

If Franklin had rebutted Lucretius on the cause of lightning, young Volta clearly saw Franklin's electrical explanation of lightning as yet another step in the Lucretian fight for freeing humanity from superstition. Classical natural philosophy and the spirit of the Enlightenment merged in the secularized mind of the former pupil of the Jesuits who was turning natural philosopher.

## Literary Interests

Volta's literary interests were not confined to scientific, didactic verse. As he wrote in 1768 to the Swiss poet Salomon Gessner, one of his early foreign correspondents, he was "fond of everything poetic."[54] Among Italian poets Volta especially enjoyed Tasso, Chiabrera, Frugoni, Metastasio, and Parini. Among foreign poets he admired John Milton, whose *Paradise Lost* he first read in French translation; Edward Young, whose *Night Thoughts* he would strongly recommend when superintendent at the schools in Como; and Friedrich Gottlieb Klopstock. Volta also liked Gessner's poetry, and especially his epic poem *The Death of Abel*.[55] If scientific prose and epic poetry were one favorite, oratory, classical and modern, was another. Among classical authors Cicero was the most beloved and unreservedly commended. Among contemporaries, Volta had particular regard for Francesco Algarotti and Giovanni Battista Roberti, a Jesuit and polymath. Tragedy also appealed to Volta. Together with French scientific prose, French tragic authors such as Racine and Corneille encouraged young Volta to learn French. Volta's life in Como was also conducive to a lighter genre of literature. Several of Volta's extant poetical compositions were of an occasional nature: their ambition did not go beyond the celebration of a wedding, or a ceremony for a woman becoming a nun.[56]

Volta enjoyed writing poetry throughout his life. To celebrate De Saussure's climbing of Mount Blanc in 1787, he wrote a 199-verse composition.[57] It conveyed effectively (though in unconvincing poetic style) Volta's view of science: the achievements of contemporary natural philosophy, as he saw them, were indissolubly part of culture, just as poetry was. Indeed, he regarded recent natural philosophy as the greatest contribution his age could make to the universalistic culture he had assimilated through classic and modern literature.

The growing ambition of contemporary natural philosophy was perceived within Volta's circle of friends, including those who were mainly interested in literature. Count Giovio, a lifelong friend of Volta's, paid homage to the new figure of the natural philosopher ("the physicists") in the following terms:

> We will not renounce being physicists either, even if there are no barometers or thermometers clanging round our necks. We will not completely forget those special words, like granite, limestone, . . . and we too will build the world as we please, [by evoking] . . . memories of Moses, phials of inflammable air, electric bursts, and the modern airs.[58]

Together with an amount of jealousy on the part of the "letterato" toward the successful physicist, Giovio's homage revealed that in Volta's circle of friends natural philosophy was fashionable: any decent "letterato" felt he *had* to be conversant with its concepts, vocabulary, and rituals.

### Views on Religion and Secularization

Volta's upbringing and family life exposed him to religion in all its forms, affecting his sensibility, formal practice, and his philosophical and theological thinking. The orthodox part of his religious education probably culminated with the reading of *Imitatio Christi*, the classic devotional book attributed to Thomas à Kempis, which the Jesuit philosophy professor lent him during the summer of 1761.[59] Volta enjoyed the book. For a while, the professor hoped that he had won his young pupil over to the Jesuits, and the Voltas feared that the boy was set to join the Society of Jesus.[60]

The fears of the relatives were partly caused by the vicissitudes the Jesuits underwent after their recent expulsion from several European states. Other fears must have risen in the family from concerns about the destiny of the riches they enjoyed as a result of the 1756 deed of trust, should even Alessandro fail to marry and leave male descendants as the deed required. To these mundane concerns, others of a philosophical nature accrued in the mind of young Alessandro himself. Some unorthodox views sprang from his reflection on the very philosophy he had been taught at the Jesuit College.

The part of the philosophy course devoted to animastics, as previously mentioned, led young Volta to believe that animals too had a soul. As for physics, the orthodox Aristotelian philosophy to which he had been exposed did not prevent him from adopting a thoroughly Democritean worldview, documented by the poem discussed above. Nor was this adhesion to a materialistic philosophy a short-lived, juvenile departure from orthodoxy: Volta subscribed to a similar brand of philosophy well into his thirties.

In 1777, while crossing the Alps on his first trip abroad, he reflected upon the theories of the earth put forward by Buffon and other naturalists.[61] In his report Volta repeatedly mentioned the "extreme old age of our planet," the "unending long series of centuries, hardly conceivable to the mind," necessary for the Alps to have attained their present, grandiosely ruined shapes. As for the factors involved in the making and transforming of mountains, Volta referred both to successive, violent, natural events or "convulsions," and to slow, gradual changes over time. Volcanic phenomena, earthquakes, and natural elements were all he had recourse to in his simple geology.[62] He made no mention of the flood, nor of events reminiscent of the Bible.

A similar worldview is sketched out in a juvenile fragment, written by Volta in French, containing reflections on a prima facie theological topic: the resurrection of bodies. The fragment, conceived as a letter to an unnamed "Rev[erend] Père," purports to explore the views expressed on resurrection by Charles Bonnet and Leibniz, and claims to accept an explanation closer to the Revelation than Bonnet's.[63] Here Volta indeed adopted a chronology compatible with the Bible, which he had not mentioned in his report of the 1777 trip across the Alps. In fact, in the undated fragment he outlined a Stahlian theory of the perennial cycle of life, a theory, he said, he had conceived well before reading Bonnet.[64]

In the fragment Volta assumes, with Buffon, that living matter cannot originate from inanimate matter. He further assumes with Stahl that human beings, like animals and vegetables, ultimately derive the matter they are made of and their means of subsistence from other human beings, as well as from animals and plants. As Volta puts it, the same living matter or substance is "common to a tree and an animal," it moves from "living machine" to "living machine," it gets dispersed and is united again in thousands of different shapes. Being the same basic substance, it has the appearances of everything that grows, vegetates, and embellishes this globe. The same measure of matter is "at once tulip, eagle, apple, caterpillar, dog, man!" Having posited this, Volta then asked himself what initial amount of living substance was needed to account for all the living beings (vegetables, animals, and humans) "from the first days since the Creation" until the end of the world. Would an original layer of vegetable matter, a hundred feet deep and covering the earth, suffice to account for the "milliard human bodies" involved? It would be able to, Volta argued, if one admitted—as he was inclined to—that the *same* amount of matter had contributed to the subsequent making of numberless different bodies. There was a difficulty with humans, however, since revelation required that each human being is resurrected from *his or her own special body*. To solve the dilemma Volta introduced what he regarded as an argument taken from, or compatible with, Leibniz's philosophy. Unfortunately, because the fragment ends abruptly we do not have Volta's detailed argument.[65]

Whatever the conclusions he might have drawn, it is clear that Volta, in the 1760s and 1770s, was well versed in the mid-eighteenth-century literature circulated by the *philosophes*. He was eager especially to submit religious issues to the tests provided by recent speculations on the earth's history, nature's equilibrium, and the endless number of forms inanimate and living beings have assumed in the course of time. If a connecting thread can be found in the fragmented evidence available, one might say that Volta, in the 1770s, continued in the same Lucretian mood documented in the poem he composed in the early 1760s. But how did all this fit in with Volta's own commitment to the Catholic religion in which he had grown up?

A guarded wariness in disclosing his personal attitude toward religion remained a constant preoccupation with Volta throughout his life. The only extant, public reference to his early departure from Christianity and adhesion to the "materialistic philosophies" of the eighteenth century is the report of a conversation he had late in his life, sometime between 1814 and 1820, with a younger friend. The piece was published only after Volta's death, in 1834. The report took the form of a poetical composition, simulating a dialogue with Volta.[66] During the conversation both the interviewer (the friend, Silvio Pellico) and the interviewed (Volta) displayed eagerness to denounce their own earlier disbelief. The "interview"—considering the peculiar circumstances—confirms that Volta adhered to the secularized views of the *philosophes* during an earlier part of his adult life.

Despite the formal homage Volta always paid to Catholic religious practice, rumors about his unorthodox views occasionally circulated during his lifetime. Not surprisingly, suspicion spread during the troubled French revolutionary and Napoleonic periods, Volta's extremely cautious public demeanor notwithstanding. The need to dispel such suspicion befell him soon after the restoration of Austrian power in Lombardy in 1815. The episode is documented by a declaration Volta wrote on his own religious views in that same year.[67] The declaration, apparently, had been requested by a canon, Giacomo Ciceri, some time before January 1815. The timing of the request speaks for itself: the era of French influence in the affairs of Lombardy had then come to an end, and it was common practice to scrutinize the conduct of public figures from the past regime. What is more, a few months earlier the restored Austrian authorities had nominated Volta head of the Philosophy Faculty at the University of Pavia, in spite of the well-known rewards he had obtained from Napoleon. The envy of some colleagues may well have stimulated Canon Ciceri's curiosity about Volta's religious views.

It does not come as a surprise, given the circumstances, that the 1815 declaration affirms the orthodoxy of its author. Writing the declaration, however, was a troublesome process, as the revisions in the minutes display. After second thoughts, Volta added a sentence mentioning not only his sins (as any Christian would), but also his "disorders," which may allude both to the philosophical doubts confessed in the interview with Pellico, and to the many love affairs which marked the first, unwed half of Volta's life.

In any case, the most likely source of the suspicions surrounding Volta's views on religion was neither his adhesion to secularized natural philosophy nor his private life but, rather, the stance he took in the 1780s and 1790s on the controversial issue of the relationship between church and state—a stance some judged to be too favorable to the state to be inspired by genuine Catholic belief.

As a professor at the University of Pavia from 1778, Volta became acquainted with several leading figures of Italian Jansenism active in the university, who supported an evangelical reawakening within the church. Volta's correspondence testifies that, during the last two decades of the century, he was on friendly terms with Pietro Tamburini and Giuseppe Zola, professors of moral theology and church history in Pavia, whose Jansenist inclinations were well known both in Italy and abroad.[68] Their teachings, and their presence in the leading educational center of Austrian Lombardy, were in keeping with the policy of Joseph II of asserting the right of the state to regulate all religious matters affecting public administration and life. Because of a similar policy, in the years immediately preceding the French Revolution, Jansenist theologians enjoyed the support of Leopold, Grand Duke of Tuscany.

That Volta subscribed to a secularized view of the relationship between church and state is documented by many statements in his correspondence, especially in the letters written to his brother, archdeacon Luigi, in the wake of the 1790 oath imposed on the clergy by the National Assembly in France.[69] A further testimony to Volta's stance is offered by his friend Gattoni, who reports having had heated discussions on the topic with Volta in those same years. The contrast led to a temporary suspension of their friendship. Gattoni, a conservative Catholic, attributed Volta's secularized view of the rights of the state vis-à-vis the church to the influence of his Jansenist acquaintances in Pavia.[70]

On that same ground, a generally accurate biographer has credited Volta with adhesion—albeit temporary and mild—to the theological tenets of Jansenism.[71] That rumors of Volta's leanings toward Jansenism continued to circulate in Como seems to be confirmed by a sentence in Volta's 1815 declaration on religion where Volta found it appropriate to assert that, while believing in the importance of grace for salvation, he also believed in the role of "good actions," as if explicitly rejecting the Jansenist notion that divine grace alone can save humans.[72] Apart from the rumors, however, there is no direct evidence of Volta's adhesion to the strict theological principles of Jansenism. If we are to believe a perceptive contemporary observer, Marsilio Landriani, the speculations on Volta's Jansenism were of a limited circulation. Writing to Volta from Vienna in May 1800, Landriani declared that Volta was *not* regarded as a Jansenist there unlike many other professors in Pavia.[73]

What has been surmised as Volta's Jansenist inclinations would probably be better explained by two facts, only partly related to Jansenism. One was his already mentioned support for the policy of restraint over the power of the church adopted in Austrian Lombardy and in certain Catholic countries at the time. He found this policy consistent with his own ideal of an efficient and independent public administration. The second fact was Volta's deep admiration for the civic and intellectual achievements of Protestant countries, several of which he toured on different occasions, and to which—having lived most of his life a few miles from the Swiss border—he felt intellectually and geographically close. The reports he wrote of his journeys (on which more in chapter 5) bear abundant evidence of his admiration for Protestant countries. He regarded them as more populous and more active in commerce and intellectual pursuits than Catholic countries.[74] In line with a trait in his personality we shall come across often, Volta found it relatively easy to reconcile his support of the state against the church, and his admiration for non-Catholic countries with a formal, occasionally sceptical, but substantive commitment to the Catholic religion of his family and country.

The most spontaneous expression of Volta's religiosity is probably the one contained in a letter written when he was forty-five. The letter was addressed to his brother, the archdeacon, during the bitter conflicts Volta had with his family over his intention (later renounced) to make an unconventional marriage. To his brother—who urged him to think again—Volta replied that "a dejected mind and an upset heart" are hardly compatible with acceptance of the maxims of Christianity, while an unconventional marriage dictated by genuine sentiment could be. In order to perceive and pursue "the great, eternal truths" of religion, Volta argued, humans must reflect upon them "with a calm mind and a peaceful heart."[75] These words offer perhaps the best portrait of the tolerant, benign religiosity—the basic ingredients of which were an implicit appeal to natural law and a blend of classical stoicism and indulgent Christianity—to which Volta subscribed throughout his life. Apparently, he found this religiosity compatible with several tenets of the secularized philosophies of the Enlightenment, to which he also subscribed.

### From Amateur, to Expert, to Public Servant

Volta's poetical composition of the early 1760s bears witness to the early development of the natural philosopher. As shown, the premises for that development came partly from the teaching Volta had received at the Jesuit College, partly from his self-taught training in the classical, especially Lucretian, tradition, and they were further stimulated by a reading of seventeenth-century authors, who had convinced him of corpuscularism as an overall worldview. A further, decisive impulse came from his reading of a handful of classics from eighteenth-century physics, notably Newton, Musschenbroek, Dufay, Nollet, Franklin, and Beccaria. In an age and society where an amateur interest in natural philosophy was fashionable—just as literature and music had been for centuries—Volta's early amateur interest in natural philosophy does not require a special explanation. What does require explanation is his decision to get in touch with the experts in the field, a decision that was taken in 1763.

The letters that Volta, barely eighteen, sent to Nollet and Beccaria in 1763 have not been preserved, but their very existence and a few surviving excerpts are significant on several counts.[76] The letter to Beccaria, for sure, was more than an *aperçu*: Volta called it "a bulky scribbling-pad."[77] From later allusions, both letters must have contained an ambitious, attractionist theory encompassing all electrical phenomena explained by analogy with the phenomena of magnetism. The amount of daring implicit in addressing leading figures such as Nollet and Beccaria—without credentials of any kind or, apparently, any personal connections—may well be explained by certain traits in young Volta's personality, which he himself described in a later letter to Beccaria. These traits were—in the self-deprecating terms he used to win Beccaria's sympathy—"fickleness" and an inclination toward "whimsical" and "childish reasoning."[78] More is needed, however, to explain how the eighteen-year-old Volta was already constructing his own system for the entire science of electricity and trying to obtain recognition as an expert. Why—having been educated in a tradition of amateur philosophers and living in a world of amateurs—did Volta want to become an expert? And why did he choose physics as his special field of expertise?

As an ambitious cadet from the lesser nobility, who had eschewed the possibility of entering the church and had no direct control over the income he lived on, young Volta was clearly in need of cultural and social recognition. To be known as a natural philosopher—not just as one of the many "letterati" the Italian provinces were overflowing with—gave him a well-defined, little-contested role in Como, and one which would eventually earn him renown beyond the confines of the provincial setting and lesser nobility which stifled him. Literature and poetry also satisfied, to some extent, similar goals of cultural and social recognition, and indeed young Volta tried them too. Yet, the kind of rational, secularized worldview the new physics offered was a more complete expression of the ideology of a "more enlightened age" to which young Volta adhered. Electricity, on the other hand, unlike discussions over the soul of animals or the age of the earth, allowed no room for controversy over religious issues, in which Volta knew how to indulge, but he was also learning to eschew.

After several attempts at establishing a regular scientific correspondence frustrated by Beccaria (who in one case delayed his answer for about one year), in 1765 Volta finally obtained almost regular answers to the letters he kept sending to the dean of Italian electricians, twenty-nine years older than himself. By the end of 1766 the notoriously difficult Beccaria had decided to afford the young Volta a degree of cautious esteem and sent him offprints of a paper of his published in the

*Philosophical Transactions* of the Royal Society of London.[79] By June 1767 Volta felt daring enough to close his latest letter to Beccaria with a sentence stating that the letter did not, after all, require an answer.[80] In September of the same year Nollet—who had answered politely though skeptically to Volta's earlier letters—wrote again and sent Volta a little book as a present.[81] In 1768 Volta wrote again to both Nollet and Beccaria, illustrating his views on how electrical repulsion could be explained in terms of attraction. By that time Volta must have been working on the treatise in Latin that he published on 18 April 1769 in Como, under the form of a letter addressed to Beccaria entitled *De vi attractiva ignis electrici* (On the attractive force of the electric fire).[82]

The content of Volta's first publication and the mixed reviews it received are discussed in chapter 3. Suffice it to say here that the dedication to Beccaria did not make the reception of Volta's *opera prima* any easier. This fact, and the plans under way for relaunching the University of Pavia, reoriented Volta's ambitions. He dedicated his next publication in 1771, still on electricity, to naturalist Lazzaro Spallanzani.[83] Spallanzani—himself not an expert in electricity—had been recently appointed to the chair of natural history at Pavia. By choosing Spallanzani as his new patron, young Volta had shifted his target from a foreign (Piedmontese) scholar like Beccaria, and an authority in his own field, to a national patron from outside his field. The shift was connected with Volta's new ambition to become not just an expert but a professor too. In that same year, 1771, Volta learnt that the support of political patrons was no less important than the esteem of senior scholars in order to join the system of public education being established in Lombardy under Austrian rule.

It was yet another scholar, Paolo Frisi—an engineer and astronomer engaged in state committees carrying out the reforms endorsed by Maria Theresa and Joseph II—who first pointed out to young Volta what kind of patrons he would need if he was to obtain public office.[84] As a consequence, in the summer of 1771, Volta turned to the secretary of the Italian Department in Vienna, Joseph Sperges, and to Carlo di Firmian, Austria's plenipotentiary minister in Milan.[85] Frisi himself took it upon himself to approach, on behalf of Volta, Gian Rinaldo Carli, who had presided over the Supremo Consiglio di Economia—a key organ of the reforming policy in Milan—in previous years. Carli, like Boscovich and Frisi, was a former natural philosopher turned public servant: until 1765 he had taught astronomy at the University of Padua. Later, while presiding over the above mentioned committee, he had studied a plan for the reform of the University of Pavia in the fields of philosophy, mathematics, and physics.

Thus, around 1770 Volta begun to consider the prospect of joining a second generation of experts in astronomy, physics, and mathematics—after the generation of scholars like Boscovich, Carli, and Frisi—seeking employment in the reformed Austrian administration of Lombardy. When Volta entered the scene, opportunities were slowly opening up, and physics was benefiting from the situation. Outlining a reform of the University of Pavia circa 1770, Frisi, while paying homage to the tradition that still regarded physics as a topic intended for the medical curriculum, cautiously suggested that there might be two, rather than one, chairs in the field: one for "general physics," another for "experimental physics."[86] Considering that the plan foresaw twenty-two chairs for the whole range of disciplines taught in the university (including the faculties of law, medicine, theology, and mathematics), prospects seemed promising for physicists. The other Lombard center for higher learning, the Scuole Palatine in the Brera Palace in Milan, was also living with the feeling, occasionally the reality, of expansion. In the Palace of Brera, including also the Sant'Alessandro Schools, it was possible to learn logic, metaphysics, physics, theology, ethics, geometry, and astronomy at university level, though no university degrees were awarded. From 1764 the construction of a new astronomical observatory was begun in Brera under the supervision of Boscovich for the Jesuits who ran the Scuole Palatine until the suppression of the Society in 1773. Frisi thought that, for the time being, science teaching should be left to develop in Pavia, while Milan would continue to have its chair of mathematics. But he let it be understood that, once the go-ahead for expansion in Pavia had been secured from the government, plans were afoot for the expansion of science teaching in Milan as well.[87]

Frisi warned, however, that expansion continued to be threatened by the low number of students enrolling in institutions of higher education, especially Pavia. Measures aimed at making a university degree compulsory for certain professions were seen as a prerequisite necessary for ensuring that expansion was rooted on firm ground. The policy of expanding science teaching was, apparently, a measure encouraged from above—that is, by the government and its experts—rather than a development brought about by an increased need for public higher education felt in Lombard society at large. Competition for the few openings in new teaching positions, in any case, was high, and chances of success only slim. Volta, having decided to become an expert *and* a professor and/or a public servant, experienced the difficulty thoroughly.

The procedures for the recruitment of new professors were being debated in Lombardy around 1770. In his plan Frisi mentioned publications and "inventions" as the credentials to be considered. Volta complied: his second publication mentioned the construction of a new apparatus (an all-wood electrostatic machine) already in the title. He still lacked the right patron, however. His competitor for the physics chair in Pavia, Carlo Barletti—a good electrician, ten years older than Volta—had got the message first. Barletti had dedicated his 1771 publication to Carlo di Firmian.[88] In October Sperges, from Vienna, told Volta that Barletti (not mentioned) had secured the support of Firmian for the chair in Pavia. Barletti was duly appointed in 1772.[89] Sperges explained to Volta that Firmian was *the* patron to whom to apply in the future if he wanted to carry on his pursuit of the "useful sciences."

His plan for a chair in Pavia temporarily frustrated, Volta went on cultivating his circle of international correspondents, which from March 1772 included Joseph Priestley.[90] Besides electricity, which was no longer Priestley's chief interest, fixed air was the frequent subject of their correspondence. Volta also kept close ties with Spallanzani in Pavia, performing a few experiments suggested by him in 1773 on the fertilization of insects and the regeneration of salamanders.[91] Volta does not seem to have taken a great interest in such experiments, however, and he is unlikely to have seriously considered forsaking electricity for natural history. However, he was not being encouraged to continue work on electricity either. Sometime before April 1774 Beccaria—after a temporary appeasement—had invited Volta to maintain an "everlasting silence" on the subject of electricity.[92]

By 1774, when prospects for a job in the public education system at last materialized—the pace of educational reform having been boosted by the suppression of the Society of Jesus the year before—Volta had learned how to channel his ambitions. This time he applied direct to the plenipotentiary minister and asked (in fact, as the custom of the day demanded, he "supplicated") for "a Lectureship in Physics, Metaphysics, or indeed, best of all, . . . the role of superintendent to public education [*tutti gli studj*] in [the] Town [of Como]."[93] The credentials he submitted were the two Latin publications on electricity, and a plan (in manuscript form) for the establishment of "public schools in Como," in which he himself suggested that there should be a superintendent.[94] In October 1774 he learned from the Royal Delegate in Como, Lodovico Peregrini, that Firmian had granted him the role of superintendent he had planned.[95] One of the first committees Volta joined in his new capacity was given the task of restructuring the library of the Jesuit College—now state property—where his interest in natural philosophy had first developed thirteen years before.[96]

After 1774 the routine job of supervising teachers and watching over the students' work and discipline did not absorb Volta's talents entirely. He saw himself, in fact, as a reformer of public education, inspired by principles carefully drawn from Enlightenment ideals.[97] The key words contained in the first brief report he submitted as superintendent to Firmian in December 1774 were "innovation," "reform," "advancement of the sciences," "a happier culture."[98]

Volta's detailed proposals for a new plan of education were submitted to Firmian in March 1775, in reply to the minister's request for a list of "things deserving reform."[99] The surviving plan does not include, unfortunately, the part dealing with natural philosophy; still, it illustrates some important innovations that Volta thought fit to introduce in the traditional curriculum offered by primary and secondary schools.

Volta, for example, argued that the time given over to arithmetic—a useful art, he stressed—should be increased, while the teaching of the Italian language should precede in order and importance that of Latin, to which schools still gave priority. The teaching of a foreign language, and more precisely French—at the time the language of the nobility, the merchants, and of science throughout Europe—should also be introduced. The suggested innovations were in line with the Enlightenment values Volta had been cultivating from his early student years.

Volta's plan of education impressed Firmian. In May 1775 he decided that all teachers in Como, if they wished to retain their job, had to pass a special exam, Volta alone being exempted.[100] In those same months Volta made a new substantive addition to his credentials as natural philosopher: he built his first widely known and successful electrical machine, the electrophorus, which he described in a brilliant paper for the recently established journal of Milanese philosophers.[101] The electrophorus also impressed Firmian. In a letter to Volta he described the machine as a "superb and extremely useful discovery, doing honor to your Country, and to all of Italy, Mother of the Sciences and the Arts."[102] On the first of November 1775 Volta was appointed—without exam—to the chair of experimental physics introduced on the occasion in the schools of Como, while at the same time retaining his previous job as superintendent. The new chair carried with it a yearly stipend of 800 lire, which was added to the 1,200 lire Volta earned in his previous capacity.[103]

The story told in this section can be summed up briefly. When Volta first focused on natural philosophy in his search for cultural expression and identity, he was following his own individual choice, but within the range of options contemplated by the Jesuits' system of education, further sustained by a fashionable trend within culture at large. When, about twenty-four years old and already an expert in natural philosophy, he joined the cadets from the lesser nobility and bourgeoisie seeking a job in the slowly expanding staff of the "enlightened" state administration, he was following in the tracks of an earlier generation of natural philosophers active in Lombardy and Piedmont, including scholars like Beccaria, Boscovich, Frisi, Carli, and Barletti. The personal development Volta underwent passing from amateur natural philosopher, to expert, to public servant, was inspired, and in turn sustained, by the efforts of the Austrian administration of Lombardy to establish a new, state system of education in which the natural philosophers had an important place.

Volta, in any case, had both the culture and personal ambition to allow him to play an active role in the shaping of the new, public figure of the expert in natural philosophy that was then in the making. In this he invested qualities and traits that stemmed also from his intense emotional life.

## Emotional Life

In his Jesuit teachers' plans, the education young Volta had received in matters concerning body, love, and sex was strict. While still hoping to attract the sixteen-year-old Volta to the Society of Jesus, his philosophy professor had instructed the boy to avoid wasting time in "bodily pastimes" that were common during the long vacations at the villas the Volta family went in for, and which were deemed dangerous for the spiritual health of the boy.[104] Advice on women was also severe. The recommendation given to Volta at sixteen was to avoid exchanging glances with women and to immediately turn away and lower his eyes should it happen.[105] Whatever the immediate results of the recommendation, the long lasting effects fell well short of the Jesuit teacher's expectations. Alessandro's behavior as an adult was in keeping with the joyous lifestyle adopted by Lombard nobility in those days.

During the first half of his long life Volta was, or saw himself as, a peculiar variety of the *cicisbeo* species, which he himself called the "cicisbeo errante"—a cicisbeo not permanently committed to serve a single lady.[106] That married women from the nobility and the rich bourgeoisie often had a "cavalier servente" was one of the traits of eighteenth-century Italian society that at once piqued the curiosity and attracted the sarcasm of local and foreign observers. The custom prospered in the relaxed lifestyle of many noble *palazzi* and *villas*, as portrayed in the poems of a well-known contemporary of Volta's, Giuseppe Parini.[107] As to Volta himself, we already know of his early exposure to the rituals of carnival and feast. This exposure had long

1.2 Alessandro Volta. Engraving by Raffaele Morghen after a portrait by Luigi Sabatelli; from Alessandro Volta, *Collezione dell'opere*, edited by Vincenzio Antinori, 2 vols., (Florence: Piatti, 1816), vol. 1, front ends (*Biblioteca Universitaria, Bologna*).

lasting effects. *Conversazioni, villeggiature*, Arcadian groups, concerts, operas, feasts, and hunting trips in and around Como, Milan and Pavia were frequent and time-consuming activities that continued well into Volta's adult life.

At twenty-five he threw himself wholeheartedly into such entertainment, as evinced from the sketch he drew of a semi serious Académie des Canapés (from the name of a then fashionable sofa) to which he belonged. Membership of the Académie included "Nymphes," as well as Armidas and Rinaldos, from the characters of Torquato Tasso's *Jerusalem Liberated*.[108] During the Carnival of 1772—three years after his debut as a natural philosopher—he was still investing vast amounts of energy in dress and masks. At that stage, however, his enthusiasm was tempered by a dissatisfaction with his hometown, expressed in philosophical jargon in the following letter:

MAKING OF A NATURAL PHILOSOPHER       **35**

Carnival is vanishing. Worse, it keeps me moving about looking for cos-
tumes and masks while everything is drenched by rain. This would be
enough to convince me that ours is not the best of all possible worlds. It
is hard to preserve intact a faith in Leibniz in Como.[109]

Dissatisfaction with his hometown kept Volta in search of "the best of
all possible worlds." During the summer of 1788 he attended *conversazioni* and poetic groups in Verona, Mantua, Cremona, and Milan, a
habit that was to endure into his forties. The "Muses" attending those
gatherings were a high point of the tour.[110] Women played a prominent
part in Volta's life. Love was more deeply rooted in his worldview than
any adhering to the rituals of Arcadia or of the cicisbeo genre would
suggest. When, in 1775, he drew up his educational plan as superintendent to the public schools in Como, he devoted several paragraphs to
the subject of love literature and its use in education. He strongly recommended love poetry and drama, arguing that an acquaintance with
this literature was the best introduction for the younger generations
to the "passion" that played such an important role in life. The new
superintendent—contrary to long-established tradition—thought that
the syllabus should *not* eschew love.[111]

In eighteenth-century Lombardy women from the nobility and the
upper bourgeoisie played an important role in social and cultural gatherings. If the rituals of Arcadia were what was most often talked about,
great store was also set by genuine friendship and lively intellectual
exchange between women and men. One such friendship Volta entertained was with Donna Teresa Ciceri. A member of a well-to-do family
from Como's nobility (in 1774 three Ciceris sat in the nobles' local assembly), five years younger than Volta, energetic and lively, with a series of cultural and practical interests, Teresa Ciceri had the qualities
Volta looked for in the woman he would eventually marry.[112] However,
his decision to marry was to be postponed until he was forty-nine,
while she married at 20. What had been perhaps a love affair, became
a deep, lifelong friendship with occasional episodes of fondness and
veiled jealousy. After his brother Luigi, though at a much deeper level
of trust and intimacy, Teresa Ciceri remained Volta's closest confidant
throughout his life. The combination of friendship and intellectual affinity that linked the two culminated in the campaign Volta launched
to obtain from the Patriotic Society of Milan a prize to celebrate Teresa
Ciceri's achievements in the field of country crafts.[113] The prize, finally
awarded in 1785, was given in recognition of Teresa's invention of a
technique that used the stems of lupines (instead of linen or hemp) to
make ropes and rough clothes.[114] The "industry" that Donna Teresa had

started, Volta proclaimed, had spread throughout the country to the benefit of the whole rural economy.

If we trust the description Volta himself gave at a turning point in his sentimental life, his bachelor days were characterized by a degree of "disorder," "dissipation," and "intemperance."[115] Jealous colleagues entertained more radical views on the matter.[116] Of better documentary worth than his colleagues' gossip is perhaps the fact that one night in Paris in November 1801—when the battery was successfully presented to the Institut and Bonaparte—Volta and Brugnatelli together took the pain to count as many as 112 whores on the streets of the French capital in one night.[117] Lichtenberg, an apparently sympathetic observer, gave the following description of the gallant side of Volta's personality, as he had observed it during Volta's European journey of 1784: "He [Volta] is a nice man, and during several extremely free hours we spent revelling after dinner in my house till one A.M., I realized that he understood a lot about the electricity of girls."[118]

Among all the revelry and entertainment, real love lay in wait, and at least once Volta was entirely won over by it. On that occasion he came close to quitting his job as professor, breaking with his brothers, and entering into a secret marriage. The woman who brought him to contemplate such a step was an opera singer, named Marianna Paris. The affair kept Volta in an exhilarated if troubled mood for about four years in his midforties, from 1788 to 1792, and caused serious tension in his private and public life.

Marianna Paris had been introduced to Volta by a common friend, a countess De Salazar from Milan, in December 1788. The friend asked Volta to support and protect the young singer during her stay in Pavia for the Carnival season—it was part of the patronage system that noble families bestowed their protection on an artist. The countess presented the singer to Volta in rather unflattering terms. In the countess's opinion, however, Marianna as a woman rated better: she had received an honest education, and her character and human qualities were deemed uncommon among theater people, who at that time rated very low in the hierarchy of social respectability. Her beauty, on the other hand, was described as no more than "average," or "even below average."[119] Marianna's activity as a singer is documented from 1785 to 1800. Her repertoire included operas by Paisiello, Cimarosa, and Salieri. A few years before, Mozart had played in Milan for Volta's patron, Firmian.[120] There are records of Marianna's singing in theaters across northern and central Italy. At the peak of her relationship with Volta, in 1792, she called off her touring.[121] Our view of what was probably Volta's deepest passion in his life is based on the letters he wrote on the subject to his best friend, Teresa Ciceri, to his brother Luigi and, as a public servant, to people higher in office, including no less a personage than the emperor himself. To all of them he appealed in the hope they would in the end bestow their approval upon the unorthodox marriage he was, at least for a while, planning to contract. This side of the story may tell something about Volta's attitude toward habits, tradition, and innovation in morals, an attitude which might in turn shed light on his views on tradition and innovation in general.

The strategy Volta adopted in his effort to make his proposed marriage acceptable to relatives and superiors turned around a few key arguments. One emphasized the fact that the moral qualities of the woman he loved "fully compensated for" the inequality of their social status. This argument, often addressed to his archdeacon brother, contained an appeal to Christian values that should, he thought, prevail over mundane concerns. Volta liked to point out that the "spiritual adviser" he had consulted on the matter—and who has been identified by some as the above mentioned Jansenist Pietro Tamburini—recognized the power of the argument far more than his brother, implicitly accused of holding passé religious views. Volta's argument that a true deep love, albeit unconventional, was "a higher good" than a marriage devoid of love but approved by society, ran along similar lines.[122] Another argument used by Volta appealed to what he perceived as changing social habits. Marriages among people of unequal social status, he stressed, were becoming increasingly common, and—especially in large towns like Milan—people no longer paid much attention to them.[123] However, in a small university town like Pavia—where Volta had moved in 1778, and where students were constantly told not to get involved with women from the theater—a university professor married to an opera singer was a cause for scandal. Hence Volta's plans for a secret marriage (a not uncommon measure at the time), or a change of job for something less visible in Milan, or else the possibility of resigning from his professorship and living modestly with his wife on the government pension he had a right to as a civil servant. From about the spring of 1789 to May 1792 Volta pondered such drastic measures, wavering between a determination to go ahead with his plans and a desire not to break completely with his relatives or depart from what was expected of him as an upright civil servant.[124]

The high points of the conflict were as follows. In the autumn of 1789, in view of the apparent determination of Volta to carry out his plans, his brother Luigi brought pressure to bear on Austria's new plenipotentiary minister in Milan, Wilzeck, to intervene. In consequence Wilzeck sent a firm reminder to Volta explaining that the intended marriage was "a risky step," and indeed an unacceptable disorder.[125] One week later Volta complied in a letter (not preserved) which must have amounted to a complete surrender: in the next few days Wilzeck wrote back praising Volta's wise decision. The minister (that is, the Volta family's ally in the cabinet) went so far as to explain to the rebel natural philosopher that in choosing a wife he should comply with his family's and colleague's sentiments, even if they clashed with his "natural feelings."[126]

But Volta's capitulation was merely temporary. One year later, in October 1790, he went back to pondering what his "real good" in the affair might be.[127] In the following spring (May 1791) he made what was to prove his boldest move: he informed the new emperor, Leopold II, of the matter. The petition was written in the old-regime spirit, which accorded the emperor the right and duty of intervening in the private life of his subjects. The petition was also written on the assumption that the Paris family would be known to the emperor. Volta opened with a moving description of the feelings that tied him to Marianna and ended by asking for two concessions: that he might exchange his job in Pavia with one in Milan, and that the emperor provide some special annuity to Marianna's family.[128]

The emperor's answer took a long time in coming. In the meantime, in September 1791, a new step was taken. Unable to obtain his brother's consent, Volta drew up a declaration, to be signed by Luigi himself, in which his brother—while refusing Marianna access to his house, and condemning the marriage as "unequal" and leaving an "indelible blemish on the family"—promised to refrain from further opposition, and "to leave them [Volta and Marianna] in peace." Though not stated, the declaration was also perhaps intended to circumvent the possibility that the disapproved marriage might affect Volta's (and his future descendants') inheritance of the Stampa deed of trust administered by Luigi. The declaration was never signed.[129] Finally the emperor's answer, negative on both counts, reached Volta toward the end of 1791.[130] Volta's attempt to bypass the opposition of his family and Minister Wilzeck by appealing directly to the emperor had failed. It took another five months for Volta to finally abandon his plans of marrying Marianna. Before making the final decision, he negotiated with his brother a sizable donation to be made to Marianna's family as compensation for the entire affair.[131]

Volta's final renunciation and the arrangements he made for Marianna did not settle the dispute with Luigi immediately. His brother continued to believe that Volta's change of heart was not sincere, dictated rather out of necessity than conviction: only by marrying someone else would Volta satisfy him that he had abandoned his plans regarding Marianna. That was to take another couple of years, but this time Volta managed to combine a respectable marriage with an advantageous financial settlement, securing him half of the rents of the Stampa deed of trust administered by his brother and a good dowry.[132] Teresa Peregrini, whom Volta married on 22 September 1794, was the youngest of eight daughters of the "royal delegate" in Como, who had a part in Volta's first appointment as a civil servant twenty years earlier.[133] Teresa was preferred over other matches for her "knowledge, experience, and great ability in running domestic affairs." She was nineteen years younger than Volta, had received a good education, read French and German, and could converse knowingly on topics that included literature, geography, the natural sciences, and matters of "taste."[134] All this, according to Volta, compensated for her lowly rank of nobility.

Her qualities certainly enabled Teresa to share part of her husband's cultural and scientific interests. The few letters they exchanged, however, contain little evidence of Teresa's involvement in Volta's scientific and intellectual life. Perhaps the most significant piece in this regard is Volta's report from Paris to his wife on meeting French scientists and Bonaparte in 1801.[135] Even on that occasion, however, the liveliest reports were addressed to his brother Luigi, rather than to his wife. Indeed, in the first ten years following their marriage in 1794, circumstances often kept Volta away from home. He taught in Pavia (when the university was not closed on account of political troubles) while Teresa and their three sons, born from 1795 through 1798, lived in Como. He was also often in Milan, which he clearly preferred to Pavia. Volta's expedition to Paris and the Consulta at Lyon, with its long Swiss appendage, lasted almost eight months, from 1 September 1801 to 22 April 1802. There is evidence, however, that Volta took a personal interest in the early education of his three sons, teaching them himself—whenever he could—the basic notions of grammar and arithmetic.[136]

## Investigative Style

It was mentioned that Volta's emotional life might be of some help in trying to get a clearer picture of his scientific personality. Controversial though such speculations are among historians of science, the present case offers grounds, I think, for a few remarks.

One such remark concerns the parallelism that is detectable in the attitudes Volta displayed toward tradition and innovation in his private life and his work. It is easy to discover, on both sides, a discernible *tension* between his impulse to break with convention, to risk disagreement and conflict, and a more cautious, realistic attitude guiding him toward consent. In Volta's scientific work, as we shall see in the following chapters, one such episode revealing this tension was his disagreement with Beccaria over the ambitious attractionist theory young Volta had worked out for the entire science of electricity. Another such example was the mature Volta's great dispute with Galvani. Such major episodes of controversy have rarely been credited by biographers with the power of altering what has often been depicted as Volta's conformist scientific personality—a picture apparently corroborated by the evidence of Volta's own carefully plotted career, and by his earnestly sought-after and well-advertised achievements. My point is that an understanding of the tension between tradition and innovation running through Volta's private life well into his mature years helps to detect the otherwise easily overlooked tension that runs through his scientific work as well.

There are other ways in which an understanding of Volta's emotional life may help elucidate his scientific personality. In reviewing Volta's character it is readily observable that his rich social and emotional life intermingled with two of his best-known qualities as a "natural philosopher." First, his aptitude as a lecturer and a performer of electrical experiments clearly owed much to the pleasure he took in social life in general, and to his passion for theater and music especially. Volta's sociable and brilliant character clearly helped him and his science to win consent in the lecture hall, among his peers, in fashionable salons in Lombardy and elsewhere in Europe during his frequent trips abroad. Lichtenberg's succinct testimonial is revealing on this account: "Volta [is] rich in knowledge, and knows how to show it."[137]

Second, Volta's great ability in carrying out delicate operations with electrical apparatus, and especially his skill in detecting weak electricity by means of his own sense organs, owed something to his rich human temper. The way in which, for example, he performed some of his experiments on electricity seems to reflect his frequent contacts with music and musicians.[138] To perceive the affinity between the performances of the eighteenth-century natural philosopher and those of the musician, one needs only to bring to mind Volta's early electrical experiments on silk strings mounted on a violin stick.[139] Indeed, when going through Volta's experiments and laboratory notes one gets the impression that he engaged in experiments body and soul. Remember, in this connection, Volta's independent discovery of the Sulzer effect, produced by two coins of different metals developing electricity detected by the tongue.[140] Or remember the later experiments performed by Volta on electricity obtained from various metals put into contact with different parts of his own body.[141] A joyous exercise of body and mind, further sustained by the emphasis on sensations proper of Enlightenment philosophies, seems to have characterized Volta's private life as well as his investigative style.

Controversial though the establishment of any direct connection between emotional life and scientific work is in the eyes of many historians of science, it is hard to deny that an understanding of Volta's private life gives a better insight into his investigative style and scientific personality.

## Conclusion

The shaping of Volta as a natural philosopher went through several stages. Each stage was characterized by a different cultural model that Volta both partly conformed to and yet partly reacted against in the molding of his individuality.

When he was about sixteen, the models he was presented with were naturally those offered by his family, and those taught by the Jesuit College he attended. He conformed to the models closely enough to raise expectations of his joining the Society of Jesus, following a pattern of service to the church that was rooted in his family tradition. Yet, at the same time he displayed a degree of resistance to that tradition when he chose, for his early philosophical readings, classical and Enlightenment authors whose ideas were at odds with the religion he was brought up in. In the end, he disappointed his Jesuit teachers by deciding not to enter a religious order and by writing a poetical composition in which he exalted the ideals and the natural philosophy of the Enlightenment.

Volta thus entered the ranks of the literary and philosophical amateurs which counted many adepts in the Italian provinces. And yet, already in his unpublished juvenile poem he chose topics (chemistry and electricity) and a genre (the didactic poem) that revealed his determination to occupy a distinctive place among the amateur *letterati*. The process of self-education as an amateur went ahead quickly, or so it seemed to Volta: dissatisfaction with the amateur stage of his interest in natural philosophy and a drive to imitate the experts ran parallel with the frustration he felt as a cadet member of the lesser nobility living in a narrow provincial setting, and having no direct control over the rents he lived on. While conforming to the habits and lifestyles of his rank and acquaintances in Como and its surroundings, he soon started to look beyond. Still only eighteen he began to contact the leading experts in the field of electricity in Turin and Paris, in an effort to achieve recognition as an expert in that branch of natural philosophy.

The aspiring-expert stage of his life lasted much longer than the amateur stage had. In Volta's mind, his amateur status must have drawn to an end when he realized, however uncertainly, that his views on electricity were obtaining the attention of experts like Nollet and Beccaria around 1763 or 1764. The next stage lasted until 1769, when Volta published his first, ambitious essay "on the attractive force of the electric fire," to be discussed in chapter 3. With this publication he qualified—at least subjectively and, to some extent, also in his peers' opinion—as an expert in the theoretical issues of the science of electricity, however little recognition the essay itself received from electricians.

Soon after the 1769 publication, Volta felt ready to start a career in the state education system launched by the enlightened administration of Lombardy. His aspirations to move from expert to professor and public servant was prompted by many factors, individual and social. Past generations of Voltas serving as "magistrates" in the local administration provided a model. Volta's own relatively subordinate role within his family, with its precarious finances, provided him with another reason for searching for a recognized social position and an independent source of income. In pursuing his ambition to take up public office, Volta was following in the tracks of an earlier generation of natural philosophers who had joined the public administration of Lombardy. Volta's commitment to the state reform movement was in keeping with the Enlightenment spirit he had first assimilated through his independent readings as a student. As in the case of that earlier generation of natural philosophers, Volta's commitment to public office matched the state administration's proclaimed support of natural philosophy and the "useful arts." However, it took about six years, and a considerable amount of self-adjustment, to convert Volta's ambitions and ideals into reality. In the process, Volta learnt that merely qualifying as an expert, enjoying the support of his peers, and sharing in the ideals of the Enlightenment were not enough to win a position in the state administration: the patronage of the appropriate political authorities was also needed. When in 1774 Volta finally obtained office in the state educational system it was not the chair he had hoped for at the celebrated University of Pavia. It was, instead, a job in Como, which he finally managed to exchange for the chair in Pavia in 1778. From then on Volta—together with renowned colleagues like Lazzaro Spallanzani, Antonio Scarpa, and Carlo Barletti—conformed to, and at the same time helped set a standard for, the model of the zealous university professor and public servant of Austrian Lombardy. With the example his scientific work, correspondence, and travels set, and with the support of his government, Volta circulated that model throughout Europe.

At no stage did Volta's acceptance of the models and career patterns he was pursuing go without a degree of conflict, reservation or dissent on his part, as is well exemplified by his views on religion and his private life. A tension between tradition and innovation, a moderate criticism of accepted habits and a desire to innovate, can be perceived throughout Volta's life, if one looks beneath the surface of his public attitudes and behavior. Volta's emotional life is especially instructive in this regard. It throws light on parts of Volta's personality that are otherwise easily overlooked. Sociability, a joyous lifestyle, deep passions—all contributed to the making of the rich personality that colleagues like Lichtenberg saw in Volta the man *and* the natural philosopher. His skills as lecturer, talker, and brilliant performer of experiments were naturally seen by contemporaries as woven of the same stuff as his personality. Indeed, the emotional side to Volta's personality helps understand how closely body and mind, passion and intellect interacted in the operations of experimenters on electricity, in ways reminiscent of the best musicians and skilled artisans of the late eighteenth century.

At each stage in the making of Volta as a natural philosopher, the tension between his willingness to conform to a particular model and yet reject parts of it was a key factor in the shaping of his individuality. At a later stage, as we shall see, a tension between Volta's own ambition to be regarded as a theoretical natural philosopher and the inclination of many contemporaries to regard him rather as a brilliant inventor of electrical machines will similarly affect the shaping of Volta's science and the assessment of his merits within the community of late eighteenth-century electricians.

# Enlightenment Science South of the Alps

## THE ITALIAN SCIENTIFIC COMMUNITY
## IN THE AGE OF VOLTA

The vast amount of scholarly work produced in recent decades on eighteenth-century Italy has kept natural philosophy and scientists at the edge of the stage. The center has been occupied by issues and personalities that are supposed to have been more directly involved than "natural philosophers" in the reform of the public administration carried out in several Italian states in the age of the Enlightenment. Historians have displayed a predilection for the intellectuals who, having a background in political philosophy, law, history, or literary studies, joined in the modernizing efforts of enlightened governments midway in the course of the century.[1] Studies devoted to scientists and science, on the other hand, have focused mainly on single personalities and institutions,[2] or on major trends in the history of scientific ideas, such as the heritage of the scientific revolution, the diffusion of Newtonianism, the controversy over Galvanism, and developments in special fields like mathematics, astronomy, chemistry, geology, or the life sciences.[3] No overall assessment of the people and resources involved in scientific activity south of the Alps in the age of the Enlightenment has been attempted so far.[4]

One reason for avoiding overall assessments of the kind mentioned has been the awareness that, strictly speaking, no Italian scientific community existed in the eighteenth century. Due to fragmentation of the peninsula into several states, often under foreign rule, Italian natural philosophers were members of, or depended for their work on, ruling classes with markedly different traditions, and they worked in institutions modeled on a variety of different patterns. Yet, those same scholars shared a common language. Many studied, traveled, and worked successively in different states of the peninsula. They often contributed to the same scientific journals, and they occasionally proclaimed themselves part of an ideal community of Italian scientists, as exemplified

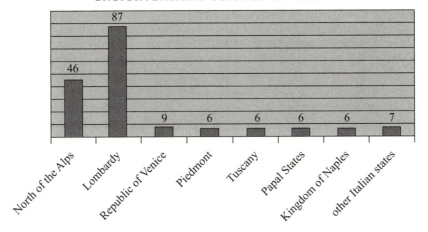

2.1. Volta's correspondents, 1761–96.

by the Italian Society of the Sciences, established in the 1780s and grouping forty scientists from the different states. Awareness of, and interest in, a common Italian cultural heritage—in which language, literature, religion, science, and the arts all played a role—was indeed occasionally expressed. In the case of natural philosophers, that awareness nurtured and was nurtured in turn by the frequent exchange of ideas, books, and instruments which accompanied scientific activity. However, allegiance to the local government and local interests remained predominant for the natural philosophers, just as for other citizens, throughout the century.

Under these circumstances, and given the presence of a dozen capitals and at least as many centers of scientific activity scattered throughout the peninsula, the exchange of knowledge, people, and instruments associated with scientific work could conform to a wide variety of patterns. Though Volta's case cannot be regarded as typical without qualification, it can offer a first glimpse into a complex scenario. The example is appropriate to remind us within what narrow limits an Italian scientific community—as distinguished from the social settings offered natural philosophers by the state they belonged to on the one hand, and by the wider, cosmopolitan Republic of Letters on the other—existed in the period under consideration.

A survey of Volta's extant correspondence[5] allows us to make a comparative estimate of the amount of attention a scientist like Volta paid to scientific work carried out in his own country—the Duchy of Milan under Austrian rule—in the other Italian states, and north of the Alps.

In the period of political stability from the 1760s to the spring of 1796, about two-thirds (87 out of 127) of Volta's Italian correspondents were from the Duchy of Milan (see fig. 2.1). The rest of the Italians (40) were almost equally distributed between the Republic of Venice (9 correspondents), the Kingdom of Piedmont-Sardinia (6), the Grand Duchy of Tuscany (6), the Papal States (6), and the Kingdom of Naples (6). One or two additional correspondents came from each of five other small states of the peninsula.

Comparing these numbers with those of Volta's correspondents from north of the Alps in the same period, one finds that Volta's non-Italian correspondents (fig. 2.1, first column) outnumbered those from the Italian states other than the Duchy of Milan by 46 to 40. Indeed, in terms of the quality and quantity of the letters, few Italian scientists ever matched up to Volta's foreign correspondents such as Priestley, Saussure, De Luc, Senebier, Magellan, Lichtenberg, van Marum, or Banks. Apparently, Italian scientific circles outside the Duchy of Milan did *not* occupy the forefront of Volta's concerns, a fact further confirmed by the limited extent of his travels in the peninsula compared to his extensive European tours.

On the other hand, Volta *did* build on the work of Italian scientists from outside the Duchy of Milan on crucial occasions throughout his career. The examples of Beccaria from Piedmont in Volta's young years, and that of Galvani, from the Papal States, for his mature age, will suffice in this connection. Furthermore, historians agree that the kind of support Volta's work enjoyed at the University of Pavia was the result of an expansion in facilities for education and research that, though more prominent in Austrian Lombardy than elsewhere, affected other Italian states as well. Thus, when setting Volta in his proper context, a survey of the scholars involved in scientific activity in the Italian states and the institutions and resources they relied on, is important, as is the assessment of Volta's own roots in Austrian Lombardy and of the European network of his correspondents discussed in chapter 5, below.

The present chapter will attempt a collective portrait of the people engaged in scientific activity in Italy in the period from 1770 to 1795. This picture is based on the study of a sample of seventy-four authors, selected from among those who published original papers in the eleven academic proceedings and scientific journals that came out in the peninsula during the period under examination. Publication in these periodicals—which had attained by then a significant, recognized degree of specialization within intellectual circles—can be regarded as the safest ground for the identification of the people engaged in scientific activity at a time when scientific work had not yet attained independent, professional status. Information on the social background, education, positions, and careers of the scholars selected will be given. This information confirms the patchy professional and social identity of the people engaged in scientific work during the period. It also offers hints, however, on some trends that were under way when the Napoleonic wars came to reshuffle educational and scientific, as well as political, institutions.

Judging from our sample of authors, in the Italian states the figure of the expert in natural philosophy—the specialist *fisico* ("physicist") and the generalist *filosofo* ("philosopher") in eighteenth-century Italian phraseology, who were to become the major components of the collective *scienziato* ("scientist") in the following century—was emerging from within social strata and institutions that were molded overwhelmingly by the concerns of the state administrations. As far as original research was concerned, the impulse given by Enlightenment culture to the recognition of the expert in natural philosophy as a new, public figure, bore fruits above all within the state-run universities, the technical corps of the public administrations, and the traditional professions like medicine. The wider audiences that were being attracted to natural philosophy by the circulation of Enlightenment literature, were apparently unable to find outside the state administrations new opportunities and careers enabling them to practice scientific work on a regular basis. A persistently stagnating economy and long-term neglect of popular education seem to be the most plausible causes for explaining a situation that, though uneven, applied to all the states of the peninsula.

To add breadth to our survey of the Italian scientific community in the age of Volta, a number of additional insights into the situation will be offered along with discussion of our group of seventy-four authors. One insight is the view of an insider, mathematician and engineer Paolo Frisi (whom we already met as an adviser of young Volta in chapter 1, above), reporting on the situation of Italian (scientific) "literature" in 1771. Because of the broad, synthetic views it contains, Frisi's report will be used to provide a general introduction to the more detailed discussion of our sample group of authors. This latter will be followed in turn by a survey of the private libraries of three distinguished personalities of the time, documenting the circulation of the major works of Enlightenment natural philosophy south of the Alps. A number of outsider's views of the scientific work carried out in the peninsula will conclude the chapter. They are taken from the reports that foreign natural philosophers, traveling through the Italian states and visiting their scientific institutions, wrote in the period examined.

The material discussed in this chapter cannot provide the long overdue and still unfinished overall assessment of scientific activity in Italy in the age of the Enlightenment. We hope, however, to have pointed out some seldom-tried lines of investigation that can bring that goal closer.

## The Soil and the Institutions

The treaty ending the War of Austrian Succession, signed in Aix-la-Chapelle in October 1748, established a settlement that, as far as Italy was concerned, lasted until the mid-1790s. The treaty ensured "the longest period of peace known in modern Italian history."[6] During that period Italy comprised a dozen states, including two kingdoms (Piedmont, under the French-speaking House of Savoy, and Naples and Sicily, under the Bourbons of Spain, Sicily having its special viceroy), one grand duchy (Tuscany, under the House of Lorraine, linked to the Habsburgs), three duchies (one being Milan with Lombardy, under Austria; another grouping Parma, Piacenza, and Guastalla under a Bourbon; the third being Modena, close to the Habsburgs), three independent republics (Venice with its region, Genoa, the lilliputian San Marino), and the Papal States. The long period of peace came to an end with the arrival of the French troops led by General Bonaparte in April 1796. The event marked the beginning of a period of political unrest, institutional reform, and cultural ferment, which affected scientific circles and ended only with the restoration following the Congress of Vienna in 1815. When considering the Italian scientific community in the age of Volta, it seems appropriate to treat the period from the beginning of Volta's public activity, around 1770, up to 1795 separately; subsequent developments will be discussed in chapter 5, below.

How Italian scientists perceived the interconnection of political fragmentation and cultural activity in the peninsula in the early 1770s is well illustrated in a manuscript report written in December 1771 by Paolo Frisi.[7] A leading mathematician, engineer, and consultant himself in several Italian states, and a foreign member of the academies of Paris, Berlin, Saint Petersburg, and the Royal Society of London, Frisi was then in charge of many educational and technological committees in Milan, and was one of Volta's early correspondents and patrons.

Frisi's report treated Italy as a cultural unity and Italian literature in the sciences as a single body. He could not, however, avoid stressing the extreme fragmentation of cultural and educational institutions in the peninsula. With its dozen states, seven major universities, and correspondingly high number of academies, astronomical observatories, natural history collections, physics cabinets and botanic gardens, Italy offered more—in terms of sheer variety—than other countries of comparable size. Yet, the lack of a center able to attract the best minds and energies of the peninsula and to support ambitious initiatives was to Frisi, who had lived in Paris and Vienna for a while, a serious drawback for Italian science.

Following Frisi's survey, the most promising initiatives for the development of scientific institutions were those taken in Austrian Lombardy by Frisi's own patrons, Vienna's Ministers Kaunitz and Firmian. Frisi described such initiatives as constituting an actual "revolution," giving new and unprecedented support to literary, scientific, and educational pursuits. The turning point of the revolution, as Frisi viewed it, consisted in having put the scholars themselves in charge of educational and scholarly institutions. Historians agree that that trend was part of a policy, carried out by Vienna and imitated by other governments, meant to diminish the power of the local nobility and the clergy, and to transfer it to officers directly responsible to the ministers appointed by central governments. It was part of that policy to attract leading or promising intellectuals from other Italian states to Lombardy: Frisi mentioned Boscovich, who had come from Rome, Spallanzani, from Modena, and himself, from Tuscany. As evidence of the results attained, Frisi mentioned the flourishing of the University of Pavia, the well-known works of Boscovich and Spallanzani, the new astronomical observatory in Milan, and the recent research of Volta and Barletti on electricity.

Compared to Milan—where Frisi had met with the best rewards for his achievements—other Italian centers fared less well in his report. Turin had greatly suffered because of the departure of the mathematician Giuseppe Luigi Lagrange, who had left for the Berlin Academy in 1766 never to return to Turin, and because of the death of the naturalist Vitaliano Donati in Bassora during a scientific expedition. The efforts of the Turin Private Society, later the Turin Royal Academy of the Sciences, and of individual scientists like Beccaria and Cigna seemed to Frisi less than adequate for the restoration of Turin's prestige as a leading scientific center.

In the Venetian Republic, which included the old University of Padua, Frisi singled out the mathematician Anton Maria Lorgna from Verona, for his contributions to algebra and hydraulics. Lorgna was also promoter of the already mentioned Italian Society of the Sciences. Frisi especially approved of the initiatives of the Mantua Academy that had established prizes able to foster the competition of scientists from northern Italy. In the case of Venice he mentioned the traditional printing industry, quoting among its recent productions the *Dictionary of the Arts and Crafts* by Francesco Griselini.[8]

Bologna, with its old university, the Academy of the Sciences established in 1714, the astronomical observatory ("the oldest in Italy"), the anatomical theater, a rich library, and a fine-arts academy, fared well in Frisi's account. He regarded the eight-volume proceedings of the Academy as the major scientific work published in Italy during the century. He spoke highly of the ninth volume then being printed, perhaps because it included a paper by himself. He valued, above all, the contributions to hydraulics by mathematicians and engineers like Eustachio and Gabriele Manfredi, who continued the tradition laid down by the celebrated Domenico Guglielmini. Frisi, a Barnabite who nonetheless supported the campaign of enlightened governments for greater independence from the church, liked to depict Bologna as being the one place in the Papal States where scientific pursuits could prosper, *unlike* in Rome.

Tuscany, where Frisi had taught from 1756 through 1764, and which had let him go, received cautious treatment. The University of Pisa had recently lost (besides Frisi himself) a long list of professors, only partly compensated for by the appointment of Giuseppe Antonio Slop, who was then launching Pisa's astronomical observatory. Florence boasted an important collection of machines kept in the physics cabinet of the grand duke, under the supervision of Felice Fontana. The collection could "be regarded as one of the finest in Italy." Attached to the cabinet a number of professorial chairs had been established, and the mathematician and engineer Leonardo Ximenes had distinguished himself among the holders. In Siena, the local Academy of the Sciences had published four volumes of proceedings, including an important memoir on irritability by Felice Fontana.

Frisi, in his brief survey, portrayed the scientific pursuits carried out in southern Italy as focusing mainly on the study of the geological phenomena and volcanism that characterized the region. He paid special homage to Sir William Hamilton, Britain's ambassador in Naples, who contributed to the geology and archaeology of the region.

The highlights of Frisi's survey were, basically, the universities, and the measures taken by enlightened governments in their effort to limit the role of the church in education. According to present-day historians, Italian universities in the eighteenth century—the seven major ones being Turin, Pavia, Padua, Bologna, Pisa, Rome, and Naples—were unevenly recovering from the ills of provincialism, low enrolments, and inbreeding which had affected them in the course of the previous century[9] as a result of competition from private (especially Jesuit) colleges and the declining number of foreign (non-Catholic) students and professors. A first wave of measures aimed at a reform of university education had taken place during the first half of the eighteenth century in Piedmont, the Venetian Republic, and the Papal States.[10] A second wave of reforms was enforced by enlightened governments in midcentury.[11] The new reforms implied, basically, the adoption of more rigorous criteria in the appointment of professors, a tighter control over teaching duties as well as over students' curricula, and the enforcement of more severe rules for access to the professions. All the measures were meant to reinforce the role of the state in education, and to provide trained individuals for the slowly increasing ranks of public servants needing technical expertise.

Where the governments' new provisions were implemented effectively, recovery of the university could be impressive as in Pavia, where the number of students grew from 150 around 1750 to 1000 in 1788.[12] In Turin graduates increased from 60 in 1730 to 150 in 1786.[13] Achievements were less impressive elsewhere. In Bologna, where the reforming effort at the beginning of the century had concentrated on the Academy of the Sciences rather than the university, (estimated) enrolments declined from 600 in 1740–41 to 300 in 1796–97.[14] However, the degrees awarded annually in "medicine and philosophy"—the curriculum closest to scientific pursuits—rose in Bologna from an average of 6 in the 1720s to 11 in the 1780s.[15]

The scantiness of data available on other universities discourages generalizations of a quantitative kind. There is evidence, however, that the reforming trend did spread to other towns and states. In the 1770s even small universities like Modena and Ferrara were able to attract good professors in the sciences.[16] Initiatives aimed at reform of the university and the establishment of new educational and research facilities were also taken, with varying degrees of success, in Naples during the same period.[17]

Side by side with the universities—with their appendages of observatories, cabinets, botanic gardens, and museums—another source of income and achievement for the professors trained in the sciences was given pride of place in Frisi's report: their role as experts in technical undertakings in which governments increasingly engaged in to favor agriculture, commerce, and manufactures. Relying on his own experience, Frisi mentioned road development and hydraulics above all. As examples of these developments he cited the new road crossing the peninsula from Modena to Tuscany through the Appennines, and the large number of studies on the water systems of the Po valley in which several experts from the north and the center of the peninsula were engaged throughout the century.[18]

The ferment accompanying the government initiatives mentioned so far affected audiences beyond the professors and the officers directly involved in education and public administration. As Frisi made plain, the spread of Enlightenment literature and a wider circulation of books were themselves contributing factors to the scene outlined in his report. We will revert to this when surveying a number of private libraries, below. It is now time to illustrate our prosopography of the natural philosophers who sustained the "revolution" sketched in Frisi's report.

### The Scholars: Provenance and Fields of Interest

Selecting a number of scholars and the literature relevant for the study of a scientific community always carries with it a degree of arbitrariness. This is especially true when dealing with an epoch characterized by a relatively loose organization of research, the proliferation of the press, and the emergence of the independent writer as a new, professional figure, often displaying an interest in the sciences. At a preliminary stage of my investigation, I went through fifteen periodicals, published in Italy between 1770 and 1795, all of which dealt with scientific issues at one time or another. It soon became clear, however, that the periodicals divided into three distinct categories: one included the proceedings published by the academies; another comprised a few, recently established, independent (i.e., not linked to an academy) journals, specifically devoted to the sciences and/or medicine; the third included the magazines devoted to "literature, sciences, and the arts," often made up of news, reviews, and articles taken from other (often foreign) journals. As it turned out, only the periodicals falling under the first two categories published *original* papers as a norm. Considering that the proceedings of the academies and the new independent journals had by then attained the status of major publication and communication channels among natural philosophers, I chose them as a basis on which to proceed in the identification of a significant group of authors who could be regarded as established members of the Italian scientific community in the age of Volta.

The periodicals thus selected include eleven titles.[19] Six are proceedings of the major academies scattered throughout the peninsula (Turin, Padua, Mantua, Bologna, Siena, Naples); one is the periodical published by the Italian Society of the Sciences, based in Verona; the remaining four are independent journals that all came out—significantly—in Austrian Lombardy. Some of the periodicals excluded because they did not publish original papers would themselves, of course, deserve consideration in a survey of eighteenth-century scientific culture.[20] They would not, however, add significantly to the sample of natural philosophers contributing original papers to the field we are interested in.

The further criteria guiding the choice of scholars contributing to the selected journals, and included in our sample, should be stated at the outset. Some of the periodicals in our list were published during only part of the period 1770–1795; others came out regularly in a large number of issues; still others came out irregularly. No simple, equal representation criterion offered itself for selecting the most active authors. Thus, when choosing the scholars for the sample, a flexible threshold for admission was adopted. When a single or a few volumes were published, all the contributors were included; when many volumes came out, only authors who contributed two to five articles (depending on the number of volumes published) were included. The adoption of this compensating measure was aimed at safeguarding the geographical representativeness of the sample. One consequence of the measure, however, is that no quantitative assessment of scientific production in the different Italian states can be made on the basis of the sample.

Fields of competence or disciplines posed other problems. The periodicals of the time were rarely devoted to single, special fields. In our list, only two independent journals explicitly covered relatively limited fields of study (medicine, and chemistry coupled with natural history). All other periodicals were encyclopedic in coverage, and the articles they published were seldom divided into separate sections. When there were sections, they were rarely the same in different journals. At the cost of including some authors who dealt with topics in the natural sciences (in the modern sense) only occasionally, I have refrained from applying modern disciplinary boundaries in selecting the scholars for my sample. This means that the scholars included belonged to the wide spectrum of competence cultivated by the typical late-eighteenth-century academies.

One selection criterion adopted was age. The aim of the study being a collective portrait of Italian scientists in the age of Volta, I have excluded scholars whose age put them outside the span of about two generations preceding and following Volta's own generation. Among the scholars present as authors in our eleven periodicals, and in line with other inclusion criteria, the few who were born before 1715 or after 1765 have been excluded (the major exclusion being Boscovich). The rationale for this is the assumption that scholars in the generation of Volta (who was born in 1745) were likely to interact, in the course of their working life, with those in the age group twenty through seventy.

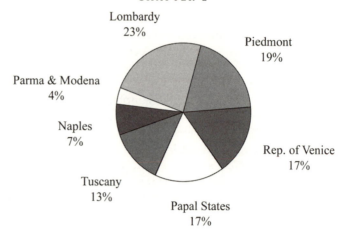

2.2. States where the scholars in our group spent most of their lives.

The resulting sample is a group of seventy-four scholars,[21] 23 percent of whom was born in the 1740s like Volta. The group is large enough to avoid the limitations of the traditional "great men" approach. Lagrange, Spallanzani, Beccaria, Frisi, Galvani, and Volta himself—the authors most often referred to by historians who adopt such an approach—are of course included in the sample we shall now discuss; but they are not isolated figures. They emerge, rather, from a wider network of scholars whose social background, education, interests and careers can add to our understanding of the making of Enlightenment science in Italy.

Considering the states where the authors in our group spent most of their life (see, however, the remarks on movement across the states, below, and the caution about productivity, above), they represent eight Italian states (fig. 2.2 above): 23 percent came from Lombardy, 19 from Piedmont, 17 from the Republic of Venice, 17 from the Papal States, 13 from Tuscany, 7 from Naples, 4 from the duchies of Parma and Modena.

While a discussion of social backgrounds, education and careers is reserved for the next section, it is worth taking a glimpse at the fields of interest represented in our group (in fig. 2.3).

Bearing in mind that no strict disciplinary boundaries in the modern sense existed, and that cross-field mobility was the norm rather than the exception, it is interesting to note that a large proportion of our scholars (31 percent) were active in the traditional area of competence ("mixed mathematics") which included mathematics, astronomy, hydraulics, engineering, and geography (fig. 2.4, p. 56). A large percent-

| | Astr. | Math. | Eng. | Chem. | Phys. | Galv. | Med. | Bot. | Agr. | Nat. Hist. | Econ. | other |
|---|---|---|---|---|---|---|---|---|---|---|---|---|
| Aldini | | | | | • | • | | | | | | |
| Allioni | | | | | | | | • | | | | |
| Amoretti | | | | | | | | | | | | • |
| Arduino | | | | | | | | • | • | | | |
| Balbo | | | | | | | | | | | • | |
| Barletti | | | | | • | | | | | | | |
| Bartalini | | | | | | | | • | • | | | |
| Beccaria | | | | | • | | | | | | | |
| Bonati | | • | | | | | | | | | | |
| Bonioli | | | | | | | • | | | | | |
| Bossi | | | | | | | | | | | | • |
| Brugnatelli | | | | • | | | | | | | | |
| Brugnone | | | | | | | • | | | | | |
| Brunelli | | | | | | | | • | | • | | |
| Buniva | | | | | | | • | • | • | | | |
| Buzzi | | | | | | | • | | | | | |
| Cagnoli | • | • | | | | | | | | | | |
| Caldani | | | | | | | • | | | | | |
| Caluso | • | • | | | | | | | | | | |
| Canovai | | • | | | | | | | | | | |
| Canterzani | • | • | | | | | | | | | | |
| Carburi | | | | • | | | | | | | | |
| Carradori | | | | | • | • | | | | | | |
| Casali | | • | | | | | | | | | | |
| Cavolini | | | | | | | • | | | • | | |
| Cesaris | • | | | | | | | | | | | |
| Chiminello | • | | | | | | | | | | | |
| Cigna | | | | | • | | | • | | | | |
| Cotugno | | | | | | | • | | | | | |
| Dana | | | | | | | • | • | | | | |
| Fabbroni | | | | • | | | | | | | | |
| Fergola | | • | | | | | | | | | | |
| Fossombroni | | | • | | | | | | | | | |
| F. Fontana | | | | • | | | | | | | | |
| G. Fontana | | • | | | • | | | | | | | |
| Fortis | | | | | | | | | | • | | |
| Franceschinis | | | • | | • | | | | | | | |
| Frisi | | • | • | | • | | | | | | | |
| Galvani | | | | | | • | • | | | | | |
| Giobert | | | | • | | | | | • | | | |
| Giovene | | | | | • | | | | • | • | | |
| Girardi | | | | | | | • | | | | | |
| Lagrange | • | • | | | | | | | | | | |
| Lorgna | | • | | | | | | | | | | |
| Malacarne | | | | | | | • | | | | | |
| Malfatti | | • | | | | | | | | | | |
| Matteucci | • | | | | | | | | | | | |
| Mondini | | | | | | | • | | | | | |
| Morozzo | | | | • | | | | | | | | |
| Moscati | | | | | | | • | | | | | |
| Mozzoni | | | | | • | | | | | | | |
| Napione | | | | | | | | | | | • | |
| Oriani | • | | | | | | | | | | | |
| Pini | | | | • | | | | | | • | | |
| Poli | | | | | • | | | | | | | |
| Presciani | | | | | | | • | | | | | |
| Riviera | | | | | | | • | | | | | |
| Robilant | | | | | | | | | | • | | |
| Rosa | | | | | | | • | | | | | |
| Rossi | | | | | | | | | | • | | |
| Rubini | | | | | | | • | | | | | |
| Saladini | | • | | | | | | | | | | |
| Saluzzo | | | • | • | | | | | | | | |
| San Martino | | | • | | • | | | | | | | |
| Scarpa | | | | | | | • | | | | | |
| Scopoli | | | | • | | | | | | | | |
| Spadoni | | | | | | | | | • | • | | |
| Spallanzani | | | | | | | | | | • | | |
| Stratico | | • | • | | | | | | | | | |
| Toaldo | • | | | | | | | | | | | |
| Vasco | | | | | | | | | | | | • |
| Venturi | • | • | | | | | | | | | | |
| Volta | | | | | • | • | | | | | | |
| Ximenes | | • | • | | | | | | | | | |

2.3. Fields of interest. Please note: Only the main fields of interest for each scientist have been considered. "Agron." means agronomy; "Econ." means economics in its broad sense; "Eng." (engineering) includes hydraulics, technology in general, and geography; "Galv." means galvanism; "Med." (medicine) includes veterinary and zoology; "Nat. Hist." (natural history) includes mineralogy and geology; "other" applies to the few journalists in our group.

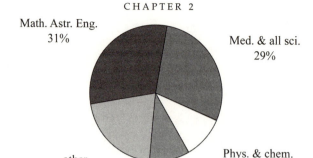

Math. Astr. Eng.
31%

Med. & all sci.
29%

Phys. & chem.
10%

Nat. Hist.
10%

other
21%

2.4.  Fields.

age (29 percent) devoted themselves to another traditional field: medicine and the allied sciences. On the other hand, the two areas including physics and chemistry, and the earth sciences and natural history—areas undergoing promising developments in the period under consideration—attracted no more than about 10 percent each of our scholars. The numbers point to the conservatism of the educational system and of the institutions supporting science throughout the peninsula, a fact I shall return to later.

### Prosopography

Prosopography is not an easy pursuit in the Italian context. The kind of fascination with biography and the "life and letters" genre that affected Victorian Britain won few converts south of the Alps. The Italian equivalent of Leslie Stephen's *Dictionary of National Biography* was started in 1960. Invaluable though the *Dizionario Biografico degli Italiani* is,[22] and despite the 57 volumes published, it does not as yet go beyond letter G. Earlier attempts, though complete, are uneven in quality. Moreover, Italian biographers seem to have disregarded some of the obvious but essential information that any British or American biographer would include, such as detailed information on social background and education received. In what follows I have tried to make the best out of the biographical repertoires, monographs, articles, and obituaries available.[23] However, in some cases (which have been pointed out) percentages refer only to part of the seventy-four scientists making up the group.

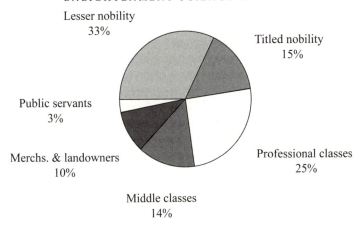

Lesser nobility
33%

Titled nobility
15%

Public servants
3%

Merchs. & landowners
10%

Professional classes
25%

Middle classes
14%

2.5. Social background.

### Social Background

Considering the social status of the families in which our scientists were born, the nobility still constituted the lion's share (fig. 2.5): putting together the titled nobility and the lesser nobility they made up 48 percent of the individuals whose father's status is known (59 out of 74). However, the titled nobility itself was only 15 percent. Moreover, the lower end of the lesser nobility (itself amounting to 33 percent) included people who had earned a degree in law that, should their income and status decline as a result of the frequent divisions of the family estate, could be turned into a new source of income, bringing them into the professional classes.

The professional classes themselves figured significantly in the social background of the group, a fact linked to the high number of towns scattered throughout the peninsula. Many physicians and lawyers, together with a few architects, pharmacists, and accountants, made up 25 percent of the fathers of our scholars. The merchants and the few landowners with no ascertained nobility, on the other hand, made up 10 percent together, while the military and the state public servants totaled a meager 3 percent. The remaining 14 percent included the lower middle classes and a few artisans.

The most common social background of our scientists therefore was the lesser nobility and the professional classes. The very limited number of public servants in the sample requires comment: in the old system of noble magistratures prevailing in the Italian states, most of the positions connected with public administration were held by the nobles, as we have seen in the case of the Voltas. On the other hand, some

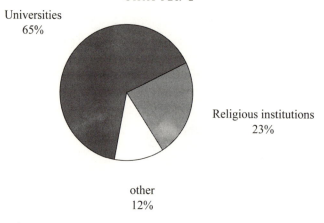

Universities
65%

Religious institutions
23%

other
12%

2.6. Education.

of the professional people held permanent or temporary positions in the courts scattered throughout the peninsula, their status coming closer to that of public servants rather than professionals.

In any case, the data seem to indicate that, in the generations preceding those active in the period from 1770 to 1795, employment by the state, independent of the old system of noble magistrates, was not common among the well-to-do south of the Alps. This changed over the next few generations, the natural philosophers themselves contributing to the change.

## Education

The higher education received by the members of our group (data being available for 65 out of 74) indicates that the universities represented the most frequent educational background for the scientists active in Italy in the period under consideration (fig. 2.6). The percentage of people in our group who attended university, in most cases earning a degree, is 65 percent, the most popular faculties being medicine and law. The seven major universities already mentioned educated more than half of the scholars in the group.

The second most popular kind of higher education was that offered by religious institutions, in which 23 percent of our scholars were educated. These institutions included a variety of schools and colleges run by religious orders or directly by the church. Given the decline of the Jesuits in the 1760s and their suppression in 1773, the number of scholars entirely trained within Jesuit institutions was very small (only two cases documented) compared to earlier periods. However, several (six cases documented) of those who attended university received at least part of their earlier education from the Jesuits.

Other religious orders fared better than the Jesuits at that time. The schools of the Piarists—a teaching order established at the end of the sixteenth century—trained, from start to finish, three of our scientists, all of repute: Beccaria, Barletti, and Gregorio Fontana. The Barnabites, another teaching order, were responsible for another three, including Paolo Frisi.

No doubt the formal and informal network of educational facilities offered by Catholic institutions still played an important part in the making of future scientists. Moreover, as we shall soon see, members and former members of the church still held many positions inside the universities. Yet, the fact that several religious orders now competed (often fiercely) among themselves in education, and that the great majority in our group were trained in universities increasingly under state control, made a significant difference.

The policy of enlightened governments was beginning to bear fruit. That does not mean, however, that the curricula offered by the universities were themselves closer to the interests and needs of students oriented towards the sciences. Should the situation have been otherwise, we would be unlikely to find some of the outstanding scientists in our group (including Beccaria, Frisi, Volta, Lagrange, Barletti, Venturi) among those who had *not* attended university. Apparently—at least until some of these men were themselves offered university positions—the universities were not offering the best education available in the sciences. In fact, when our scientists were trained, the main concern of Italian universities was (and would long remain) the traditional professions rather than the sciences.

### Positions and Careers

The great majority of the scientists in our group (72 percent) were professors, their main source of income coming from teaching in the universities or in other public or private educational institutions (fig. 2.7, p. 60). The universities alone provided jobs for 57 percent of them at one stage or another of their careers.

Compared to teaching, the other niches usually available to scholars oriented towards the sciences did not amount to much. Considering together those who, throughout their life, worked as public servants in various technical, administrative, or military capacities, or had jobs as directors and assistants in public institutions not directly devoted

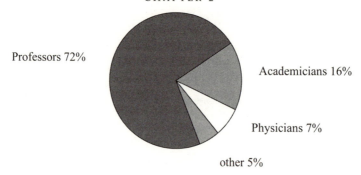

Professors 72%

Academicians 16%

Physicians 7%

other 5%

2.7. Positions.

to education, including scientific academies and astronomical observatories, they amounted to 16 percent. It should be noted, however, that many professors routinely acted as consultants in technical undertakings the public administrations were engaged in: roads, hydraulics, mines, gun and nitre production, etc. Moreover, it should be remembered that in the period immediately following the one under consideration many professors were lured into various offices in the new institutions of Napoleonic Italy.

Our survey shows that, from 1770 to 1795, by far the most frequent source of support for scientific activity was work in the educational field, variously combined with consulting activity in the service of public administrations, the medical profession (23 percent of our professors also practiced medicine) or, rarely, involvement in some independent economic initiative.

Apparently, the technical competences that enlightened governments and private enterprises were in need of in Italy in the period under consideration were not such as to require the establishment of a dramatic number of new, specific jobs, independent of the traditional business of higher education. Or, when that did happen, the new jobs were not at the level of competence proper to scholars publishing original papers in scientific journals and likely to be included in our group.

The professions, especially medicine (when not combined with a university position, as above), provided additional opportunities (7 percent) for indirect support of scientific research. Other opportunities were offered by journalism, in which five of our scholars were engaged full-time during at least part of their life.

The indirect support of scientific activity offered by religious orders and the church requires a comment. Those in our group who were members of the church or of some religious order throughout their life, plus those who were members (at one level or another) and active in the educational institutions run by the church for part of their life make up a considerable 32 percent of our group. Considering their careers in detail, however, it turns out that only three (4 percent), undistinguished scholars (Canovai, Franceschinis, Giovene) found their main source of income in the church in their lifetimes. All the others earned most of their living working in institutions run by the states, or acting as consultants, temporary officers, writers, etc. in the variegated setting sketched above.

Apparently, when recruiting new manpower, enlightened governments and the publishers exploiting the new market for education, journals, and books still had to rely heavily on people who were or had been attached to the church for part of their life. Yet, the very fact that the overwhelming majority of the people earned most of their living from sources other than the church was clearly bringing about, in the period under examination, the shift in allegiance pursued by enlightened governments. The shift was soon to be confirmed by the adhesion of many *abati* to the new institutions of Napoleonic Italy.

### Mobility between the States

Inbreeding has been highlighted by historians as being a major cause of the decline of Italian universities in the seventeenth and eighteenth centuries.[24] A limited movement of students and scholars across the Italian states, a drastically reduced power to attract students and scholars from north of the Alps (compared to earlier periods), and the consequent trend to recruit professors from local people alone, were certainly among the factors responsible for the decline of Italian universities. The data emerging from our prosopography suggest that scientists were not the major culprits of inbreeding, nor were they a major exception.

The number of those in our group who held jobs successively in different Italian states is not negligible (24 percent). By adding those who had studied in a state other than the one where they worked (14 percent), and those who either acted as consultants in different states or traveled extensively abroad at their employer's expense (12 percent), we obtain a figure of 50 percent having substantial experience of different Italian and/or European states. Needless to say, the rest belonged to the majority of locally oriented, inward-looking academics and professionals. Among the latter, however, there were also scholars of international repute like Galvani, or scholars who had a vision of an Italian scientific community like Lorgna.

Judging from the biographies of the people in our group, three factors contributed to keeping movement alive across Italian states and,
to a lesser degree, throughout Europe. The first factor was the cosmopolitan and traveling habits of the nobility. The second factor was the
transnational organization of the church and the religious orders, to
which a number of our scholars were attached during part of their
lives. The third factor was more directly connected with the competence of the scientists—the competitive spirit nurtured by the educational and technical corps of the states, which kept alive the tradition
of trips abroad, in which figures like Volta eagerly participated.[25]

The uneven quality of the biographical sources used in our study
advises caution in drawing conclusions. The information discussed,
however, allows us to outline some features that can be regarded as
typical of the scholar in our group of authors that included the natural
philosophers engaged in scientific work. He (no woman is represented
in our group[26]) was typically a professor, at a public university or some
other state-run institution of higher learning. As such he occasionally
had additional incumbencies in the technical corps of the public administration when, rarely, this was not his main occupation. His social
and cultural background was overwhelmingly aristocracy and/or
from the church, or, less often, the professional classes. He usually represented the first generation in his family earning a living in the administration of the state. Less frequently, he could be a physician or some
other kind of independent scholar, having the leisure and motivations
needed to contribute to the kind of periodicals that have been used to
select our group of authors.


### The Circulation of Enlightenment Literature

The scholars discussed in the previous section were apparently the
chief protagonists of the peaceful "revolution" Frisi felt involved in
when writing his 1771 report. These scholars were also the main,
though certainly not the only, audience sustaining the circulation of
Enlightenment literature south of the Alps. The taste for books and
journals, Frisi noted in his report, had grown steadily in Milan in the
1760s. Literature with a scientific bent enjoyed an increasing popularity
everywhere in the Italian states. Frisi, who admired the *philosophes* and
had become acquainted with them in Paris, mentioned above all the
success of the *Encyclopédie*. When Frisi wrote, in December 1771, a first
Italian edition of the *Encyclopédie* (in French), published in Lucca, had
sold out all or most of its expensive folio volumes.[27] By that time the
third volume of a new Italian edition (also in French) published in Leg-

horn had just been printed and the growing number of subscribers had called for a new printing of the first volume.[28] By 1775 the publishers of the Leghorn edition claimed to have sold 1,500 sets. As Robert Darnton has noted, "despite the unevenness of its cultural and political geography, Italy seems to have been a great market for the literature of the Enlightenment."[29] Apparently, the condemnation of the *Encyclopédie* by the Catholic Church was not very effective, and it was, as it were, compensated for by the church's own campaign against the Jesuits—the greatest enemy of the *Encyclopédie*—culminating with the suppression of the Society of Jesus. So it could and did happen that, "when the publishers of the Leghorn edition sent a traveling salesman around Italy, they chose a priest and gave him a 10 percent commission on every sale."[30]

A survey of the books kept in a few private libraries gives an idea of the circulation of Enlightenment literature south of the Alps. I have chosen three libraries whose owners can be regarded as representative, in their different capacities, of enlightened culture in the period under consideration. One is the library of Carlo di Firmian, Vienna's minister in Milan and a staunch supporter of Volta and other Lombard scientists. Another is the library of Leonardo Ximenes, a mathematician and chief public engineer in Florence. The third is the library of the electrician Giambatista Beccaria, physics professor at the University of Turin. Though different in intent, the three libraries offer interesting insights into the circulation of the literature of the Enlightenment. I shall focus most of the attention on the Italian translations and editions of major foreign works in the field of natural philosophy, assuming that local editions and translations are indicative of deeper trends within cultural circles throughout the peninsula.

Of Firmian a contemporary wrote: "Firmiano, quidquid erat Britannicum, id ferme omne in honore ac deliciis erat." ("Almost everything that came from Britain was held by Firmian in high consideration, and delighted him.")[31] Not surprisingly, Locke figured prominently among the philosophers in his library,[32] and Locke's books included the Italian translations, by Francesco Soave, of the *Essay concerning Human Understanding* and *Of the Conduct of Human Understanding*, published in Milan in 1775–76. The former book was in its fourth edition in 1801, while a Latin edition had been published in Naples in 1789. The sections devoted to philosophy and theology (including a special subsection for "Theologia heterodoxa") contained several works by Condillac and Voltaire, whose *Eléments de la philosophie de Newton* had been translated in Naples in 1741. Among science classics were Galileo's *Dialogue* in the Florentine edition of 1710 and the four volume collection of Galileo's works published in Padua in 1744. Many works by Newton and his followers and commentators, as well as Frisi, Giovanni Del Turco, and Jacopo Andrea Luciani also figured, side by side with "eclectic" physicists like Aimé Henry Paulian, who had attempted a compromise between Descartes and Newton. Physics textbooks included Italian editions of works by John Keill, as well as 'sGravesande, Musschenbroek—the most successful in Italy among physics textbook writers—and Franklin, in the translation of Carlo Giuseppe Campi, Volta's friend. Among the electricians—besides Franklin—figured Nollet, Priestley, Sigaud de la Fond, Beccaria, Barletti, and Volta.

Given the position of Firmian, several of the books in his library were gifts: they bear witness to the initiative of the authors, publishers, and translators more than to the interests of the owner. The libraries of Beccaria and Ximenes are different in that respect: they display the professional interests of the owners themselves.

Ximenes[33] had a very fine (for a private library) collection of academic proceedings, with works from Paris to London, Berlin to Saint Petersburg, along with the proceedings of Italian academies. There was a copy of the *Encyclopédie* (the Lucca edition), and Chambers's *Cyclopedia* in the edition published (in French) in Venice in 1748. Chambers had been printed in Naples as well, from 1747 through 1754. Given the interests of the owner, the mathematics section was especially rich. It included Italian editions of de l'Hospital, and several works by William Whiston. Among the many works by the Eulers, there was Leonard Euler's treatise on shipbuilding in the Italian translation published in Venice in 1776. Nollet was especially well represented among the electricians, including two Italian translations. 'sGravesande's *Philosophia newtoniana* was present in the edition published in Bassano in 1749, while Gassendi's works were represented by the Florentine edition of 1727. Descartes, Newton, d'Alembert, and Maupertuis were the authors who appeared more frequently in the collection.

The library of Beccaria was also remarkable.[34] Among the many Newtoniana, it contained a first edition of the *Principia*, of which only seven copies are preserved in Italian public libraries.[35] The *Opticks* had a wider circulation: the Latin edition printed in Padua in 1749 was reprinted in 1773. The *Arithmetica universalis* also had an Italian edition, in Milan in 1753. Science classics in Beccaria's library included Robert Boyle's *Opera omnia* published in Venice in 1696–97, and the collection of Galileo's works published in Bologna in 1656. Physics textbooks included Desaguliers, Musschenbroek (the Padua edition, 1768, of the *Introductio ad philosophiam*), Hauksbee (Florence, 1716), and Hales' *Vegetable staticks*, which had had two Italian editions, in Naples and Milan in 1756 and 1776. The works of Christian Wolf were especially well represented. Voltaire's *Eléments de la philosophie de Newton* were there in the Italian translation already mentioned. Academic proceedings were numerous, and so too, of course, were the books on electricity which included Aepinus's *Tentamen*.

Significant though the Italian editions and translations are in revealing a deeper interest in, and wider circulation of, their authors, it is clear that—given the predominant use of French and Latin in eighteenth-century scientific circles, and the fact that these languages were familiar to every educated person in the peninsula—no major language barrier hindered the circulation of scientific literature. Books in German and English, on the other hand, were less well represented in libraries south of the Alps. In the period in question, however, the influence of German culture was itself spreading through the Austrian or Austria-related rulers governing several of the Italian states, as exemplified by Volta's own experience. As for English, the anglophile sentiment to which people like Firmian, Volta, and many others subscribed—a sentiment that was itself part of Enlightenment culture on the continent—brought with it some fluency in written English too. Works in English and German, though, were mostly known through the French translations. The mediation apparently did not reduce their impact, as the popularity in Italy of authors like Franklin and Priestley shows overwhelmingly. French remained, in any case, the lingua franca of Italian scientists when corresponding with or meeting colleagues from other countries.

A remarkable receptiveness of intellectual circles south of the Alps to the major products of Enlightenment culture and natural philosophy is mirrored by the circulation of scientific books. Together with the reform of educational institutions and public administration carried out by several governments, the circulation of Enlightenment literature was creating conditions that were potentially favorable to scientific activity. The process, however, must have been very slow, and other circumstances much less favorable, if we have to accommodate into our picture of Italian science in the age of Volta the opinions expressed by a number of foreign visitors.

### Views from the Outside

The Grand Tour tradition kept a considerable number of foreigners traveling throughout the peninsula in the eighteenth century. Not a few among them were scholars who combined a traditional interest in the historical and artistic relics of the past with an interest in the individuals, institutions, and collections connected with science. Their reports are a useful, seldom exploited source of information for historians of science. Some of our "scientific travelers," indeed, seem to have been so overwhelmed by artistic and historical monuments, as well as by the pleasures and pains of eighteenth-century traveling, that they came to devote comparatively few pages of their reports to the state of the sciences in Italy. Though often sketchy, these pages deserve careful consideration. While one should avoid assuming that views from the outside were, per se, less biased than views from the inside, it is clear that evaluations by foreigners acquainted with scientific institutions abroad may contain useful clues for the historian interested in comparative assessments of scientific activity.

The travelers considered here occupy a reasonably wide spectrum of competence: they are the astronomer Joseph-Jérôme Le Français de Lalande, the physicist and "mechanic" Adam Walker, the botanist James Edward Smith, and the agriculturalist Arthur Young. Lalande, astronomer and *associé* of the Paris Academy, traveled though Italy in 1765–66. In order to update the second edition (1787) of his seven-volume account of the trip, he set up a network of correspondents who enabled him to add interesting remarks on recent developments in Italy. Smith's journey took place in 1786–87. At the time Smith, who had been a medical student in Edinburgh and London, had just acquired Linnaeus's herbarium; soon after his return he was to launch the Linnean Society in London. Walker traveled to Italy in 1787–88, and was an independent "lecturer on natural philosophy," based in London. Young, author of many works on agriculture, traveled through northern and central Italy in 1789.[36]

Lalande considered "la littérature italienne" (a phrase which, in his use, included the sciences) to be a cultural unit. He noted, at the same time, how little known it was in Paris and stated his intention to contribute with his work to that literature being more widely known.[37] Though rather guarded when assessing the merits of the individuals and institutions he visited (and, occasionally, pointedly generous), it is not difficult to understand what he recognized to be the Italian centers of excellence in the sciences. Among astronomical observatories, he regarded only Milan (laid down by Boscovich) and Pisa (then being relaunched by Slop) as being sufficiently well equipped to compete with observatories abroad, while noticing that the best instruments used in those observatories were made in Paris and London. The observatory in Bologna he judged considerably inferior to those of Milan and Pisa, though superior to other Italian observatories. Among physics cabinets, he commended the one in Padua (established by Giovanni Poleni and then being run by Simone Stratico) and that of Bologna. In Padua, he similarly bestowed praise on the chemistry laboratory recently established by Carburi. Curiously, Lalande failed to mention the physics museum in Florence: he celebrated the merits of Ximenes (who was attached to the museum), but not those of Felice Fontana, its director. At the time of Lalande's journey (1766), he found facilities at the University of Pavia unsatisfactory or nonexistent. Through his contacts there (which included Volta) Lalande recorded, in his second edition, the great improvements achieved during the 1770s and 1780s.

Interested as he was in the administrative side to scientific institutions, whenever possible Lalande made careful note of the money spent in running them. He was impressed by the disparity in the endowments of different institutions. The University of Pisa, with its forty-two professors (each paid between 840 and 2,800 French livres), had a budget of 90,000 livres, that Lalande regarded as generous, provided by the Grand Duke of Tuscany. The celebrated Institute or Academy of the Sciences of Bologna, on the other hand, had a budget of only 10,667 livres (one-tenth of which devoted to the library), administered by the local senate. Lalande remarked that, in Bologna, "le zèle des professeurs tient lieu de richesses."[38] In fact, when visiting Bologna he praised the natural history and physics collections, and showed little interest in the professors.

Commenting on the Academy of Siena, Lalande noticed that its functioning depended entirely on the meager "pension" accorded to the Academy by the Grand Duke of Tuscany, which had allowed the Academy to publish four volumes of its acts. Only from 1767 had the government granted a separate stipend for the secretary of the Academy. In fact, direct funding from the sovereign or his government—with all the uncertainties and dependence that that entailed—was the rule in Siena as in Turin, Milan, Padua, Bologna, Florence, Pisa, Rome, and Naples. The direct relation with the sovereign or government affected the universities just as it did the academies and the other scientific institutions.

Higher education, though a major concern of the best among enlightened governments, was not their only reason for supporting science. Prestige, and the amount of show and appearance connected with scientific collections, was another. The fact did not escape the attention of foreign travelers, especially of those among them who were less inclined to cherish symbols of prestige and power. As Adam Walker noted on visiting the Institute of Bologna, the place looked as it had been conceived "for inspection, rather than instruction."[39] Walker's vivid account of the apparatus kept in Bologna, like that of the museum in Florence by Arthur Young that we cite below, deserves to be quoted at length. Here is Walker on the chemical apparatus in Bologna:

We were then led into the School of Chymistry—here we found a great variety of retorts, cucurbits, stills, &c. hung up for shew; but neither these, nor the various ill-contrived furnaces seem to have ever been used. No apparatus for catching the glasses that fly off from heated substances— and which make so very material a part of modern chemistry—nor one process of any sort going forward.

The School of Philosophy ... I was particularly wishful to see—sorry that in this I was also disappointed. Here we found a number of ill-made instruments (mostly on the plan of Gravesand's) nothing new, except a pretty large plate electrical machine. A new air-pump was just trying, with the two cylinders horizontal, and open in the old way, very clumsy, which extracted very slow, yet seemed tolerably staunch.[40]

Walker's severe comments on scientific apparatus (exempting only the electrical machine: the one used by Galvani?) are in keeping with his general comment on "mechanics" south of the Alps: "They are wretched mechanics in Italy, never saw a clock with a minute hand to it, in the whole country."[41]

Arthur Young, himself not interested primarily in natural philoso- phy, commented on the collections in Florence in a similar key:

To Signore Fabbroni, who is second in command under il cavaliere Fon- tana, in the whole museum of the Grand Duke; he shewed me, and our party, the cabinets of natural history, anatomy, machines, pneumatics, magnetism, optics, &c. which are ranked among the finest collections in the world; and, for arrangement, or rather exhibition, exceed all of them; but note, no chamber for agriculture; no collection of machines relative to that first of arts; no mechanics, of great talents or abilities, employed in improving, easing, and simplifying the common tools used by the hus- bandmen or inventing new ones, to add to his forces, and to lessen the expense of his efforts! Is not this an object as important as magnetism, optics, or astronomy? Or rather, is it not so infinitely superior, as to leave a comparison absurd?[42]

On visiting the Institute in Bologna Young made other perceptive, caustic remarks (and some useful comparisons) on laboratories in- tended for show rather than research:

A well arranged laboratory, clean, and every thing in order, in a holy-day dress, is detestable; but I found a combination of many pleasures in the disorderly dirty laboratories of Messrs. de Morveau and la Voisier [sic]. There is a face of business; there is evidently work going forwards; and if so there is use. Why move here [in Bologna], and at Florence, through rooms well garnished with pneumatical instruments that are never used?

Why are not experiments going forwards? . . . Half these implements grow good for nothing from rest; and, before they are used, demand to be new arranged. A prince, who is at the expence of making such great collections of machines, should always order a series of experiments to be carrying on by their means. . . . [The cabinets in Florence and Bologna] would be better employed than in their present state, painted and patched like an opera girl, for idle to stare at. What would a Watson, a Milner, or a Priestley say, upon a proposal to have their laboratories brushed out clean and spruce? I believe they would kick out the operator who came on such an errand. . . .

Here is a chamber for machines applicable to mechanics; and the country is full of carts, with wheels two feet high, with large axles; what experiments have been made in this chamber to inform the people on a point of such consequence to the conduct of almost every art? . . .

Bologna may produce great men, but she will not be indebted for them to this establishment.[43]

Institutions specifically devoted to agriculture and the arts made a similar impression on Young. On attending a meeting of the Società Patriottica in Milan, he remarked:

I looked about to see a practical farmer enter the room, but looked in vain. A goodly company of i Marchesi, i Conti, i Cavalieri, i Abbati, but not one close clipped wig, or a dirty pair of breeches, to give authority to their proceedings.[44]

No doubt part of the harshness in these comments came from Young's outspoken committment to liberal political views, which made him appreciate only Tuscany among the Italian states. Similarly, in the case of Adam Walker, one should discount his strong anti-Catholic feelings, to which his report bears frequent witness. However, it would be mistaken to disregard their testimonials as irrelevant on such grounds: James Edward Smith seems to have suffered from neither of those biases, and his judgments about natural history collections and botanic gardens were equally severe. Of all the botanic gardens Smith visited in Italy, only the one in Pavia appeared to him to be well run, and "as well furnished as most out of England."[45] The garden in Pisa was decently arranged and kept.[46] In contrast, the one in Turin was "in its infancy," though promising thanks to the efforts of "the very able Professor Dana."[47] The botanic gardens in Rome, Bologna, and Milan he found equally "indigent," "poor," "small," and their keepers generally ignorant of the Linnean nomenclature. Nor did Smith enjoy the botanic garden in Florence. But he did like the natural history collection of Pier Antonio Micheli better, which gave him the opportunity to comment on working, as distinct from mere exhibition collections, in terms similar to those used by Young and Walker:

> The minerals and corals of this collection are very numerous, and, being ticketed by Micheli and Targioni, have that peculiar value which renders the original museum of a working naturalist so far preferable to those of Emperors and Princes destitute of such authority.[48]

The impression of the state of the sciences emerging from the reports of foreign observers traveling through Italy in the 1780s can be easily summed up: a few, well-known, and generally overrated princely institutions (the museums in Florence and Bologna), continued amid a variety of other, more recently established facilities, all still weakly rooted, some quite good, some poor but promising, many badly or very badly run. Unlike insiders like Paolo Frisi, the foreign visitors valued achievements above intentions and good will. While one should not overrate the reliability of the sharp comments by foreign scholars quoted above, it seems clear that they did point to some real problems. Apparently, the recent investment of enlightened governments in educational and scientific institutions was not, and could not, bear fruit as quickly as the scholars engaged in carrying out the reforming measures expected. The more so because (as more than one foreign observer remarked) the recently relaunched educational and scientific institutions retained many characteristics typical of their courtly, pre-Enlightenment ancestors—a fact probably linked to the frequent, traditional aristocratic and/or religious training of the staff in charge of the institutions, and one that the slow pace of economic and technological change in society at large (noted especially by British observers) was doing nothing against.

## Conclusion

When assuming that no interest in real scientific work or real technology could develop side by side with the "show" of the cabinets and stately collections, foreign observers were perhaps rushing to premature conclusions. The amount of scientific work and technological consulting in which some of the "professors" discussed in our prosopographical study engaged was impressive. Yet, British observers were probably right to stress the relative separateness in which Italian scientific institutions seemed to live with respect to the real world. Isolation was the outcome of several factors. One of these was the fact that these institutions (even when attached to the universities) were still chiefly intended to be ornaments of the courts, symbols of achievement for courtly culture and for the public servants in charge of them. This carried with it functions and rituals which the professors—often nobles themselves, and directly dependent on government patrons—subscribed to all too easily, further reducing the margins for autonomy accorded by governments. One result could be the well-painted, though deserted, laboratories and the inactive machines which attracted the sarcasm of British visitors.

At first glance, the situation of scientific institutions just referred to seems to conflict with that wide circulation of Enlightenment literature south of the Alps mentioned earlier in this chapter. As noted there, the intellectual circles of the peninsula *were* receptive to the literature of the Enlightenment. This literature emphasized—in principle at least—the alliance between philosophy and the sciences, and between the sciences and the arts. Its circulation in Italy too implied that wider audiences would become interested in the sciences, and this was itself—as Frisi hinted in his report—among the impulses promoting the creation of new facilities for higher education and the sciences, and supporting the expanded market for scientific books. What, then—other than the persistence of the courtly mentality lamented by some foreign observers—retarded or prevented the scientific institutions from benefiting from the spread of the new literature?

One obvious but very likely cause was the all-too-recent and still uneven effort to improve higher education. The effort, though begun vigorously and successfully in states like Lombardy and Tuscany, was only beginning to bear its fruits during the 1780s, and it was soon to be disturbed by the vicissitudes of the Napoleonic age. In such conditions, and given the proliferation of scientific institutions throughout the peninsula, well-trained and highly motivated staffs continued to be rare commodities. That neither Volta nor Galvani established a research school, despite their long careers and their deep roots in the universities of Pavia and Bologna respectively, is a telling example.

The fact that the state universities continued to be both the main reservoir of new manpower for the sciences, and also, together with the state academies, collections, and observatories, its most frequent destination suggests another circumstance that may explain the faltering life of scientific institutions south of the Alps in the period under consideration. The apparent monopoly that the state administrations had over the sciences revealed the fact that agriculturalists, manufacturers, skilled artisans, and independent teachers for their part either did not have a keen interest in the sciences or did not have a say in matters of education. Both circumstances probably applied and related to what was—to different degrees in different Italian states—a persistently stagnating economy and a long term neglect of popular education.[49]

Apparently, neither the spread of Enlightenment reforms and literature, nor the recent, exceptional achievements of a handful natural philosophers like Volta could possibly overcome the effects of those long term trends.

## CHAPTER THREE

# The Electrophorus

## THEORY, INSTRUMENT DESIGN, AND THE SOCIAL USES OF SCIENTIFIC APPARATUS

Announced by Volta in 1775 as "the perpetual carrier of electricity," and still depicted by the *Encyclopedia Britannica* twenty-two years later as "the most surprising machine hitherto invented,"[1] the electrophorus seems for historians of physics to have retained some of this fascination. The machine is still cherished as "the most intriguing electrical device since the Leyden jar,"[2] and has gained an exemplary role in recent historiography. The electrophorus and its inventor could well be seen as representing the recent inclination of historians of science to emphasize experiment and apparatus over theory, instrumentalism over deep philosophical commitment, efficacy and public recognition over private achievement.

All such ingredients figure prominently in the story of the electrophorus, an apparatus of intriguing simplicity, whose basic parts are a metal "shield," a resin "cake," and (optional) a second metal disc under the cake (fig. 3.1, p. 74).

The easiest way of operating the machine, as described by a contemporary, is as follows:[3] Holding the upper metal shield *AA* by its insulating handle *E*, the resin cake placed on top of the metal dish *CC* is rubbed. The shield is then lowered and put on top of the cake. The experimenter now touches the shield with a finger, while he touches the inferior metal dish *CC* with another finger. On lifting the shield up again, and approaching a finger or any conducting body to it, a spark is drawn. The operation can be repeated "a hundred times," obtaining as many sparks without charging the cake again.

The story of the electrophorus is the story of an important segment of the science of electricity in the last quarter of the eighteenth century. The efforts made after 1775 by dozens of electricians around Europe in order to come to terms with the new machine and with its curious exploits—its apparently endless source of static electricity and the puz-

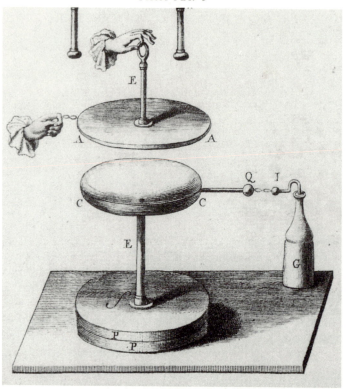

3.1. Volta's electrophorus, here combined with a Leyden jar used to "revive" the electricity of the apparatus when—rarely, Volta emphasized—needed. From *Scelta di Opuscoli* (1775) (*British Library, London*).

zling operations that made the trick possible, later explained on the principle of induction—are regarded as having been the most powerful impulse in drawing the electricians away from the old and shaky notion of "electric atmospheres." By casting doubts on that notion, the electrophorus was instrumental in discarding the last seriously misleading tenet of the effluvial interpretation of electricity. It thus freed electricians from the major shortcoming of Franklin's otherwise successful theory of electricity and opened the way to the crucial distinction between the electrical matter and its field. It also revived an interest in Aepinus's work, which was to culminate in the formulation of Coulomb's celebrated law of electrostatics; a law to which—to add spice to the story—the inventor of the electrophorus refused to recognize any kind of general applicability.[4]

In view of the impressive impact of the electrophorus—convincingly described in the most authoritative account of the history of electricity in the eighteenth century, and virtually unchallenged (with the possible exception of the part concerning Aepinus[5])—it is surprising how little is known about the invention of the apparatus, about the context in which it took place, and the ways in which its inventor managed to bring the machine to the attention of physicists throughout Europe. What follows is an attempt at filling the gap, while showing Alessandro Volta at work in his chosen field of expertise during the first successful part of his career.

One point of the present chapter is that Volta, when building the electrophorus, knew about similar devices described by Johan Carl Wilcke and Franz Ulrich Theodosius Aepinus several years earlier. The circumstance, so far unnoticed by historians, sheds light on the process of Volta's invention and on the assessment of his achievement by his contemporaries. Rather than detracting from the interest of the case (or dissolving it into a search for precursors), the new evidence invites a closer analysis of the ways in which theory, instrument design, and the social uses of scientific apparatus interacted in Volta's early investigations, and in the assessment of scientific innovation in eighteenth-century science. A second, central point of the chapter is that, in Volta's path to the electrophorus, the instrument itself—whose features were partly shaped by the social uses for which it was conceived—led the conceptualization of the principles able to explain its functioning.

Our story is based on three main sources: Volta's early publications, displaying his theoretical ambitions and frustrations, as well as his talent for unconventional apparatus; a reconstruction of the printed sources on the science of electricity available to him; and Volta's local and international correspondence, which he was learning to cultivate—as he would for the rest of his life—as a form of very special, private academy. Our reconstruction cannot rely on Volta's laboratory notes. Of the numerous such documents preserved among the Volta papers in Milan, none can be related with certainty to the invention of the electrophorus, the first apparatus to win him wide recognition. If we trust Volta's own testimony, eighteen months before announcing the electrophorus he had not yet made a custom of taking regular notes of his observations.[6] On the other hand, soon after the electrophorus, Volta's preserved laboratory notes become relatively numerous. It is tempting to think that, just as the occasion of announcing and defending the electrophorus allowed Volta to develop new skills in making use of his publications and correspondence to serve his science and ambitions, so that same occasion taught him how to refine his working style and laboratory habits. The lack of laboratory notes prior to 1775, in any case, means that the reconstruction that follows is circumstantial.

Around 1775, as we know from chapter 1, Volta's efforts as an electrician, a science teacher, and a public servant were gaining momentum on all fronts. As we shall see, Volta's path to the electrophorus was inspired by many converging goals. One goal he followed out of a certain disenchantment with the theoretical ambitions he had nurtured as a young natural philosopher. Another was associated with his growing success as a performer of electrical demonstrations and as an inventor of electrical machines. The third was dictated by his firm determination to conquer a place in the local and international communities of physicists. All these goals were further connected— though often at variance—with the notions and experiments on electricity available in principle to any physicist of the time through learned journals and books.

As will become clear, in order to map and represent the complex goals that led Volta to the electrophorus, the historian is obliged to adopt, as a stage for his story, a multi-dimensional space: a space of a kind that is not usually provided by studies focusing exclusively, either on invention as a strictly individual and intellectual enterprise, or on ideas and their supposedly impersonal development, or on scientific practice conceived merely as the outcome of social negotiations among the experts and with their audience.[7]

### Fire, Magnetism, Electricity

As mentioned in chapter 1, the earliest recorded steps taken by Volta to enter the community of physicists go back to 1763. At the age of eighteen, he thought that submitting a very general theory, embracing electrical as well as magnetic phenomena, all explained through a universal principle of attraction, would be the best way of entering that community. Volta had left the philosophy class at the Jesuit College in Como two years earlier and was now a boarder at a seminary in the same town. However, he felt confident enough to address letters to Nollet and Beccaria expounding his bold speculations. The letters have not been preserved. We know of their content through summary descriptions given by Nollet in 1763 and by Volta in 1764 and 1769.[8] The fact that the two latter descriptions were addressed to the same correspondents, or intended to be read by them, allows us to regard them as giving reliable evidence about Volta's views in 1763, though perhaps revised and updated to some extent.

Volta was then focusing on the two varieties of electrification known as resinous and vitreous. Dufay had called attention to these in the 1730s. In the early 1760s, experiments made by Robert Symmer—centered on his curious observation that two, differently colored silk stockings worn for some time one over the other attract each other, and on the notion of two distinct electric fluids—were being interpreted by Nollet as offering new support for his own two-stream theory of electricity; a theory which others regarded as superseded by Franklin's one-fluid theory, which had been circulating during the previous decade.[9]

We do not know for certain whether Volta knew of Symmer's work in 1763. The possibility that he had been involved in Nollet's efforts at circulating Symmer's results cannot be excluded. By the end of 1762 Nollet had printed a French translation of Symmer, made by Dutour and including Nollet's comments, and distributed it privately to electricians he hoped could be convinced to abandon Franklin's system. Cigna in Turin was one of the recipients, and made what Nollet regarded as good use of Symmer's translation.[10] What we know is that Volta and Nollet exchanged books side by side with letters in those years.[11] In any case, the fact that Volta addressed his 1763 speculations to Nollet *and* to Beccaria—who was one of Franklin's most committed supporters—indicates that he perceived what was at stake. Judging from the theory Volta himself propounded, he was hoping to win the attention of both the Nolletists and the Franklinists by suggesting a higher level of explanation, in which both groups could be interested. Not surprisingly in those years, it was the notion of attraction which seemed to offer the higher level of explanation he sought after.

Acknowledging receipt of Volta's now lost letter of 1763, Nollet described Volta's theory as explaining the causes of electricity by means of Newtonian principles. Volta's goal appeared to the Parisian electrician exceedingly ambitious, and very probably unattainable:

> I will consider with much pleasure your new system concerning the causes of electricity, when you will publish it: it would surprise me to see the physics of these phenomena explained by Newtonian attraction, as you plan to do; it seems to me that, within the laws attributed to that kind of virtue, it is very difficult to account for the main facts [of electricity]; so far nobody dared to initiate that; it would bring you glory to have done it successfully.[12]

The evidence Volta provided in support of his theory came mainly from experiments on magnets.[13] He assumed that the visible behavior of a magnet interacting with metal plates and metal filings suggested

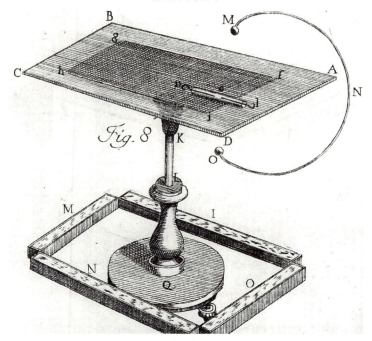

3.2. A Franklin square, from G. B. Beccaria, *Elettricismo artificiale* (Turin, 1772) (*British Library, London*).

something about the behavior of the electric fluid in a condenser, for example a Franklin square (fig. 3.2).

This was by then a common piece of apparatus among electricians, and was equivalent to the classic Leyden jar. The "square" (a parallel-plate condenser) consisted of a sheet of glass to which two metal coatings were applied on both sides, acting like the two coatings in the jar (the shape of the glass and coatings had been demonstrated to be irrelevant to the functioning of the apparatus).[14] Through the analogy with the magnet, Volta hoped to outline a general theory of the phenomena displayed by Franklin squares.

In this connection, Volta's key experiment was the well-known experiment on the magnet Newton had introduced in his *Principia* to give an idea of how attraction might act in fluids.[15] It consisted of suspending an iron body from a loadstone, the iron being the heaviest the magnet could hold. When a second iron body was brought near the magnet, the first dropped. Newton (and Volta) suggested that the forces of the magnet, now exerted toward the second body as well, were no longer sufficient to sustain the first. To develop the analogy with the Franklin square, Volta varied the experiment as follows: He used a magnetized steel plate, to the surfaces of which he applied varying quantities of iron or iron filings. When the two surfaces were holding as many filings as they could, he found that by bringing additional filings near surface $A$, a corresponding amount of filings was dropped from surface $B$. And vice versa, when removing some filings from surface $A$, surface $B$ attracted additional filings brought near to it. Volta conceptualized the results by saying that, for any given loadstone or magnetized steel plate, there was a certain amount of filings by which its attractive forces were "saturated." By analogy, he argued that something similar happened in a Franklin square with regard to the ability of the two surfaces of its electrified glass to attract or repel additional electric fluid. Having extended the analogy to include the many combinations obtained by adding glass panes to the square, and by joining and separating the metal coatings, Volta tried to develop a general "theory of squares."[16]

The details of this theory can only be inferred from Volta's first publication, six years later. There is, however, some other evidence which offers hints on the broader context in which Volta's first theory of Franklin squares had been conceived. This is the poetical fragment in Latin composed in 1764, and mentioned in chapter 1, above.[17] In that long fragment Volta developed a series of reflections on fire, combustion and heat. Clearly inspired by the Descartes-Boerhaave tradition in the study of heat,[18] Volta asserted that the combustion of bodies depended on the presence of small particles of fire (ignicula), hidden inside them. When the ignicula were excited by the addition of fire from the outside, they acquired a greater power and produced the phenomena perceived as heat and flame. The different susceptibility of bodies to heat and combustion was explained in terms of the internal texture of different substances, which was assumed to affect the ignicula through forces of attraction active at the microscopic level. Volta used this model of explanation to account for the phenomena displayed by gunpowder, fulminating gold, and will-o'-the-wisps. He also tried to establish a scale of different bodies according to their inflammability, with "sulphura and pingues materiae" at the top.[19]

In 1764 Volta explicitly ruled out electricity from the topics to be discussed under the heading of fire.[20] Recent discoveries, he noted, had shown that phenomena like lightning and other "ignita meteora," traditionally associated with inflammable substances, belonged to the specific domain of electricity. Accordingly, as we know, Volta planned to devote a separate poetical composition to them. However, there were similarities in the ways Volta dealt with heat, magnetism, and electricity in those years; similarities resulting from his effort to ac-

count for all of them in terms of corpuscular interactions. Hints at possible connections between the "electric fire" and "common fire" were
found in Franklin's works known to Volta.[21] Indeed, the tendency toward exchange explanatory models between the fields of heat, magnetism, and electricity was widespread among mid-eighteenth-century
physicists.[22] In 1763–65, in any case, Volta was not committed to his
general theorizing ambitions to the point of not finding the time and
leisure to pursue other fruitful lines of investigation. A letter he sent
to Beccaria in the spring of 1765 shows that his penchant for electrical
apparatus was another important source of inspiration.

In April 1765 Volta was developing a line of investigation focusing
on the electricity of different kinds of silk, and on resinous and vitreous
electricity in general.[23] He had built a "scale" of the electrical properties
of different substances, as they resulted from rubbing them against
each other, and then checking whether they became electrified plus or
minus.[24] The implicit premise was that the two known kinds of electricity—plus and minus in Franklin's phraseology, assumed to correspond
to the vitreous and resinous electricities of Dufay and Symmer—were
found to be present in different substances to different degrees. An additional premise was that there were "gradual" shifts in natural products and, accordingly, in the distribution of the electric fluid intrinsic
to each substance.

The simple experimental set-up, and the substances tested, were
probably those described in the undated table drawn by Volta and reproduced in figure 3.3. The table does little more than describe a series
of tests carried out by rubbing different substances against each other.
The results presented to Beccaria in spring 1765, however, incorpo-
rated a higher degree of conceptualization. Volta assumed that resin
and glass occupied the two extremes of a series, the electric fire being
extremely plus in resin and extremely minus in glass. Conducting bodies were equally distant from the extremes, being nonelectric per se
and susceptible of being electrified by either kinds of electricity simply
by contact. Bodies that were electric per se belonged either to the class
of resinous or to the class of vitreous substances, each occupying a relative position determined by the susceptibility to being electrified plus,
minus, or either when rubbed together with other substances in the
series. Volta's conceptualization[25] can be illustrated by figure 3.4 on
page 82.

To take an example, let us consider silk. Silk belongs to the class of
vitreous bodies: it can easily get charged minus when rubbed together
with glass. However, it will get charged plus when rubbed together
with a conducting body or with resin. The relative position of silk in
the class is determined by the fact that silk is more inclined to "give"

3.3. Volta's undated laboratory table describing systematic observations on triboelectricity. From VMS, I, 3 (*Istituto Lombardo, Milan*).

| RESIN | +++++ |
| ... | ++++ |
| sealing wax | +++ |
| ... | ++ |
| ... | + |
| INDIFFERENT BODIES (electrifiable by communication) | 0 |
| nail, horn, wool, flax, hemp | - |
| silk | -- |
| hair | --- |
| animal hair (cat, dog, horse) | ---- |
| GLASS | ----- |

3.4. Triboelectric series. Diagram of the triboelectric series described by Volta in the letter of 12 April 1765 to Beccaria, *VO*, III 6–8.

electric fluid to glass (more distant in the scale) than to "receive" it from indifferent bodies (closer in the scale). In other words, substances distant in the scale, when rubbed together, become electrified more easily than substances closer in the scale.

Research on triboelectricity was not a novelty at the time.[26] Moreover, Volta admitted that his knowledge of the series of resinous bodies was less accurate than that of vitreous bodies (see the gaps in the diagram). However, some of Volta's results and their interpretation are worth noting. Volta found that by rubbing glass and silk they became electrified at a higher degree (that is, they produced more distinct electrical signs) than when rubbing glass and a conducting body, like the hand. In other words, silk—despite being itself a vitreous body—"gave" to glass more fluid than conductors did. This could not have been predicted on the basis of the series above and forced Volta to suggest a specific explanation. It turned around the notion that different parts of the same body act differently in the process of rubbing. He assumed that the portions of the rubbing body not directly subject to friction "claimed back" from glass part of the fluid that other portions of the body gave to it. Silk being less active than conductors in claiming the fluid back from glass, friction between glass and silk was more effective than friction between glass and a conductor in producing the imbalance of the fluid responsible for electrical phenomena.[27]

Taking a step we will meet time and again in Volta's subsequent investigations, in 1765 he tried to apply his notions on triboelectricity to the construction of a piece of apparatus. This was a new version of the classic electrostatic machine (fig. 3.5), in which a silk wrap sustained by a wooden wheel took the place of the glass globe (right), and a piece of glass took the place of the rubbing cushion (mounted on a flexible

3.5. Nollet-type globe generator, 1760s, Museo di Storia della Scienza, Florence (*Franca Principe, Istituto e Museo di Storia della Scienza, Florence*).

brass strip facing the globe).[28] The arrangement turned out to be impractical,[29] but the machine was the first of a long series of devices Volta conceived with the explicit purpose of displaying at work, as it were, the notions he was developing at the conceptual level. The machine is evidence of the close link, in Volta's investigations, between the process of explaining electricity and that of displaying it. In the realization of the silk electrostatic machine, conceptualization and explanation, apparently, guided the realization of the displaying apparatus; as we shall see, in the case of the electrophorus, ten years later, the machine was to guide conceptualization.

## "Vindicating Electricity"

Beccaria was sparing of comments on Volta's letter of April 1765.[30] He expressed reservations on some of Volta's points,[31] but he was interested enough to send some of his recent publications to him.[32] In them Beccaria described a series of experiments that, as far as the friction of different substances and their respective inclination to "give" or "receive" the electric fluid was concerned, Volta judged similar to those in which he himself was engaged. Beccaria, however, was then also

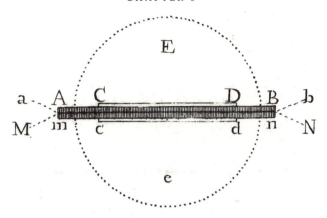

3.6. Beccaria's diagram of glass and metal plates used in experiments on "vindicating electricity." From G. B. Beccaria, *Elettricismo artificiale* (Turin, 1772) (*British Library, London*).

developing a related line of inquiry, started by a famous experiment devised by the Jesuits in Peking in 1755 and reported in Europe by Aepinus three years later.[33]

The Peking experiment was the object of frequent reflections by electricians in the 1760s. In it, a glass plate electrified by rubbing was alternately placed on and removed from the glass lid of a compass box. When the electrified glass was placed on top of the glass of the compass box, the compass needle rose to the top, stayed for a while, and then fell. When the upper glass was removed, the needle rose again, to fall when the glass was brought back. The effects lasted for hours, without rerubbing.[34] By varying the experiment, by combining it with the line of investigation indicated by Symmer, and by using basic Franklinist concepts, Beccaria developed the theory of "vindicating electricity," which he first described in an essay dedicated to Franklin in February 1767.[35] Volta heard of Beccaria's essay that same year,[36] and the theory of "vindicating electricity" became one of Volta's major concerns up to the realization of the electrophorus.[37]

Beccaria's theory[38] applied to electrified plates of different materials (fig. 3.6), joined and then separated. In its most general formulation, Beccaria's principle of "vindicating electricity" asserted that (a) when an insulator is joined to another insulator of opposite charge or to a conductor, the electricities of the two bodies become null, and that (b) the same two bodies regain ("vindicate") their respective electricities when they are separated again.[39]

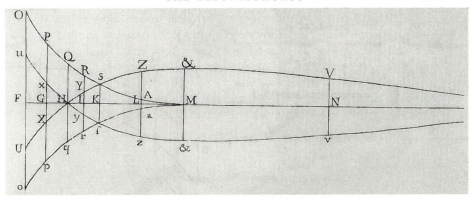

3.7. Beccaria's diagram of "vindicating electricity." From Beccaria, *Elettricismo artificiale* (Turin, 1772) (*British Library, London*).

Beccaria tried to establishing a sophisticated law of such phenomena and illustrated it by a graphic representation (fig. 3.7). Keeping in mind the apparatus illustrated in figure 3.6, where *ABab* and *MNmn* are glass plates (dielectrics), while *CD* and *cd* are metal foil coatings (conductors), figure 3.7 describes the changing quantities of electricity present in the glass plates *ABab* and *MNmn* once the apparatus has been charged, and metal coatings are repeatedly disjoined from and rejoined to the glass plates. After charging the apparatus from the chain (plus) of an electrostatic machine, face *MN* of the *MNmn* pane will have the plus charge *OF*, and face *mn* the minus charge *oF*. Pane *ABab* is assumed throughout to have electricities which are equal and contrary to those in *MNmn*. Disjoining a metal coating a first time (by means of silk threads), *MN* will lose portion *uF* of its plus charge. By joining it again, *MN* will regain ("vindicate") some of the charge, so that its residual charge will be *PG*. At disjoining again, *MN* will lose *xG*. It will then have the residual plus charge *QH*, no further diminished by subsequent disjoinings because in *H*, Beccaria claims, "vindicating electricity" is changed from positive to negative. Thus, at a subsequent disjoining, when residual plus charge is *RI*, the pane will have regained portion *Iy* of its lost charge. *MN* will subsequently turn out to be charged *KS* in *K*, *LA* in *L*, zero in *M*. *MN* is therefore regaining growing portions of its lost plus charge, indicated by *Ks*, *Lz*, and *M&*. The portion of lost plus charge, regained on subsequently disjoining and joining, will slowly decrease (*Nv*), and the residual charge (*NV*) will decrease accordingly.[40]

Beccaria's sophisticated account was admired but did not attract adherents. Presented as being a satisfactory compromise between the classic, Franklinist one-fluid approach and the troublesome reality of Symmerian phenomena, it convinced young Volta only to a limited extent, and even that only temporarily. Judging from Volta's first publication, a reason for Volta's dissatisfaction with "vindicating electricity" was that Beccaria did not even try to base his notions on what was going on at the microscopic level, and did not mention attraction.[41] Apparently, young Volta continued to regard these as prerequisites for any convincing explanation of electrical phenomena.

### Attraction and the Atmospheres

Volta's first publication, a theoretical essay written in an idiosyncratic Latin and published in 1769, was composed with the intention of bringing the nexus between electricity and attraction to the center of attention. The title read *De vi attractiva ignis electrici, ac phaenomenis inde pendentibus* ("On the Attractive Force of the Electric Fire, and on the Phenomena Dependent on It").[42] Written in the form of a letter to Beccaria, the essay is the first systematic account extant of Volta's efforts at reconciling electrical phenomena and mechanical principles, efforts in which he had been engaged since 1763. To understand the 1769 essay it is essential to keep in mind Volta's main sources and terms of reference. These were, basically, Franklin and Beccaria on electricity, and Newton, the British and Dutch Newtonians, and Roger Boscovich on physics in general and attraction in particular.

Franklin's attitude toward the notion of attraction and its possible role in explaining electrical phenomena was admittedly a complex one. As Volta knew through Dalibard's French translation of Franklin's *Experiments*, the American admitted an attractive force active between the particles of ordinary matter and a repulsive force active between the particles of the "electric fire," which he conceived of as being a different substance from common fire.[43] Historians seem to agree that Franklin did not specify clearly whether microscopic forces could also account for the observed, macroscopic phenomena of attraction and repulsion between electrified bodies.[44] What is certain is that, in order to explain the latter, he also had recourse to the notion of "electrical atmosphere." By this he intended a kind of layer of electric fluid, surrounding any positively electrified body and affecting other bodies which might enter it.[45]

Some of Franklin's followers seem to have been more concerned than he was with the problem of reconciling these notions with classic, mechanistic concepts.[46] Beccaria, especially at the beginning of his career, had been one of them. This concern for mechanistic models of explanation, which he had displayed in his early works, may well have influenced young Volta.[47] Beccaria, however, had increasingly refrained from using similar principles in his subsequent publications. Franklin and Beccaria, in any case, were not the only and perhaps not even the main source of Volta's interest in attraction. Judging from the literature Volta quoted in 1769, he was following above all in the steps of the eclectic brand of "Newtonianism" circulated by British and Dutch physicists, by their Italian imitators, and by the Jesuits.[48] John Keill, John Freind, and Pieter van Musschenbroek were Volta's acknowledged sources for the rich repertoire of short-range attractions mentioned at the beginning of the 1769 memoir.[49] Keill was especially popular in physics textbooks published in Italy during the eighteenth century.[50] Volta judged that the attempts of Keill and Freind to include short-range attractions in the framework of a thorough mechanical model of explanation had been "not without success."[51] Musschenbroek, on the other hand , supplied him with a list of examples of short-range attractions drawn from optics, mechanics, hydraulics, and chemistry.[52]

As historians have pointed out,[53] this literature often pursued, side by side with updated, "Newtonian" lines of investigation, older Cartesian concerns.[54] Volta's 1765 notion of a corpuscular "texture," characteristic of each substance and affecting, through attraction, its properties with regard to electricity, seems to have stemmed precisely from this eclectic tradition. In 1769, however, he was pursuing a more strictly attractionist ideal. Having found that the same substance could both "receive" and "give" the electric fluid, he convinced himself that a satisfactory explanation should rest, not on properties specific to each substance, but on general attraction principles, valid for any substance. In shifting him toward this conclusion, *A Theory of Natural Philosophy* by Roger Joseph Boscovich had played an important role.[55]

Boscovich was in those years active in Milan, supervising the construction of the astronomical observatory at the Brera Schools run by the Jesuits.[56] In the early 1770s, Volta was to be a frequent visitor to the scientific circles that had developed around Brera. Though there is no evidence of personal contacts between Volta and Boscovich, the presence of the Jesuit scientist in Austrian Lombardy was clearly significant in those years. It was very likely that Volta could not grasp all the details of Boscovich's mathematics and physics, but he liked the general goal and the main thesis expounded in Boscovich's masterpiece. The thesis was, to use J. M. Child's effective phrase, that "matter is composed of perfectly indivisible, non-extended, discrete points," and that "there is a mutual *vis* between every pair of points, the magnitude of which depends only on the distance between them."[57] A characteris-

tic addition to the thesis—in which a Newtonian methodology mingled with Leibnizian assumptions[58]—was the notion that the same
basic *vis* accounted for both attraction and repulsion phenomena. As
Volta put it, by adopting Boscovich's system it was possible to explain
how frequent changes from attractive to repulsive forces occurred "in
the smallest possible space."[59]

Boscovich himself had not developed the implications of his theory
for electricity in depth. However, he had stressed that his system was
compatible with the theories of Franklin and Beccaria.[60] He had also
called attention to what he regarded as the most significant link between his theory and Franklin's. This was the notion of the "relative
saturation" (*respectiva saturitas*) of bodies with regard to "the fiery substance"; a concept, Boscovich declared, that applied to the electric fluid
as well. As he put it, "relative saturation" explained why, for example,
"when the fluid, under the action of a mutual force, passes from one
substance to another, it is readily seen that those bodies, of which the
particles attract the fluids to themselves although with unequal forces,
must also attract one another."[61] Boscovich had mentioned, and had
tried to explain, the functioning of Franklin squares along the same
lines, accounting for the fact that the apparatus performed appropriately only when the glass pane was sufficiently thin.[62] In considering
magnetism from a similar perspective, Boscovich had asserted that "all
phenomena with regard to it reduce to a mere attraction of certain substances for one another"; a position also echoed by Volta in his 1769
essay.[63] Volta, however, had by then a command of electric theory and
manipulation superior to Boscovich's. Accordingly, his discussion in
*De vi attractiva ignis electrici*, though inspired by similar, general goals,
attained new, interesting results.

Volta started from the assumption that all the motions observed in
electrical phenomena could be reduced to a simple law: any two bodies, having unequal amounts of electric fire (and other conditions being
equal), attract each other with a force proportionate to the imbalance
between the two amounts of fluid.[64] Adopting Boscovich's concept of
attraction changing into repulsion, and insisting on the notion that the
force acting between any two bodies, including the particles of air, depends on the unequal amount of fluid they contain, attraction was
thought to explain the phenomena of repulsion as well.[65] As Volta
viewed things now, the amount of electric fluid present in a body did
not depend exclusively on properties intrinsic to each substance. It depended, more generally, on "mechanical forces" determined by the internal, corpuscular texture of the body as well by its relations with surrounding bodies.[66] Mechanical forces, Volta maintained, could explain
why friction, percussion and "chemical motions" affected the electrical

properties of a body.[67] They explained also why, by adding or subtracting electric fluid to a body (for example with the aid of an electrostatic machine), the attractive forces of the body's particles toward the
fluid changed: diminishing when they were shared with an increased
amount of fluid, or increasing with a decreased amount of fluid, and
affecting in turn the body's attitude to "give" or "receive" additional
fluid. Thus, the attitude toward the fluid was affected by the contact
or proximity of other bodies. Echoing Boscovich, Volta stressed that
the *saturitas* of a body with regard to the electric fluid was always *respectiva*, that is, it depended on the attractive forces resulting from
other bodies and their fluid interacting with the first body.[68]

Such notions were applied by Volta to develop a theory of Franklin
squares and of Symmerian phenomena somewhat different from the
theory he had sketched in 1763. The key concept was now that of *applicatio*. This concept also clarifies Volta's stand with regard to the tricky
Franklinist notion of electrical atmospheres. Volta called *applicatio* any
situation in which the electric fire external to a body affected the body's
electric condition *via attractive forces*.[69] As Volta exemplified, when a
body is brought near the charged chain of an electrostatic machine, it
will emit part of its internal fire not because the fire of the chain actually invades the body, but because the *attraction* exerted by the fire of
the chain will diminish the body's attraction toward its own internal
fire. Similar notions applied to negatively charged bodies as well.[70]
Thus, no actual exchange of electric matter took place in such cases,
and the effect traditionally ascribed to the fluid contained in the "electric atmospheres" surrounding the bodies was ascribed to attraction
instead. According to Volta, on the other hand, there is a minimum
distance between electrified bodies below which the fire can actually
pass from one body to the other depending on the amounts of the electric fluid and on the attraction forces involved.[71]

In 1769, therefore, Volta did not abandon the old notion of electrical
atmospheres entirely. In fact, he would continue to use the expression
for years to come. His appeal to attraction, however, had the effect of
drastically limiting his use of the old notion, as well as curbing some
of the ambiguities it carried with it.

Volta still hoped, in 1769, that his approach could coexist with Beccaria's notion of vindicating electricity, which, he claimed, was compatible with his own theory.[72] Some striking differences, however, were already in place, and were brought about by the two different
approaches Volta and Beccaria were pursuing. By focusing on conjec-
tures about underlying microscopical events, Volta drew from the
same observed phenomena conclusions which were different from
those of Beccaria, who now refrained from speculations on microscopi-

cal events. Let us consider, for example, the simple case of a metal coating joined to and then separated from an electrified glass pane. After joining, and having touched the coating with a finger, the glass and the coating do no longer produce electric signs and remain attached to each other. Beccaria interpreted this by saying that electricity actually disappeared at that point, only to be "vindicated" by glass when glass and coating were separated again. Volta suggested, instead, that attractive forces due to the internal texture of glass "retained" the electric fluid for some time, perhaps for hours, and were responsible for the firm adhesion of the coating to the glass pane. This last phenomenon, then called "electrical cohesion," had already been observed and tentatively quantified by Symmer.[73] In Volta's interpretation, by joining glass and coating a temporary equilibrium was created, responsible for the cessation of electrical phenomena, to which no actual disappearance of the electric fluid and of attractive forces corresponded.

Unwilling to emphasize the differences from Beccaria's theory in 1769, in the following years Volta developed a growing and explicit dissatisfaction with it.[74] Indeed, he seems to have developed a certain dissatisfaction with his own theoretical ambitions as well.

## Disenchanted Theorist

The reception of Volta's first effort at an ambitious contribution to the theory of electricity was, to say the least, cool. In 1768, before publishing, Volta had made known to Beccaria some of his ideas. Beccaria—with whom Volta had been trying to keep a correspondence alive for some five years—reminded young Volta that he himself had used the notion of attraction to explain observable, electrical phenomena in a book published fifteen years earlier, and that he (Beccaria) had since abandoned the idea.[75]

The next recorded reaction of Beccaria was a refutation of Volta's interpretation of vindicating electricity in Beccaria's big treatise on *Elettricismo artificiale*, published in 1772. Beccaria alluded to Volta, but he did not even mention Volta's name and work. He insisted that, in the mentioned experiments on glass panes and metal coatings, there were no signs that glass retained its electricity throughout.[76] In April 1774, perhaps after another fruitless exchange of letters, Volta understood that Beccaria had invited him "to keep silent forever" on matters concerning electricity.[77] Volta let Beccaria know that he had got the message, and complied with the prohibition, writing to him, instead, on small-pox inoculation.[78]

Volta had also anticipated some of his ideas to Nollet[79] with whom, as we know, he had occasionally corresponded since 1763 and to whom he had very probably sent his 1769 essay. No reaction on the part of Nollet is recorded (he died in 1770). However, it is hard to imagine anything better than a cool response to the work of an author, like Volta, who was perceived as staying in the camp of the Franklinists and who had dedicated his first publication to Beccaria, notoriously adverse to the system of Nollet.

Consolation, and solid advice, came to Volta from Paolo Frisi, another correspondent he had wisely included in his network and who, working in Milan and having many connections with the Lombard nobility of which the Voltas were part, was much closer to him as a model and an adviser. In July 1771 Frisi gave Volta to understand that he attached very little sense, if any, to Beccaria's notion of "vindicating electricity."[80] Implicitly, he was advising Volta to desist from his efforts to place the notion on firmer theoretical grounds. More positively, Frisi suggested that Volta, when advertising his credentials as a young natural philosopher, should place as much emphasis on scientific instruments as on (controversial) theory. Second, he put a translation of Priestley's history of electricity in Volta's hands.[81] The suggested reading and the advice—intellectual and strategic—set Volta on the road to the electrophorus.

## Scientific Instruments and Their Social Uses

Frisi's advice fell on fertile soil. Volta had long shared an interest in the sort of eighteenth-century collection in which physical machines found their place side by side with all kinds of natural and artificial objects. The unfinished Latin poem of 1764 showed Volta enjoying such semidomestic arcane phenomena as fireworks, and more sophisticated chemical marvels like fulminating gold. The use of Latin verse reflected the gentlemanly, Arcadian character of the undertaking; but it did not conceal that Volta enjoyed the practical business of preparing his own fulminating gold.[82] There are signs that Volta turned increasingly to the practical side of natural philosophy after the cool reception accorded to his theoretical essay of 1769. Volta's second published memoir, in 1771, already promised a "novus ac simplicissimus electricorum tentaminum apparatus" in the title itself.[83]

The 1771 essay showed Volta's undiminished commitment to a corpuscular and mechanical model of explanation. It included new speculations on the internal makeup or "texture" of different materials, affecting their electrical properties via attraction.[84] The notion of "applicatio" was reasserted and developed to include repulsive forces active in bodies which are electric per se, or dielectrics.[85] The survey of the electrical properties of dielectrics was extended. New observations were reported on the propensity of metal foils to "give" the electric fluid to different dielectrics.[86] The essay also aimed at accounting for the change some materials underwent from the condition of conductors to that of dielectrics when submitted to special treatment. Emphasis was placed, above all, on the description of two, new electrical apparatuses that—following a pattern we have already met when considering the silk electrostatic machine of 1765—displayed at work the principles Volta invoked in the theoretical part of his essay.

One instrument[87] was built like a classic electrostatic machine, except that the usual glass globe was replaced by three wooden discs. To produce electricity by friction like glass, the wooden discs had to be heated, either by frying them in oil (and then drying them carefully), or by baking them in a oven. Under these conditions, wood, often acting as a conductor, was found to act as a dielectric. The other unconvential apparatus Volta described was a Franklin square in which wood took the place of the usual glass, and with which he tried successfully the Peking and similar experiments.[88]

Soon after the 1771 publication, Volta discovered in Priestley's *History of Electricity* that not everything in his experiments on wood was new: a certain Ammersinus had tried similar experiments to an extent Volta was apparently unaware of.[89] Notwithstanding this, Volta's machine was perceived as curious and new. The fact that it simulated the classic electrostatic machine—based on glass, the dielectric par excellence—using wood instead, was an especially striking illustration of how much physicists still had to learn about conductors and dielectrics.

Volta used his all-wood electrical machine to promote his work and his ambitions as a young physicist. He addressed the 1771 memoir to Lazzaro Spallanzani, professor of natural history at the University of Pavia, and sent him a version of the electrostatic machine and the wood Franklin square together with the printed essay itself.[90] At that time, as we know, Volta nurtured hopes that he was a likely candidate for a chair of physics soon to become vacant in Pavia.[91] He probably also sent a machine to Frisi in Milan, who in turn suggested that he send the printed essay and the machine to Carlo di Firmian, Austria's plenipotentiary minister in Lombardy, who was to become Volta's powerful patron in a few years time.[92]

That Lombardy was receptive ground for the kind of physics Volta was beginning to practice in the early 1770s, is shown by many circumstances. Volta himself attended meetings arranged in Milan by similarly inclined amateurs, science professors, and instrument makers active in the Brera schools.[93] Until 1773 a college run by the Jesuits, during the construction of the observatory under the direction of Boscovich and again after the suppression of the Society of Jesus, Brera was a lively center for science teaching and research. Brera's physics cabinet and astronomy observatory attracted a considerable number of instrument makers, or *macchinisti*, the presence of whom in Milan would continue to be a subject of envy for Volta during his subsequent career in Como and Pavia.[94]

In frequent trips to Milan, Volta enjoyed the evening *conversazioni* arranged by an accomplished physicist, Carlo Giuseppe Campi, the Italian translator of Franklin's *Experiments*, and the first to present the electrophorus in Milan the following year.[95] Volta also had frequent contacts with Marsilio Landriani, a nobleman interested in physics and chemistry, who was soon to be professor of physics at Brera.[96] Campi and Landriani were the sort of people to whom, in commissioning some new physical apparatus, one would turn for the supervision of the work of the Milan instrument makers.[97] Another Milanese in close touch with Volta at the time was Giovanni Francesco Fromond, a canon employed by the government to work in the physical cabinet in Brera, and known for his skill in polishing prisms.[98]

In those same circles, the project to start a new scientific journal was slowly gathering speed. It was especially designed to bring the works of foreign natural philosophers to a public south of the Alps, and it was focused on chemistry, physics, and natural history. Volta had already been invited to contribute to the journal in 1773.[99] The publication was finally launched in 1775, and Volta was to publish the first notice of the electrophorus there.[100] The chief editor of the new periodical was Carlo Amoretti, later secretary of the Patriotic Society of Milan, established in the 1770s with the purpose of promoting the "useful arts," especially agriculture, among the landed classes of Lombardy.[101] As Pietro Verri, the leading personality among enlightened intellectuals in Lombardy, was to put it in 1778, members of the Society were expected to act as "mediators between the educated physicist and the artisan machinist."[102]

Education, however, was still the chief source of support for the kind of physics Volta and his friends were cultivating. The extent of the support it obtained from the Austrian government is further illustrated by developments at the University of Pavia. The physics professor there was Carlo Barletti, a distinguished electrician and another of Volta's frequent correspondents in the early 1770s.[103] The examples of Milan and Pavia made Volta hope that he could encourage and exploit similar developments in Como. His initiatives to this effect were closely bound up with the discovery of the electrophorus.

After lobbying in Como, Milan, and Vienna, exploiting his family connections to find appropriate patrons, Volta was appointed superintendent to the secondary schools in Como on October 1774.[104] This success came just eight months before he publicly presented the electrophorus. The machine was conceived by Volta having just taken up his responsibilities as a public servant, in charge of surveying the slow start to the educational reform launched in Como by the Austrian government. Volta's duties included that of lecturing should a particular professor be unavailable.

He relished the change in his status from that of none-too-wealthy amateur to that of an official enjoying public responsibility and resources. Two months before announcing the discovery of the electrophorus, he had sent the detailed plan for the reform of education in Como (discussed in chapter 1) to Firmian.[105] Four months after the announcement he was appointed to the chair of physics in the secondary schools of Como, a post he combined with that of superintendent.[106] Thus well placed, he started making plans for the impressive development of physics facilities he had been contemplating for some time. These plans envisaged: a collection of physical instruments, for research as well as for teaching and public demonstrations; an assistant, to be employed during demonstrations, and when otherwise needed; a substantial addition to the school's library; a new, fittingly grand lecture hall for the physics professor; money for prizes to be offered to the best students.[107] If realized, Volta's plans would have amounted to little less than a revolution in the teaching of physics in Como. As it turned out, the Austrian government was unwilling to allow such a development in a small town like Como. In the end it was decided to send Volta to Pavia instead. The plans Volta conceived around 1775, however, offer a clue to the scope of his ambitions as physicist, teacher, and public servant at the time he was developing the electrophorus.

If education was the main, it was not the only, source of support and encouragement for the kind of physics Volta was increasingly drawn to in the early 1770s. The local nobility, from Archduke Ferdinand of Lorraine, the highest representative of Empress Maria Theresa in Lombardy, down, had a reputation for being fond of physical machines and experiments.[108] When Volta made inquiries to see whether a good *meccanico*, Marco Saruggia, would occasionally work for him, he found that the artisan was too busy to accept new commissions, having to attend almost daily the archduke's physical collections.[109] As a teacher and a superintendent to the schools in Como, Volta soon learned how to please his students' parents with public demonstrations of physical experiments, which, accompanied by refreshments, offered fashionable opportunities for the local nobility to gather.[110] In this as in other pursuits, Como imitated Milan, just as Milan imitated Vienna. Sending a big electrophorus to Vienna was to be one of Volta's first priorities when fighting for recognition as inventor of the machine.

In Como, as in Milan and Vienna, physical instruments were thus being used in social transactions in which the shared values of education, the enlightened "pursuit of knowledge," as well as the notion of "public utility," all played a role. As Volta was by this time well aware, the physicists' cosmopolitan Republic of Letters also had a role to play in these social interactions. As he had come to see things in the years 1771–75, the Republic had its most effective spokesman in Joseph Priestley. Accordingly, Volta turned to him for advice, for recognition, and in an effort to bypass the jealousy of authorities closer to home like Beccaria.

## The Path to the Electrophorus

Frisi lent Volta the French translation of Priestley's *History of Electricity* in summer 1771, as already mentioned. Some time later, certainly by early December 1771, Volta had assimilated enough of it to find that he could not claim priority for his experiments on fried or baked wood.[111] By that time, however, Volta had also developed a new version of his unconventional electrostatic machine. He now used a disc made of baked cardboard, instead of wood. Early in 1772 he felt confident enough of the new apparatus to describe it in a letter to Priestley.[112] In writing directly to Priestley he was treading a path he had already followed, not very successfully, in his approaches to Nollet and Beccaria. Priestley was more encouraging, however. He wrote back to Volta:

> The idea of your machine, made of cardboard, impressed me. That is why I had it built for me, and I was much surprised to see its effects, however inferior they are compared to the ones produced by our glass globes. I can well imagine that, should the machine be built in a better manner, its force would be greater; especially by adding several sheets of cardboard or wood (what would be quite easy to do), with one friction pad acting on two sheets at a time.[113]

This was the first, authoritative recognition Volta had won in his career as an electrician. It touched his role as a philosophical instrument maker, rather than as a theory maker. The circumstance, together with Frisi's earlier advice, throws light on Volta's subsequent strategy when dealing with theory and instruments.

Soon afterwards, in May 1772, Volta sent some new reflections on conductors turned into dielectrics to Priestley.[114] On this occasion Volta reinstated his mechanical model of explanation. He also showed a keen interest in those special circumstances under which the phenomena of so-called vindicating electricity were seen to persist for some time, and in the materials most likely to be conducive to the effect. From his experiments he had inferred that resin preserves its electricity for a longer time than glass.[115] He explained this by arguing that glass, being more "solid"—that is, having a microscopic texture composed of a higher number of particles as compared to resin—exerts less attraction than resin on the electric fluid and it lets it go quite easily.[116] Wood and resin seemed to Volta comparatively "lazy" or unwilling to free their electric fire.[117]

For some time, empirical investigations of the kind just described went on side by side with his plan to write a new theoretical essay on "electric atmospheres." However, in the summer of 1773, describing his intentions to Carlo Amoretti, who was pressing him to publish, Volta was cautious about the practicability of a theoretical memoir on the subject.[118] Before writing, he declared, he needed to consult several publications not available in Como. These included the *Philosophical Transactions* from 1770, Beccaria's large treatise of 1772 on artificial electricity, and Barletti's book, also published in 1772, which he asked Amoretti to obtain for him. This was not a difficult thing to do in Milan, and Amoretti is likely to have satisfied the request.[119]

Of these works, Beccaria's new treatise was certainly the one in reading which Volta felt most engaged. Volta's disappointment must have been accordingly acute. As we know, Beccaria had rejected Volta's interpretation of the theory of vindicating electricity. Beccaria's persistent hostility toward Volta's theoretical ambitions must have reinforced Priestley's and Frisi's positive encouragement to focus, temporarily at least, on instrument and experiment rather than on theory. Moreover, in works Volta certainly had in hand at that point—in Priestley's *History* and in Barletti's *Physica specimina*—he found the description of several experiments, in the repeating of which he could extend his line of investigation on unconventional dielectrics.

The most intriguing of such experiments were those of Wilcke and Aepinus on melted dielectrics. As Priestley reported,[120] Wilcke had established that melted sulphur, placed upon conductors and then cooled, was found to be strongly electrical when removed. Wilcke had varied his experiment ingeniously. He had found that melted sealing wax poured into glass acquired a negative electricity; when poured into sulphur, on the other hand, it acquired a positive charge and left the sulphur negative.

3.8. A diagram of Aepinus's sulphur and cup apparatus, from Aepinus, *Tentamen theoriae electricitatis et magnetismi* (St. Petersburg, 1759) (*Biblioteca Nazionale Braidense, Milan*). Volta, apparently, did not know Aepinus's figure when building the electrophorus; but he certainly knew the descriptions given by Priestley and Barletti. Compare Aepinus's sulphur and cup apparatus with Volta's first illustration of the electrophorus, fig. 3.1, above.

Even more intriguing were Aepinus's experiments as reported by Priestley and Barletti.[121] Aepinus had poured "melted sulphur into metal cups, and [had] observed that when the sulphur was cool, the cup and the sulphur together showed no signs of electricity, but showed very strong signs of it the moment they were separated."[122] Aepinus had also noted that "electricity always disappeared when the sulphur was replaced in the cup, and revived upon being taken out again." In such circumstances the cup acquired a negative and sulphur a positive electricity. But, Priestley's report continued, "if the electricity of either of them had been taken off while they were separate [for example, by drawing a spark from them with a finger], they would both, when united, show signs of that electricity which had not been taken off."[123]

Comparing Priestley's and Barletti's reports of Aepinus's experiments with the sulphur and cup apparatus (fig. 3.8) and Volta's early descriptions of the electrophorus (fig. 3.1), it is impossible to miss the convergence of lines of investigation. Through Priestley and Barletti, Aepinus's sulphur and cup experiment played a key role in Volta's path to the electrophorus; a point so far unnoticed by historians.[124]

The convergence, however, did *not* point necessarily toward the realization of an apparatus like the electrophorus. In fact, not even the many who, unlike Volta, knew Wilcke's and Aepinus's descriptions of their experiments (published in 1757 and 1759, respectively) at first-hand, were led to build a machine of the kind conceived by Volta. Judging from Volta's early descriptions of the electrophorus, one suspects that an early intersection between the lines of investigation of Volta on one hand, and that of Wilcke and Aepinus on the other, resulted, not in the idea of building a new apparatus for displaying electricity, but in experiments on fusions of different dielectrics, intended to test the electrical properties resulting from the combination.

This line of investigation, documented in Volta's work for a period following his announcement of the electrophorus, should probably be regarded by historians in the way biologists regard "rudimentary organs" in living beings: that is, as evidence of some significant past step, important in the process of evolution and irrelevant to future stages of the process. As it turned out, the dielectric of the electrophorus, as recommended by Volta, could be made of many materials. For example, in an early version built by George Adams the younger in England—who followed for the rest Volta's instructions—the dielectric was made of glass, as in an ordinary Franklin square.[125] Another "rudimentary" characteristic of the electrophorus as described by Volta was the presence of *two* metal plates, one on top of and the other supporting the dielectric, as in a Franklin square. As soon became apparent, the latter metal plate was superfluous to the functioning of the machine.[126]

In any case—going back to Volta's path to the electrophorus—Priestley's and Barletti's reports of Wilcke and Aepinus appealed to Volta above all because of their implications for "vindicating electricity," or "spontaneous electricity," as Wilcke-Priestley put it. In this regard, the procedure described by Aepinus-Priestley—consisting in removing some electricity by drawing a spark from the sulphur and the cup when separated, finding the two still electrical when united—was particularly interesting. The procedure certainly reminded Volta of an intriguing maneuver he was familiar with, originally suggested by Giovanni Francesco Cigna in Turin in 1765, and to which Priestley had called attention in his *History*.[127] The procedure consisted in charging a Leyden jar by a series of operations involving an electrified body like a stocking, a plate of lead, and the jar. In favourable conditions (dry weather, etc.) the jar could be charged without subtracting much electricity from the stocking. Cigna proceeded as follows: While the charged stocking was brought near the plate of lead, he took a spark from the opposite side of the plate with the hook of the Leyden jar. Having removed the stocking, he took another spark using his finger or any earthed conductor. He then approached the stocking again, taking sparks with the jar and with the finger in the same manner. After repeating the operation a number of times, the jar was charged.

As Priestley reported, the effect was explained by saying that, when the stocking was brought near the plate, the electricity of the stocking—unable to actually enter the plate—drew the fluid away from the part of the plate facing the stocking, pushing it to the opposite side. This side of the plate, being overcharged, shared its electricity with the hook of the jar whenever the jar was approached under similar conditions.[128] In fact, we would now say, the trick was done when earthing the metal plate, so that it became charged by induction without significantly affecting the charge of the stocking.

At this stage, one might say that to build his "perpetual electrophorus" Volta had only to combine four ingredients that were all at his disposal in summer 1773. These were:

- Volta's view that certain dielectrics had the specific power to retain electricity for a long time.
- Volta's notion, still imperfectly formulated and controversial among those who had paid attention to it, according to which no real transfer, nor a regeneration of the electric fluid was to be postulated when dealing with "vindicating" electricity.
- Aepinus's suggestion, encountered via Priestley and Barletti, of an apparatus in which melted sulphur and a metal cup interacted in an intriguing manner, became electrified and preserved electricity for a long time.
- Cigna's sophisticated way of maneuvering an electrified dielectric and a conducting plate to charge a Leyden jar.

If we want to avoid reading history backwards, however, the four ingredients mentioned should be seen merely as cognitive and operational strategies somehow interacting in Volta's investigation from 1773 onward. No one of these strategies may be said to have pointed at the realization of a specific, separate machine of general applicability like the one Volta announced to Priestley on 10 June 1775.[129] The various types of apparatus used up to that point in research on "vindicating" electricity had been strictly *experimental* devices. They were used, basically, to test, and possibly to offer new clues on, the laws hidden in the amazing variety of such electrical phenomena. Even Aepinus's cup and sulphur apparatus did not deviate from the pattern, and as such it remained little noticed for more than a dozen years. In order to realize the electrophorus, while combining the four ingredients mentioned above, Volta had to add another, important ingredient. This was

- conceiving and directing his energies toward the goal of realizing a simple, easily replicable, portable machine to be used in displaying electricity, and enabling a lecturer to give a full course on it, while at the same time conspicuously testifying to the ingenuity of its inventor.

The social context we have sketched above, and Volta's personal ambitions, were instrumental in both shaping the goal of building such a machine and helping Volta to achieve it. In focusing on the goal of creating such a device between 1773 and 1775, Volta was putting to good use Priestley's encouragement and Frisi's advice. If all the five ingredients mentioned so far were indispensable in concocting the new machine, it is the latter goal which explains more than anything else what made *the difference* between the electrophorus and the many apparatuses that had been used in experiments on "vindicating" electricity for some twenty years before 1775.

### Instrument Design

The goal Volta had at last conceived interacted with his theoretical notions—by then well-rooted, though still volatile and controversial—in the shaping of the new instrument he finally had ready in spring 1775. Yet, the early descriptions of the machine bear witness to the fact that, when building it, Volta was focusing above all on the machine itself rather than on the principles involved: in the letter to Priestley of 10 June 1775 Volta did not even mention his beloved concept of attraction. The emphases and the omissions used there tell us something about the process of discovery as well.

The instrument Volta slowly developed between 1773 and the spring of 1775 (fig. 3.1) had many characteristics in common with a Franklin square (fig. 3.2) and with Aepinus's sulphur and cup apparatus (fig. 3.8), as suggested by the "rudimentary organs" already mentioned.

Improvements on the models were the result both of theoretical notions and refinements by trial and error. An instance of the role of the former was the use of resin instead of glass or sulphur for the dielectric—in fact, Volta had also tried wood.

Instances of trial and error refinements in the development of the apparatus recur time and again in Volta's descriptions. They range

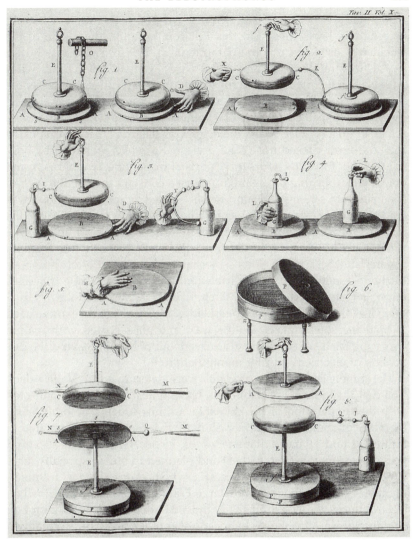

3.9. Volta's repertoire of experiments for the electrophorus. From *Scelta di opuscoli* (1775) (*British Library, London*).

from the choice of the metal for the lower dish (tin or brass, fig. 3.1, *CC*), and its shape (with a raised border, one line deep, it performed better), to the material and shape most appropriate for the upper metal dish or "shield" (fig. 3.1, *AA*). For this Volta first tried a simple metal plate. As it turned out to be too heavy and impractical for repeated manipulations, he used a wooden disc covered with gold leaf instead, or a hollow brass disc.[130] In this last configuration the inside of the shield was used to carry the paraphernalia the electrician needed for his demonstrations: silk and metal wires, a small Leyden jar, wires with cork balls fitted to them for testing the electricities. One essential feature of the "shield" was its gently curved borders, to avoid electricity getting dispersed by way of "points," as it was well-known to be the case.[131] The choice of the dielectric was itself the result of repeated trial-and-error procedures. Volta preferred a "cake" obtained by boiling together for some hours three parts of turpentine, two of rosin, and one of wax (burning of the mixture was a constant threat). But he had tried sulphur, sealing wax, and mastic as well.[132] Special attention was paid to avoid the cake getting cracked when cooling, in which case the instrument would not work properly. The thinness of the cake—thought to be essential because of the analogy with the Franklin square, and because of Volta's mechanistic notion of short-range attractions—soon proved to be irrelevant, following a simple trial-and-error procedure performed by Archduke Ferdinand in Milan.[133] Equal care was taken to devise the composition and shape of the handle used to hold the shield: glass, sealing wax, and glass coated with sealing wax were tried, all conveniently curved and polished to avoid the dispersion of electricity during manipulation.[134]

The dimensions of the apparatus were determined by Volta's decision that it should be portable, and by technical limitations connected with the preparation of the cake. Several instruments he built had a diameter of about one foot, occasionally two feet. Those illustrated in figures 3.1 and 3.9 had a diameter of five inches. However, he also tried pocket instruments (two inches) and planned to build especially large versions. Clearly inspired by Beccaria, who had described a "fulminating table" built in the form of a large Franklin square, he imagined an electrophorus the size of a table. Indeed, extending the technique of charging one electrophorus by means of another, he imagined that two such tables could be built, which would act on each other: supposedly, an especially impressive challenge to Beccaria's scepticism about young Volta's abilities.[135]

For the largest instruments, Volta and several others—including Joseph Klinkosch, Marsilio Landriani, Erasmus Darwin and William Nicholson—were later to suggest that a mechanical device be used for raising and lowering the shield.[136] The arrangement, combining electrical and mechanical apparatuses, had a special appeal for physicists and instrument makers between 1775 and 1800. In one version of a mechanically operated device, suggested by Volta in 1775, a pendulum was supposed to act alternately on two electrophoruses and a Leyden jar, producing "effects multiplied more than tenfold."[137]

Volta was apparently keen to suggest some kind of "useful application" for his machine. The only field in which there were applications of electricity at that time (the lightning rod excepted) was medicine. By electrifying parts of the body and entire people in various ways, some physicians had declared that they had obtained striking (and highly controversial) results, especially in the treatment of paralysis and deafness. Priestley, in his *History of Electricity*, had suggested that hospitals should have especially equipped rooms for electrical therapy.[138] Volta thought that the electrophorus was a particularly convenient device for the purpose and imagined gigantic electrophoruses— big as "lightning clouds"—hanging from the ceiling of the hospital rooms recommended by Priestley.[139]

The goals pursued by Volta in the realization of his instrument, and what distinguished the electrophorus from other similar apparatus, were further marked by the striking name Volta "dared to attach" to his machine: "the perpetual carrier of electricity."[140] Impinging on needs, as well as fancies shared by the community of the electricians at large, the name soon attracted special attention and helped to overcome the reluctance to recognize that the machine represented an important new development. Naming the machine, and choosing an impressive name for it, was not minor a step in Volta's achievement. Suggesting a rich repertoire of experiments to be carried out with the apparatus was another step to which Volta paid considerable attention, as can be seen from the table that accompanied the first description of the machine (fig. 3.9).

Demonstrations could be varied in many ways. The electrophorus lent itself easily to the function of being the main attraction in the kind of popular courses on electricity to which, early in 1776, Rozier's journal invited amateurs in Paris, "chez M. Sigaud de la Fond, rue St. Jacques, près St. Yves, maison de l'Université."[141] For the amateur with some background in physics, as well as for the physicist, the appeal exerted by the machine was to a great extent associated with the troublesome question: Where does the electricity producing all those sparks come from when nobody is rubbing the dielectric or turning the wheel of an electrostatic machine?

The question was, and was to remain for some time, unanswered. Volta himself, despite his theoretical commitments—which he had clearly embodied in the machine—felt that he had no truly convincing answer. When first illustrating the instrument, he mentioned that he was deeply interested in its *theory*. He also announced his intention of publishing a theoretical memoir on "electric atmospheres" in the near future.[142] In fact, it took another three years for him to produce something on these lines.[143] Meanwhile, the electrophorus had earned its important place in the eyes of the international community of electricians, independently of such a theory.

Another circumstance indicates the relative independence of the electrophorus from Volta's ongoing theoretical work. This involved the information Volta actually had at hand when developing the machine, and sheds additional if intriguing light on the process. Historians have assumed so far that Volta, in 1775, ignored the earlier works of Wilcke and Aepinus.[144] We have shown that, in fact, he knew of them through Priestley and Barletti. According to our reconstruction, Volta made a good deal of Wilcke's and Aepinus' experiments on melted dielectrics. Yet he apparently did not acquire Wilcke's and Aepinus' works before 1775. The circumstance requires explanation. When assuming that Volta was altogether uninformed of Wilcke and Aepinus prior to 1775, historians have mentioned that their works were not easily available. They have further implicitly assumed that Volta had trouble in procuring foreign books in the provincial setting in which he was living. Both these factors were no doubt significant. But all we know about Volta and his local and international connections at the time suggests that these did not represent insurmountable obstacles. Barletti, Volta's correspondent, had Aepinus at hand in Pavia or Milan in 1772, Fromond possessed Aepinus's book on the tourmaline in 1775,[145] and Beccaria had Aepinus's *Tentamen* in his personal library in Turin.[146] As for Wilcke, his publications were not easily available, and were written in Swedish. Yet, in the same 1775 series of the *Scelta di opuscoli* in which the electrophorus was announced, a short paper of Wilcke from the memoirs of the Swedish Academy of Sciences was published in Italian translation[147]

The fact that Volta did not acquire Wilcke's and Aepinus's works seems to confirm that, on his road to the electrophorus, he was not really focusing on the theoretical side of the investigations on "spontaneous" electricity. If Volta had had a theoretical goal in mind, acquiring such works would have been essential; and he did acquire them after 1775, when working on the long-delayed memoir on electrical atmospheres.

## Publicizing Discovery

Volta's decision to send the first description of the electrophorus to Priestley does not require any special explanation: we know the encouragement Priestley had granted Volta three years earlier and the correspondence they had exchanged in the meantime. Judging from the way Volta dealt with Priestley and with local scientific circles, he clearly expected from Priestley a declaration stating both the signifi- cance and originality of his machine; a recognition to be carefully circu- lated in his subsequent dealings in Lombardy and elsewhere. Priest- ley's answer, however, was delayed for almost one year. Meanwhile, because of information leaks among his circle of friends, which he him- self encouraged, Volta developed a parallel publicizing campaign in Milan in order to seek recognition and defend priority.[148] Not surpris- ingly, the most enthusiastic of his supporters in the campaign turned out to be people like Campi and Fromond, who had closer links with the instrument makers.[149]

"Philosophers" like Landriani and Barletti, on the other hand—who knew more about the antecedents of the machine and saw an ambi- tious competitor in Volta—were much more cautious.[150] Nevertheless, Milanese circles proved receptive ground for the first exploits of the electrophorus. By the summer of 1775 demonstrations were being car- ried out in Brera for audiences including well-known public personali- ties, professors, and many amateurs.[151] On 2 August the discovery of the electrophorus had been communicated to Carlo di Firmian, Aus- tria's plenipotentiary in Milan, who apparently attended a demonstra- tion that same week.[152] By 8 August many similar machines had been built in Milan.[153] Excited by the success, Volta was developing a larger apparatus the following autumn, to obtain "the grandiose effects that the big [electrophoruses] pompously display."[154] At that stage, wealthy amateurs like one Cavalier Litta, were anxious to order larger appara- tus from instrument makers and sought advice among Volta's friends.[155] In September and October Amoretti's journal published in Italian Volta's letter to Priestley, together with the detailed account and the table reproduced in figure 3.9.[156] On 1 November Firmian informed Volta that he had appointed him professor of physics in Como, a posi- tion he added to the position of superintendent.[157]

Local reward did not deflect Volta from seeking international recog- nition. On 13 November a box containing electrophoruses and off- prints was put in the hands of a Captain Ziegler, of the Imperial Guards, to be brought to Vienna and circulated among members of Lombard nobility there. In January 1776 a number of philosophers, physicians, and amateurs in Vienna learned of the electrophorus.[158] Among them was John Ingenhousz, then physician at the Austrian court, who went on, in 1778, to publish the most detailed account of the machine in the *Philosophical Transactions* of the Royal Society of London, unambiguously attributing the invention to Volta.[159] Ingenhousz, apparently, had learned of the machine directly from Archduke Ferdinand from Milan, who himself boasted of a high degree of competence about the machine.[160]

Despite the delay in Priestley's answer, Volta's letter to him proved very effective in circulating the news in Britain. Sometime before March 1776, George Adams the younger, "philosophical instrument maker to his majesty," showed what he proudly called "a machine for exhibiting perpetual electricity" to William Henley, who in turn published the first, short report on the electrophorus in the *Philosophical Transactions*.[161] At that stage the machine was also known to Edward Nairne, who was able to inform Henley that its inventor was a "M. Volta, of Coma [*sic*], near Milan."[162]

Priestley's answer, at last, arrived in spring 1776, when other messages and offprints sent by Volta had reached him. Priestley indeed gave Volta the acknowledgement he was seeking, although rather guardedly:

> I have very lately received the printed sequel of your letter to me, and also another letter containing a concise account of your discovery. It is indeed, a very curious one, and the study and application of it will, I doubt not, contribute greatly to improve the science of Electricity, and must do you the greatest honour with all who know the real value of it. We have begun to make the instrument in England, and succeed very well with small ones. We shall soon attempt some of a larger size. It must give you great pleasure to make so fine an experiment of your own discovery.[163]

The note of caution detectable in Priestley's statement was echoed and explained in another message Priestley sent through Jean-Hyacinthe Magellan. Writing on a copy of *Experiments and Observations on Different Kinds of Air* that Priestley had asked Magellan to address to Volta, Magellan transmitted Priestley's congratulations on the electrophorus, together with a comment by Magellan himself that recalled Wilcke's and Aepinus's experiments on melted sulphur as reported by Priestley. Magellan concluded: "Quoi qu'il en soit, c'est un phénomène fort remarquable, que la machine de M. Volta donne en perfection, et avec un avantage très agréable" ("However, it is a very remarkable phenomenon, that Mr. Volta's machine displays perfectly, and with a profit which is highly agreeable").[164]

Thanks to the frequent contacts between intellectual circles in Milan and Vienna and with Paris, the electrophorus soon reached France. Volta himself, in August 1775, had begun to subscribe to Rozier's *Observations sur la physique* through his friends in Milan.[165] In January 1776 Rozier's journal published a report on the machine sent from Vienna by Louis Sébastien Jacquet de Malzet: Volta's paternity was acknowledged in the title itself.[166] At the same time Sigaud de la Fond and his nephew Rouland were publicizing the course of lectures using the electrophorus to which we have already referred.[167]

In 1776 the electrophorus also found its way to Berlin. Franz Karl Achard reported on the machine in the *Mémoires* of the Berlin Academy for that year. Achard offers evidence of the variety of expectations that surrounded the machine at the beginning of its public career. After describing a series of experiments, Achard reported having also (unsuccessfully) tested the claim, advanced by an unnamed "celebrated naturalist," according to which a pendulum made to oscillate above or below an electrophorus kept its oscillations "within the magnetic meridian."[168] Linking the electrophorus with magnetism was, it would seem, not uncommon. Joseph Klinkosch, a physician from Prague who had learned of the machine in Vienna and who was among the first to try it out, used the electrophorus to deflect a compass needle. The experiment had the explicit purpose of refuting "the foundations on which the system of animal magnetism rests."[169]

The early success of Volta in seeking local and international recognition for the electrophorus did not prevent subsequent polemic over priority, as we shall see in the following chapter. That early success, however, did seriously inhibit later challenges. Volta himself, on the other hand, while eager to circulate the machine under his name, was extremely discreet when it came to spelling out his role in its realization. In a typical statement of 1776 he contented himself with claiming to have substituted resin for glass as a dielectric, and having found the right shape for the metal shield.[170]

Indeed, everybody acquainted with current developments in electricity, and Volta himself, saw how much the electrophorus had in common with devices that had been known for years previously. However, having pointed at the general use of the machine as a "carrier of electricity," and having improved on its design and performance, Volta could rightly claim to have achieved something genuinely new. The nature, and the limitations, of the breakthrough were in keeping with the goals Volta had imposed on himself as a result of his perception of the ways open to him in the community of the physicists. In turn, when linking the machine to Volta or when using the name he had suggested for it, electricians throughout Europe did not so much recognize the intrinsic novelty of the machine, as the new functions and the goals suggested by Volta. In regarding the electrophorus a genuine departure, as many did, they apparently subscribed to expectations and goals similar to those that had guided Volta himself in the construction of the device.

## Conclusion

It is difficult for the historian to distinguish—and very likely there never was—a complete, clear-cut separation between the theoretical, technical, social, and personal dimensions to Volta's investigation explored in the present chapter. The various goals that converged in directing Volta interacted, to varying degrees of intensity, at almost every step in the process.

The appropriate image for describing Volta's own motives and the intellectual and social contexts that, together, brought about the electrophorus, is one focusing on Volta's frequent movement back and forth between theory and the instrument-displaying-theory-at-work, in the decade preceding 1775. Volta had probably conceived, and certainly presented his silk electrostatic machine of 1765 as a striking proof of his own conceptualization of dielectric behavior. He had done the same with the all-wood or cardboard machine, and with the wood Franklin square presented in 1771. It seems that, in these cases, conceptualization did lead Volta toward the realization of the apparatus. In the years closer to the development of the electrophorus, however, things were somewhat different. The frustrations of the young theorist, the encouragement coming from many quarters to focus on instrument rather than on controversial theory, the amazing variety of the phenomena of "vindicating electricity," and the disputed explanations of them, all pushed Volta to pursue a different strategy. At this stage Volta came to concentrate more on instrument design than on theory. Certainly, in this case too he had a theoretical goal in mind, and that was the refutation of Beccaria's theory of vindicating electricity. Occasionally, Volta did present the electrophorus as being a practical refutation of Beccaria's theory: the instrument, he said, was there to show that electricity did not disappear when the metal disc was put on the resin cake. Yet, it still took years for Volta to develop a theory of the new instrument that would satisfy him: in the case of the electrophorus, the instrument itself and its potential social uses led conceptualization and theory.

When accounting for the discovery and the public impact of the electrophorus, the historian needs (as a stage for his account) to assume the multidimensional space mentioned at the beginning of this chapter; a space in which theories, instruments, social uses, and personal ambitions all have a role to play, and the paths leading from one another are two-way streets. To be sure, Volta himself was the main beneficiary of his well-publicized achievement, as we have seen in considering his rapid advancement in Como. Yet, in Lombardy and elsewhere his instrument became a starting point for many other electricians pursuing their own goals and careers, either in perfecting it or in pointing to new applications of it. Finally, if Volta's main concern in the crucial steps leading to his discovery had been the machine itself and its uses for his career, the performances of the machine soon drove him and many others to consider, above all, its implications for the theory of electricity.

Indeed, nothing less than a complex, multidimesional historiographical space seems to offer an appropriate setting for Volta's simple machine.

CHAPTER FOUR

# Volta's Science of Electricity

## CONCEPTION, LABORATORY WORK, AND PUBLIC RECOGNITION

### Reluctant Theorist

Soon after 1775, while savoring the mixed success of the electrophorus and once again following the steps of Priestley,[1] Volta entered a
new line of investigation, the chemistry of airs. From 1776 to 1791 he
devoted considerable energy to the new field and published extensively, achieving significant results, such as the discovery of methane,
and in the process acquiring wide recognition as a chemist.[2] The choice
of the new field, which he extended to include the study of heat, reflected the various facets of Volta's approach to the natural sciences. It
revealed the wide-ranging interests of the natural philosopher, as well
as his gusto for and opportunism in tackling fashionable and controversial issues. More relevant in the present context, it exposed Volta's
hesitation in carrying out the program for developing a comprehensive
system of electrical science that he had outlined in 1769, and continued
to pursue only intermittently in the 1770s and 1780s.

This chapter deals with the main features of Volta's science of electricity as it developed between the presentation of the electrophorus
in 1775 and the outbreak of the controversy over galvanism in the early
1790s, which provoked a new and momentous shift in his investigations. Volta's science of electricity will be considered both in its day-
to-day making—as the result of developments affecting concepts, instruments, and measurement techniques—and as it was perceived by
other scientists Volta encountered in his negotiations aimed at recognition. Indeed, one main purpose of the chapter is to explore the important links connecting conception, laboratory work, and public recognition in Volta's own investigations. The astounding degree to
which these different dimensions were interwoven in his work was
partly the result of Volta's own personality and education. Yet, in considering the rich documentation available—ranging from published
papers to laboratory notes, an extensive correspondence illustrating

private or semiprivate attitudes and reactions—it is tempting to regard our case as depicting some general characteristics of scientific investigation in the late eighteenth century, characteristics the historian is likely to meet whenever he or she is confronted by a similar wealth of documentary evidence. If so, this chapter may be regarded as a case study in the interaction between conception, laboratory work, and public recognition in late-eighteenth-century science.

The sudden success of the electrophorus among physicists and amateurs throughout Europe, described in chapter 3, was accompanied by frequent attacks on the legitimacy of Volta's claim to be regarded as its inventor. Though generally mild in tone, the controversy went on intermittently for a quarter of a century. The controversy affected, of course, the physicists' perception of Volta's work; more subtly, it also affected Volta's own perception of his contributions to the science of electricity. The main episodes of the controversy included the claim of Wilcke, supported by Lichtenberg, to priority in discovering the principle of the electrophorus, though not the machine itself,[3] and the veiled but severe criticism of Volta's assertion of originality in Haüy's survey of the works of Aepinus and Coulomb in 1787. Echoes of Haüy's criticism could still be detected in France in 1801, the year of Volta's triumphant presentation of his newly discovered electric battery in Paris.[4]

One consequence of the mixed impact of the electrophorus and of Volta's publications was that while some in the 1780s could depict Volta as "the Newton of electricity," Franklin being regarded as the Kepler, others continued to regard him as merely an inventor of "amusements électriques."[5] Volta himself was conscious, through his network of correspondents and frequent travels, of the ambiguity that accompanied his celebrity. Apparently as a consequence of this situation—as well as of the cool reception accorded to his juvenile theoretical essay—Volta avoided committing himself publicly to any systematic enunciation of the principles of electrical science. Nevertheless, several electricians throughout Europe were convinced by the early 1780s that there *was* indeed a Voltaic "system" of electrical science, which could valuably be built on by its author, or, should he fail to pursue the task himself, by others. Such a development engaged the attention and efforts of Jean André De Luc between 1782 and 1786. In the end, however—for reasons we will discuss—the Voltaic "system" of electrical science never materialized. Volta's contributions to the theory of electricity remained a number of midrange theoretical concepts stemming directly from his laboratory work, his other major contribution being his machines.

The task confronting the historian studying Volta's science of electricity in the period under consideration is analogous to that confronting the geologist studying a multilayer geological formation: both the elements internal to each layer and those making up the whole must be taken into account to explain stratification. Assuming that the whole of Volta's electrical science can be properly described as the work of a reluctant theorist, it must be borne in mind that several, diverse elements combined in its making. Among these elements the following figured prominently: Volta's long-lasting, underlying commitment to the microscopic, mechanical models of explanation cherished by natural philosophers; the instrumentalism typical of the practical electrician; and the pragmatism proper to an ambitious man who was trying to climb the hierarchy by which the community of electricians ranked its members. The implicit hierarchy within the community, as we shall see, included such diverse figures as the prestigious natural philosopher, the less prestigious inventor, the simple instrument maker, and the disparaged physicist juggler. The story we deal with in the present chapter amounts, in a sense, to an assessment of why Volta, who was occasionally recognized as the Newton of electricity, came to be known instead as a brilliant inventor of electrical machines. More important, our story shows how the limitations imposed on Volta's ambitions within the community of electricians contributed to the shaping of his investigative style. In several respects, Volta's best qualities as an investigator were the result of adjustment between his aspirations as a natural philosopher and the more modest role of inventor of intriguing machines that sections of the community of electricians came to assign him. That adjustment took place, ostensibly, during the negotiations Volta engaged in with his peers in his quest for public recognition. The adjustment, however, also improved his ability as an investigator to go back and forth between theory and instrument-displaying-theory-at-work, and to develop effective, midrange conceptualizations that proved seminal for the science of electricity.

To revert to our geological metaphor, while each layer making up Volta's investigation was molded by its own internal forces—the commitments of the natural philosopher, the conceptual and practical needs of the laboratory man and lecturer, the instruments themselves—it was the continuous adjustment between Volta's ambition and the response of the community of electricians that forged the special stratification we call Volta's science of electricity.

### Midrange Conceptualization and a New Machine: Capacity, Tension, "Actuation," and the *Condensatore*

The main achievements of Volta in the field of electricity in the years following the electrophorus are represented by two articles—one also

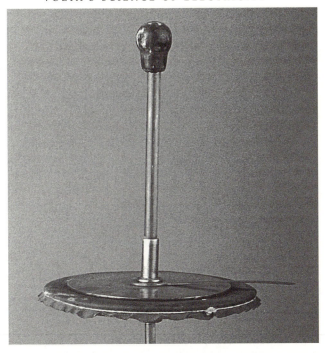

4.1. Volta's *condensatore*, reconstruction in the Tempio Voltiano, Como (*Musei Civici, Como*).

published in abridged form in the *Philosophical Transactions* for 1782—and a new machine, the *condensatore*, first conceived in 1780 and widely publicized two years later. The genealogy of the new machine, as viewed both by contemporaries and present-day historians, is much less controversial than that of the electrophorus. Yet, the origins of the new apparatus also lie in that entangled ground where conceptual developments, the inventive tinkering of the electrician, and the search for recognition intertwined. Indeed, the *condensatore* can properly be seen as emblematic of the period of Volta's investigation this chapter is concerned with. It is accordingly convenient to start with a description of it.

Volta's *condensatore* is a very simple apparatus, similar in shape and composition to the electrophorus, but having an entirely different function. It consists (fig. 4.1) of an insulating surface, such as a marble or wood plane (better if given a thin coating of copal paint), on which a metal disc lies, perfectly flush to adhere to the plane. The two parts making up the machine clearly remind one of the "cake" and the "shield" of the electrophorus, respectively. Their use, however, is dif-

ferent. If the metal disc is charged with any weak source of electricity
while lying on the insulating surface, when lifted it turns out to be
electrified to a much higher degree than if charged with the same weak
source of electricity under different conditions. The apparatus thus
makes detectable to ordinary electroscopes otherwise undetectable
amounts of electricity; hence the name "micro-electroscope" that Volta
also gave to the machine, suggesting that it might be useful especially
in the investigation of "weak" atmospheric electricity.[6]

It is worth stressing that if the composition of the *condensatore* resem-
bles that of the electrophorus, its proper use requires precisely that the
apparatus should *not* function as an electrophorus, that is, that it does
not itself develop electricity when being used to detect weak electricity
from an external source. It is also appropriate to remember that problems similar to those associated with the above-mentioned difficulty
were to affect a number of other machines introduced in the 1780s and
1790s to detect weak electricity. In fact, the difficulty and the task of
avoiding it were to play a role in Volta's development of the battery,
as we shall see in chapter 6, below.

If we take Volta's word at face value, the immediate antecedents of
the *condensatore* are to be found in the chance observation of an amateur electrician, the Marquis Pio Bellisomi of Pavia.[7] According to
Volta, at some time in 1780 Bellisomi was experimenting with an electrophorus when he noticed that, if he left the exhausted machine lying
on a table covered with animal skin, and after a while lifted the shield,
he could still draw a spark from it with a finger. As Volta tells us, having learned of the observation, he could derive from it and his own
notions on electrical atmospheres all the ingredients needed to develop
the theory and practice of the *condensatore*. Indeed, if we adopt Volta's
own published report, he could understand, explain, and build upon
that chance observation *because* he knew how to fit it into his conceptualization of the action of electrical atmospheres.

Volta thus presents us with a curious, double attribution concerning
the *condensatore*. On the one hand he credits an obscure amateur—
though conveniently noble and powerful in Pavia circles, and like
Volta fond of opera—with a significant, though merely empirical and
not decisive, step toward the realization of the *condensatore*. On the
other hand, he stresses that the new machine had been based, above
all, on his own *theory* of electrical atmospheres, a theory that, in those
same years, he was also presenting as the source of the electrophorus,
legitimizing his controversial claims to originality as inventor of the
earlier machine.

There is no easy way through the intricate plot that Volta himself
sets out. Indeed, the plot becomes more intricate if we include evidence

taken from Volta's private correspondence, where he once hinted that the *condensatore* had somehow *intruded* on his theoretical work and had diverted him from writing a memoir on electrical atmospheres.[8] The story gets even more complicated if we look at some manuscript drafts, in which Volta illustrated the operations of the electrophorus soon after its first presentation in 1775. In those manuscripts we find that he had noticed a phenomenon that, with hindsight, can be said to have hinted at the principle underlying the *condensatore*.

Both as normal practice and as a distinctive trait of his way of op-erating with the electrophorus,[9] Volta used a small Leyden jar to re-charge the apparatus when it was exhausted (fig. 3.1, chapter 3, p. 74). The operation consisted in bringing the hook of the jar into contact several times with the insulating cake of the electrophorus; as a result, the electrophorus regained its power to give strong electrical signals. While varying this operation using simple trial and error in order to find the most effective way of displaying the marvels of his machine, Volta had checked what happened if the Leyden jar was brought into contact with the metal shield rather than with the insulating cake. He had found that, if the shield lay on the insulating plane, it retained much of the electricity communicated by the jar, transmitting only a small part of it to the insulating cake and thus frustrating the attempt at making the exhausted electrophorus recover its strength quickly.[10] Under such circumstances the electrophorus clearly acted as a *conde-nsatore*. At that stage, however, Volta did not draw any conclusion from the observation other than a useful hint on how *not* to handle the electrophorus.

Brilliant performances with the electrophorus, in any case, were not Volta's only concern in those years. The attacks on his claims to be the discoverer of the machine,[11] and the advice coming from respected natural philosophers like Horace Bénédicte De Saussure and Jean Senebier with whom he had established regular contacts in the course of 1777 on the occasion of a trip to Switzerland, suggested to him that developing and publishing a convincing *theory* of the electrophorus should be his first priority.[12] Two considerations played a role here. First, such a theory would have established Volta's claim, adumbrated in the memoir presenting the electrophorus and often repeated later, that the machine originated in concepts already hinted at in his juve-nile theoretical essay of 1769. Second, a theory of the electrophorus was needed to reinforce Volta's reputation as a genuine natural philoso-pher, a reputation he was eager to add to the one he enjoyed as an inventor of intriguing machines, especially at a time when he was ma-neuvering to exchange his job in secondary schools in Como for a chair at the University of Pavia, the leading educational institution in Austrian Lombardy.[13]

The plan for a long, three-part memoir on the electrophorus, giving considerable space to a discussion of the "theoretical principles" of the machine, took shape during the winter 1777–78, partly at the instigation of Senebier.[14] By March or April 1778 a draft of the first part was ready, and Volta sent it to Senebier in Geneva.[15] It dealt with the construction of the electrophorus, and it was written with a view to publication in Rozier's journal.[16] The two other parts, which Volta informed Senebier were to be forthcoming in May 1778, were to deal with some "remarkable phenomena of the electrophorus" and with its "theoretical principles."[17] In fact, their realization underwent substantial changes in content and destination. By early August 1778 Volta asked Senebier to return the manuscript on the construction of the electrophorus, on the grounds that he now planned to address that memoir to the Royal Society, which he was approaching through George Cowper in Florence.[18] Meanwhile, he provided Senebier with some interesting new observations on conductors, revealing his plan to publish a long memoir on conductors and his decision *not* to write the two additional parts on the electrophorus which he had promised earlier.[19] As in the most creative stages of his investigation, Volta was going back and forth between theory and instrument-displaying-theory-at-work.

The planned three-part memoir on the electrophorus was slowly becoming a three-part memoir on conductors, which would include the theory of the electrophorus as a special case study.[20] The first memoir of the newly conceived series, dealing with single conductors, was in fact addressed to Saussure and published in Italian in 1778 and in French the following year. The two other parts, on conjugated conductors and on insulating panes, in turn underwent delays and changes in content and destination. Choice of destination for publication was largely a matter of policy. For example, Volta's decision to publish his first memoir on conductors in the Milanese *Opuscoli scelti* was clearly dictated by the reproach, addressed to Volta earlier that year by Volta's patron, Carlo di Firmian, to the effect that as a Lombard professor he was expected to publish in Lombard journals.[21] But here we are concerned primarily with content, and several concepts presented in the 1778 memoir deserve to be considered in detail: they represented the first intimations of Volta's system of the science of electricity to circulate widely, and they were at work in Volta's mind when the *condensatore* was conceived in 1780.

The notions in question were those of capacity, tension, and a special version of the notion of electrical atmospheres, often referred to by Volta with the word *attuazione* ("actuation"), in some way corresponding to the concept of electrical induction. Volta did not give formal definitions of his notions in the 1778 memoir, protesting that he was not writing a treatise, and that he could therefore avoid tackling the principles of the science of electricity.[22] He did, however, offer effective, working instructions to convey what he meant.

By *capacity* of a body Volta meant the amount of electricity the body could retain without letting the electricity drop off and be diffused in the air.[23] This implied something more sophisticated (and practical) than Volta's old, theoretical notion of "saturation,"[24] for he now *measured* capacity by the number of turns of the friction machine needed to make the body reach that condition, which he also called the condition of "maximum electrical tension" proper to that body.[25] He measured the *tension* of a body by counting the number of turns of the friction machine needed to make an electrometer connected to the body rise to a given degree, the electrometer being a simple flax thread with a cork ball attached to it.[26] Having made the measuring techniques remarkably reliable (considering their extreme simplicity!) Volta compared the capacity of several different conductors.

Measurement confirmed some notions already well established: for example, that capacity was proportional to the external surface, not to the mass of the body, something that Franklin and Saussure had already noticed.[27] Other results of the measuring program threw fresh light on phenomena which remained little understood, such as how best to compare the electrical performances of simple conductors with those of condensers like the Leyden jar or the Franklin square.

Using the above-mentioned measuring technique, Volta established that a conductor made of a series of wooden sticks 6 lines in diameter, coated with silver foil, arranged as in figure 4.2, page 118, and having a total length of 96 feet, had about the same capacity as a small condenser like a Leyden jar or a Franklin square having a coated surface of only 4 square inches.[28]

According to Volta, the quantitative comparison made it plain that the large shock given by the condenser was not due to the presence of plus and minus electricities on the plates, but simply to the large amount of electricity contained in the condenser because of its capacity. The capacity and tension of a single conductor being equal to that of a condenser, the conductor would give comparable shocks.

The notions of capacity and tension, and the related practice of systematic measurement, encouraged the posing of some old and more recent questions in new ways. For example: Why did coated areas as different as Volta's long conductor and a small Leyden jar have the same capacity? Why did bringing the sticks making up the long conductor closer than 3 or 4 feet apart drastically diminish its capacity?

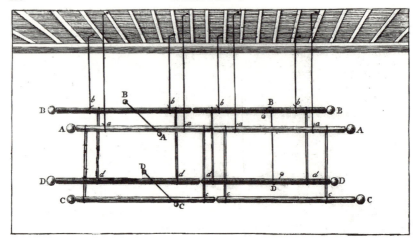

4.2. Volta's conductor as illustrated in his memoir to Saussure, from *Opuscoli scelti* (1778) (*Biblioteca Universitaria, Bologna*).

And how was the well-known fact to be explained that in a Leyden jar and a Franklin square the thickness of the glass so markedly affected the capacity of the condenser?

These questions could be addressed and answered satisfactorily, Volta maintained, by adopting his special notion of electrical atmospheres. He summarized it by saying that any electrified conductor brought close to another conductor "actuates" the latter, that is, it increases its tension and "accidental" (Volta's word) electricity, limiting its ability to acquire additional tension and reducing its capacity.[29] As Volta stressed several times, no exchange of electrical fluid took place in the "actuating" process: what was at stake was only the "sphere of activity."[30] At the same time, should an electrified body actuate another body up to the latter's maximum tension, no exchange of "absolute" or "real" (Volta's words) electricity would take place by bringing the two bodies into contact.[31]

The notion of actuation applied to the different parts of the same body as well. That was why, Volta maintained, the shape of a body, not just its surface, affected its capacity. Volta proposed that every single conductor be conceived as being divided into several bands or stripes, all "actuating" one another to various degrees.[32] Within this perspective, he maintained that it could intuitively be explained why a long thin conductor had a higher capacity than a short broad conductor with the same surface area: the individual components of the long conductor interfered less with the capacity of their neighboring parts.

Volta deferred a detailed discussion of his views on electrical atmospheres to the planned memoir on "conjugated" conductors. However, already in 1778 he had proclaimed his preference for the sort of midrange conceptualizations illustrated above, rather than for basic principles and fundamental notions. The phenomena displayed in the "actuating" process, Volta emphasized, were "a matter of fact," and the search for their deep causes could be postponed.[33] The phenomena of capacity and tension, he stressed, had to be studied through careful measurement rather than indulging in speculations on a supposed microscopic constitution, the pores of the bodies, or their "oscillations" in conductors.[34]

Yet, Volta's current preference for midrange conceptualization and measurement did not imply the renunciation of all sorts of "visualization" of what went on at the microscopic level. If the operative, measurable notion of tension played the principal role in Volta's memoir of 1778, he did not refrain from occasionally describing tension as the "effort [of the electric fluid] to thrust itself out" of the body.[35] In his mind, apparently, the physical, microscopic image of the electric fluid—reminiscent of the young Volta's speculations on electricity and attraction[36]—acted alongside the new, operative, measurable notion of tension. Certainly, Volta now refrained from speculations on attraction and on the interaction between the fluid and the particles of common matter. However, especially when explaining tension as the result of the "actuating" process—when no real exchange of fluid was postulated—the persistence of the image of the fluid could cause problems for Volta and his readers.

For his readers, other problems arose from the unusual combination of concept and measurement techniques in Volta's essay of 1778. Those who paid more attention to concept and experiment than to measurement—still a majority among the electricians at that time—failed to see anything new in Volta's essay. This is well exemplified in the reaction of an otherwise perceptive colleague of Volta: Marsilio Landriani. Landriani was on such terms with Volta that he felt able to write frankly that he saw absolutely nothing original in the paper.[37] According to Landriani, Volta's notions and discussion had been anticipated by a long series of electricians, including no less than Franklin, Nollet, Beccaria, Watson, as well as Le Monnier, Gordon, and Barbier. Landriani's challenge deserves to be considered in detail: it invites clarification of the interaction between conception and recognition that we are focusing on in this chapter.

Of the electricians mentioned by Landriani, Beccaria—whom Volta did not even quote in his 1778 essay—was the one to whom Volta owed most for the development of his own concepts. In his *Elettricismo*

*artificiale* of 1772 Beccaria had devoted a long and unusually clear discussion to the theory of "electrical atmospheres" and of "actuation,"
itself Beccaria's term.[38] Beccaria had repeated the view—which he
boasted he had been the first to propound twenty years earlier—that
no real exchange of electric fluid takes place in the process of actuation.[39] Beccaria had also repeated, and very penetratingly analyzed,
Aepinus's and Wilcke's experiments on the air condenser,[40] as well as
Richman's experiments on Franklin squares.[41] Volta had certainly read
and benefited from Beccaria's discussion of electrical atmospheres and
actuation. Yet, he was gradually, and incompletely as we shall see,
emancipating himself from a trait which remained central to Beccaria's
physics: a strict adhesion to a material interpretation of the atmospheres. Beccaria continued to regard the air surrounding a body as an
essential mediator in the process of actuation.

Problems different from those highlighted in Landriani's reaction
must have arisen in the admittedly few electricians well acquainted
and sympathetic with Aepinus's *Tentamen* and with Henry Cavendish's recent papers in the *Philosophical Transactions*.[42] These works proposed an approach to the science of electricity akin to Volta's; yet, Volta
gave no sign of being prepared to build on them.

As for Aepinus, by 1778 Volta knew the *Tentamen* at firsthand and
praised it highly; however, he insisted that he had looked at the book
only briefly.[43] At this stage, in any case, Volta is likely to have carefully
considered Aepinus's discussion of electrical atmospheres, which was
tantamount to the jettisoning of the old notion and to the adoption
of action at a distance as an acceptable notion within the science of
electricity.[44] This certainly helped Volta clarify ideas he had already expressed but not formulated clearly in his 1769 theoretical essay, and
which he had also found in Beccaria. Yet, probably with an eye on the
ongoing priority dispute over the electrophorus, Volta did not miss the
opportunity of criticizing Aepinus on crucial issues, such as the latter's
discussion of the air condenser.[45]

With respect to Cavendish, historians agree[46] that, by 1778, Volta was
acquainted with his papers on electricity of 1771 and 1776; the latter
he certainly knew and praised in 1782.[47] Though Volta did not quote
Cavendish in his 1778 essay, textual comparisons confirm that he must
have known the 1771 paper at least.[48] In that paper Cavendish, like
Aepinus, did without effluvia and admitted actions at a distance between the particles of the electrical fluid. Considering that very few
electricians at that time gave Cavendish's essays the credit they deserved,[49] the convergence of interests and approaches between Volta
and Cavendish is striking.

Indeed, around 1780, Volta's acquaintance with the best of Beccaria, with Aepinus and with Cavendish placed him in good position to attempt a systematic, updated survey of the basic notions of the science of electricity. The midrange conceptualizations Volta was employing at that time, and that he had put forward in his 1778 essay, as well as his reiterated plans to produce a long memoir on the theory of conductors, testify to his commitment to theory during the period when he developed his new machine, the *condensatore*. Yet, as in the case of the electrophorus, the new machine crystallized well before his planned theoretical memoir. By the summer of 1780 Volta was using his *condensatore* to detect atmospheric electricity during an aurora borealis, and advertising the machine to his fellow electricians. By the autumn of 1780 a number of electricians were also using the machine. On the other hand, it took another year to draw together his long memoir "on the capacity of conjugate conductors," and two more years to have it laboriously, and imperfectly, digested by scientific circles around Europe.

### Natural Philosopher or Inventor of *Amusements Électriques?*

By the late 1770s Volta's researches on inflammable air and eudiometry had won him considerable fame throughout Europe. His "letters" on the subject had been translated into French and German, and some of his experiments had been reported to the Académie des Sciences in Paris. On the other hand, his first attempt in December 1778, through Cowper and Nairne, to be admitted to the Royal Society, had failed mainly because of difficulties internal to the Society itself and to the plethora of foreign members at that time.[50] This made Volta eager to consolidate and further expand his connections abroad. Travel was the chief means to this end, and a popular one among professors in Lombardy.[51] Some time after his 1777 trip to Switzerland, Volta had submitted a new petition to his patron for a leave of absence and money to make a new journey. The stated goals included Paris, Brussels, Holland, England, and Scotland.[52] Meanwhile, in September 1780, he visited Tuscany, strengthening his contacts with Cowper in Florence, at the time his main link with the Royal Society.[53] The grand European tour, which Volta finally undertook from September 1781 to October 1782, was closely bound to the laborious reworking of his planned long memoir on conjugate conductors, with the circulation of the *condensatore*, and with the reception of his system of electrical science.

Volta had been working on the memoir on conjugate conductors, intended primarily for Cowper and the Royal Society, in the summer and autumn of 1780. It was actually sent to Cowper one year later, as Volta was setting out on his grand tour.[54] By the autumn of 1781 the memoir consisted of no fewer than 104 sections, occupying 108 manuscript pages in the English translation. It was presented as part of "A Treatise on Conjugate Conductors," and entitled "A First Memoir, Containing More Particularly an Indication of the Most Remarkable Advantages Resulting from a Mode of Insulation So Far Imperfect That It Can Scarce Be Called Such, When Compared with Perfect Insulation."[55]

The translation, provided by Cowper, was submitted to the Royal Society during the winter of 1781–82. It was read and discussed at four successive meetings, between 7 February and 14 March 1782.[56] Cavendish and Bennet attended at least this last meeting, but the reporter was almost certainly Cavallo. During those weeks Volta himself was in Paris, where he also circulated the memoir. At some point before the end of his tour he was to give a copy of the long memoir, now intended for Rozier's journal, to Lavoisier.[57] Meanwhile, on 3 May 1782, Volta had reached London.[58] Informed of the mixed reception met by the memoir—not least because of its prolixity—he prepared a much shorter and less ambitious version, focusing on the *condensatore* rather than on the general theory of conjugate conductors. This was the version finally published, in English and Italian, in the *Philosophical Transactions* for 1782. The long version was later published, in French, in Rozier's journal in 1783.[59] The reactions within the Royal Society, and the subsequent emphasis on instruments and move away from theory in the paper finally published in the *Philosophical Transactions*, are worth noting. Though mainly reflecting Cavallo's own views, the report in the "Journal Book" of the Royal Society may also retain traces of the reactions of other members, including Cavendish.

The most striking aspect of these reactions is the difficulty Volta's fellow electricians experienced in grasping the notions of tension and capacity. At a crucial step in the report on Volta's views,[60] the Royal Society reporter confessed his own difficulties and the need he felt to reinterpret Volta. As we know, Volta's implicit definitions of tension and capacity were operational and closely linked to his measuring techniques, while he abstained from speculations on the microscopic events involved. The Royal Society reporter, however, left operation and measurement aside, and reinterpreted tension as the "intenseness" or "elasticity" of the electric fluid contained in the conductor. Elasticity itself was assumed to depend on the nature of the "medium"—a perfect or an imperfect insulator—with which the conductor was in contact. Emphasis on the "elasticity" of the electric fluid was frequent in Cavallo's correspondence and in his published works.[61] On the other hand, the "medium," where the electric fluid circulates or expands, and the related notion of "resistance," also mentioned in the report,[62] were part of Cavendish's concepts.[63]

Given the difficulties at the conceptual level, within the Royal Society the safest ground for appreciating Volta's contribution remained the performances of his machine, the *condensatore*. Accordingly, the short version of Volta's paper for the *Philosophical Transactions*—that Volta prepared in London in close touch with Cavallo, Magellan, and De Luc—began with the words "A machine. . . ." Emphasis was placed, above all, on the apparatus, for which Volta proposed, once again, an attractive name, that of "micro-electroscope." The title of the paper focused on practice rather than concepts: "Of the Method of Rendering Very Sensible the Weakest Natural or Artificial Electricity."[64]

If these were the main shifts in emphasis and meaning Volta's work underwent in London in 1782, it was to experience other, significant vicissitudes in Paris. According to Volta's own report, also published in the *Philosophical Transactions*, his *condensatore* played a decisive role in allowing Volta, together with Lavoisier and Laplace, to obtain in Paris for the first time—on 13 April 1782—"undoubted signs of electricity from the simple evaporation of water, and from various chemical effervescences."[65] Lavoisier and Laplace, in fact, gave little space to Volta's machine in their account of the breakthrough.[66] Nevertheless, in Paris as in London Volta's achievements in 1782 seem to have depended more on his machine and on his performances as an electrician than on the notions developed in the long theoretical memoir that he carried with him and circulated during his tour.

That this was indeed the case is confirmed by a survey of opinions carried out by Jean André De Luc in the autumn of 1782 in London and in January 1783 in Paris. De Luc, a brilliant amateur whose inordinate interests ranged from geology to electricity to theology, had been won over to Volta's cause when they had met in London in the spring of 1782. As a consequence, De Luc planned to devote an entire treatise to illustrating Volta's system of electrical science. By interviewing electricians in Paris and London De Luc found, to his dismay, that Volta was regarded by many as a mere inventor of "amusements électriques,"[67] rather than as a genuine natural philosopher. De Luc's report is crucial testimony of the limited impact Volta's ideas made on his colleagues in Paris and London in 1781–82, and to the strategy De Luc himself thought necessary if they were to be made to reconsider Volta's views. De Luc's report is also an impressive instance of how closely conception and recognition interact in scientific activity, if its results are to influence the opinion of the peers.

As De Luc reported to Volta on the situation in Paris:

I found that, with the exception of Mr. De La Place and to some extent Lavoisier, all the others had not really listened to you [a year before, when Volta was in Paris], and the one who especially piques himself on being an Electrician, and on whom others so to speak rely on matters concerning Electricity, made fun of my enthusiasm, and almost shrugged his shoulders.

I resented the shock, and despaired. The more I felt, the less I expressed, while I pondered over the means appropriate to launch some major attack against these obstinate fellows, I announced that I would denounce the prejudices with more than just a memoir in the *Journal de Physique*.

Back in London, I reconsidered my Experiments, carried out by following your indications, as seen from the eyes that I could expect (having seen it myself) in those who especially deserved to be convinced; and I realized that they would not be impressed, and would remain indifferent. I realized that a general, fundamental Cause had to be discovered, acting amidst the several different causes that were modifying it; and while I never lost trace of that cause throughout all its modifications—and minds such as that of Mr. De La Place would certainly have figured it out—I understood that I could not assume these minds to feel like us, since you yourself had attracted so little attention when performing under their very eyes.[68]

Who was the man "who especially piques himself of being an electrician," and whose shoulders shrugged at Volta's proselytizing efforts in Paris? It has been speculated that it might have been Jean-Paul Marat, to whose house Volta was certainly admitted and where he saw the host performing electrical experiments.[69] But De Luc's hint that he was the person on whom others, including Laplace and Lavoisier, relied in electrical matters, makes it more likely that he was somebody of the standing of Jean Baptiste Le Roy, Coulomb, or Haüy. The identification of Coulomb or Haüy as Volta's antagonist in Paris would go some way to explaining the harsh tones of Volta's later rejection of Coulomb's inverse square law, as well as Haüy's veiled but severe criticism of Volta in his successful treatise of 1787, in which he supported Aepinus and, especially, Coulomb.[70]

Volta's own recollections of his stay in Paris suggests that he received a warmer reception among amateurs and instrument makers like a certain Billaum or Billaux than among people like Coulomb, Haüy, Laplace, and Lavoisier. At any event, in Volta's opinion the lack of a research tradition in "electrical atmospheres" in Paris caused serious misunderstanding. As he put it in acknowledging De Luc's report on the state of Volta's standing in Paris :

It must be noted that no French physicist has been acquainted so far with electrical atmospheres. That is why (with the exception of Mr. De La Place, this excellent mind, and Mr. Lavoisier to whom I explained my ideas more in detail), in Paris they have not listened to me, despite the fundamental experiments that I showed with my discs.[71]

Judging again from Volta's own report, in London he especially enjoyed the gatherings in which instrument makers, independent scholars, and fellows of the Royal Society met. Volta described with special vividness a meeting at Bennet's that Cavallo, Kirwan, and Walker also attended.[72] Bennet was a well-known "amateur" and instrument maker; Cavallo and Kirwan were both to some extent outsiders though fellows of the Royal Society; Walker was probably Adam Walker, a successful, itinerant lecturer in natural philosophy, and inventor of several mechanical devices.[73] It was on that occasion, in Bennet's laboratory and under the eyes of such a mixed group of people, that—as Volta reported—"the experiment on the evaporation of water, which did not respond so well at Paris, succeeded much better."[74]

Whatever the strains Volta's work underwent in Paris and London, the grand tour of 1781–82 had proved on the whole a success. The trip spread Volta's fame as a leading electrician and inventor; and if some whom he met did not see him as such, he also met people who came to regard him precisely as a first-rank natural philosopher. The latter included, as already mentioned, Jean André De Luc in London. He was not himself a prominent electrician, and his contribution to the diffusion of Volta's views proved, in the end, marginal. However, the correspondence that De Luc established with Volta while engaged in promoting Volta's approach to the science of electricity offers interesting evidence of the motives underlying Volta's approach, and of how Volta himself could bend his notions in his quest for recognition.

### Explanatory Models and Presentation Strategies: True Causes vs. Instrumentalism

In the course of several meetings in London in 1782 Volta presented De Luc what he regarded as his complete "body of doctrines" concerning electrical phenomena; a "system" of knowledge that, Volta emphasized, went well beyond what he had published thus far. Deeply impressed by Volta, De Luc gradually developed the idea, which he submitted to Volta in 1783, of publishing a treatise entirely devoted to setting out Volta's system. Volta, who, after his recent experiences in Paris and London must have become rather skeptical about his own ability to present his theoretical views in a convincing manner, welcomed the initiative and encouraged De Luc. He clearly saw De Luc's project as a good chance for seeing his ambitious and long-delayed plan of publications on the "theory of electrical atmospheres" realized. Volta's directives to De Luc, contained in a letter of March 1784, which has recently come to light, are revealing.[75]

First of all, Volta invited De Luc to read his treatise of 1769, and to present it to the public as containing, despite its juvenile faults, the germs of all his subsequent discoveries, of the electrophorus as well as of the *condensatore*.[76] In a systematic exposition of his views, Volta suggested, theory had to came first. By stressing the role of his juvenile, theoretical treatise, he was clearly also aiming at reinforcing the legitimacy of his claims to be regarded as the genuine discoverer of the machines.

With regard to theory itself, Volta stressed that the connecting thread of his work had been the principles he had laid down concerning the action of electrical atmospheres. Several electricians had preceded him in that field to be sure, but, he invited De Luc to emphasize, nobody else had brought such a large number of phenomena under the same explanatory model:

> But to my knowledge nobody ever tried to submit such a vast number of phenomena to the action of electrical atmospheres as I have done. This is perhaps my only contribution. Having deduced the theory of the Leyden charge and discharge, the Electrophorus, the *condensatore*, and the points from the same principle, the general theory of electricity—which was exposed to strong objections on the part of those who did not see the whole, and was in any case considered far from perfect—becomes the more luminous and satisfactory; and I do not doubt at all that it will turn out victorious and triumphant when treated by you, Sir.[77]

On the suggested strategy of presenting theory as legitimizing discovery, De Luc understood Volta's message very well. As he wrote back to Volta, anticipating the line of argument he would develop in his future treatise, given Volta's general theory of electricity, it will be easy to show that "you must have found the Electrophorus and the *Condensatore not by chance, but a priori*."[78]

In a similar systematic mood, in 1784 Volta offered De Luc an explanation of "electric movements," in which he revealed his old penchant for "attraction" and for the view that electrical attractions are real, while repulsions are only apparent. The general principle supporting his explanation was that any two bodies, which are not in equilibrium in respect of their electrical status, attract each other. In order to explain apparent repulsion under this principle, Volta returned to the notion that the particles of air surrounding electrified bodies—the atmospheres, as he continued to call them—are subject to the same phenomena of actuation, tension, etc., as all other bodies. That is, the particles of air gain a certain degree of tension by being "actuated" by the electrified body immersed in them. Because air particles closer to the body attain a higher tension, light threads attached to the body will depart from it, being attracted by the particles of air more remote from the body and having a lower tension.

Prompted by the same systematic spirit, in his instructions to De Luc Volta also attempted to sketch out an analogy between electricity and heat. He assumed that bodies having different degrees of heat also attracted each other. He then imagined two glowing lumps of coal placed close to each other, suspended from threads, and free to oscillate. Because the particles of air between the two lumps acquire a higher degree of heat, they will move apart, being attracted by the particles of more remote and fresher air. Adding a third glowing lump of coal, it will move away from the other pieces and from their atmospheres for the same reason. The movements caused by the electric fluid were thus assumed to be similar to the movements caused by heat, which Volta also regarded as a fluid.[79]

The arguments set out for De Luc's intended treatise show that Volta, in his capacity as a systematic natural philosopher, held notions that still bore in 1784 traces of the old, material view of electrical atmospheres. This view was sustained to some extent by the need, felt by physicists in general, to reconcile the science of electricity with the classical, mechanical models of explanation, based on attraction but not necessarily committed to action at a distance as an acceptable explanatory model.

Volta's speculations on electrical movements remained abstruse, and they were an easy target for controversy. These speculations, in fact, became the main ground for disagreement with De Luc himself, who—at least at the methodological level—was able to address a very effective lesson to Volta on the subject. The lesson was an earnest call for instrumentalism, and against the search for true causes. As De Luc wrote Volta:

> In all your speculations on the topic we are dealing with [electrical movements], you combine two things that, on the contrary, I always keep separate; that is, the *cause* of the *electrical Movements,* and the *condition* [*état*] of the bodies which are moving. In the latest letter that I had the pleasure to send you, I was alluding to the latter [i.e., the condition], and now I would like to explain to you why I think we should treat it separately.
>
> To begin with, I notice that the *Phenomenon* to be studied is nothing else than the *movements* of *free Bodies*, resulting from whatever Electrification. These *Bodies* seem to *attract* or *repel* each other, that is, they come closer

or depart according to their different *electrical conditions*; and the relevant question, as far as *theory* is concerned, consists in establishing what are the *conditions* of these Bodies when they move in one way or another. In one word, what we need to do is to establish the *Laws* of the *attractions* or *repulsions* in the Bodies *that we see moving*.

After that, no doubt, one can and even must address the question of the *Cause* of these *Movements*: but that is a *systematic* Question, very different from the former, and one that must follow after the former is assessed.[80]

Because of the disagreements mentioned, because of the disappointing outcome of the survey of opinions on Volta he had carried out in Paris and London, and because of other personal reasons, De Luc abandoned his project of writing a treatise devoted to the illustration of Volta's "system."[81] Volta in turn found De Luc's later publications full of ideas that he regarded as "too hypothetical, and too transcendental."[82] The treatise expounding Volta's science of electricity never materialized.

Volta's instructions to De Luc in 1784 clearly exemplified the role he ascribed himself as a systematic natural philosopher. This, however, did not exhaust his repertoire and ambitions. Indeed, in the course of the 1780s it was the roles of lecturer and electrical experimenter, both at the University of Pavia and in his frequent trips abroad, that came to the fore. In these latter capacities he had recourse not so much to his "system" as to the very effective midrange conceptualizations we already know. These apparently provided a more solid basis for his teaching and for explaining the functioning of his machines.

A set of notes taken in French by somebody to whom Volta had lectured during his journey of 1781–82 show the refinement those midrange notions had attained through Volta's daily practice as a lecturer and a performer of electrical experiments. In these lecture notes, which he had translated into Italian and revised for his course of 1783 in Pavia, we find, for example, one of his clearest formulations of the relation between the quantity or "action" of electricity, capacity, and tension: $Q = CT$. As he put it:

> It is convenient to conceive the action of electricity [$Q$] as being in compound reason of its intensity [tension, $T$], and the capacity [$C$] of the conductors. Thus, a small conductor strongly electrified will send a spark to a great distance, but the spark will not produce a great effect. On the contrary, a very large conductor weakly electrified will send a spark to a shorter distance, but producing a great effect. Hence it is important to distinguish carefully between the *intensity*, and the quantity of electricity. Quantity is the sum of all the electricities contained in all the points making up the surface of the electrified body. *Intensity* is the force exercised

by each one of these points, in order to get itself rid of the electricity it enjoys, and to reestablish its equilibrium.[83]

In the lecture notes, the atmospheres too received a more effective, straightforward treatment:

The electricity of a body is felt by all surrounding bodies well beyond the distance at which a transfusion of the electric fluid can take place, as one can judge from the movements of attraction and repulsion taking place within a few feet, when the aigrette or the spark cannot reach beyond a few inches.[84]

In the same lecture notes, the explanation of how the electrophorus and the *condensatore* worked was deduced from these simple principles in the enunciation of which, it will be noted, the particles of air surrounding the electrified body were not even mentioned. It is time, however, to see how the machines, the notions, and the explanatory models mentioned so far worked in Volta's own day-to-day investigations. It is time to enter Volta's laboratory, as a number of manuscript notes that have been preserved allow us to do.

## Volta's Laboratory: Measuring Electricity

The most striking feature of Volta's investigation during the 1780s, as documented by extant laboratory notes, is the vast amount of calculation that occurred in his experimental practice. This circumstance is striking because the essays Volta published in those same years bear little evidence of such an effort at mathematization. Measurement was the main connecting link between the two sides of Volta's investigation just mentioned: experimenting and calculating. However, the fact that mathematization as such found little place in Volta's published work requires explanation: To what extent did the "quantifying spirit"[85] guide Volta, and why did new machines like the *condensatore* and the battery, rather than new, mathematized laws for describing electrical phenomena continue to be the main outcome of his work?

The measuring setup illustrated in figure 4.3, page 130, is a good starting point for describing the simple laboratory techniques adopted by Volta in this period. The figure illustrates two similar Leyden jars receiving electricity from the chain of a friction machine (not illustrated). The jars can be connected, singly or jointly (in parallel), to a metal stick, to the other end of which a Henley electrometer is attached. Volta first charged jar *A* to a certain tension, then connected it to *B* and measured the resulting "average" ("*med.*" in Volta's notes) tension. After having

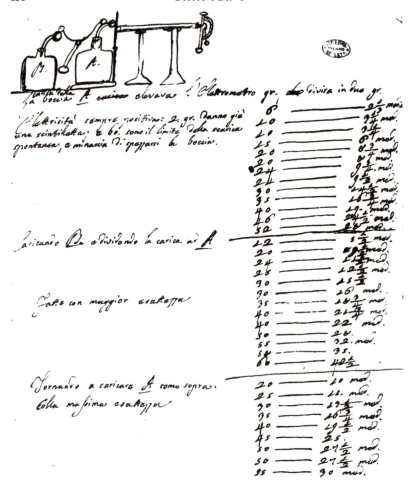

4.3. Measuring setup consisting of two Leyden jars, a conductor, and a Henley electrometer, from VMS, I, 6 (*Istituto Lombardo, Milan*).

tried it the other way round (that is, charging *B* first), he repeated the measurements following the first procedure. The results are registered in the three sets of figures forming the column to the right.

Compared with the measuring techniques described by Cavendish in his memoirs in the *Philosophical Transactions* in the 1770s, Volta's apparatus was less sophisticated. For similar purposes Cavendish used large batteries consisting of forty-nine Leyden jars, arranged in rows of seven jars each connected and disconnected according to need.[86] Equally simple (compared to Cavendish) was Volta's method of measuring the quantity of electricity communicated to a body by the num-

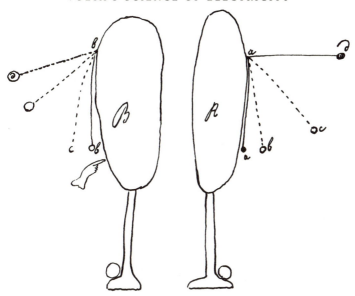

4.4. Volta's apparatus for research on "actuation," from VMS, L, 3 (*Istituto Lombardo, Milan*).

ber of turns of the friction machine. Cavendish regarded this method as "very fallacious."[87] In his experiments he preferred to administer electricity from the same jar, or sets of jars, the charge of which he had tested on several occasions using standardized procedures. Yet, what Volta lost in accuracy at each measurement, he probably gained by repeating his simpler procedures many times. Moreover, at the level of conceptualization, when it came to accounting for and comparing the electrical properties of different bodies, Cavendish's notion of "degree of electrification" was probably less effective than Volta's combined notions of "capacity" and "tension."

Volta's first steps in a systematic study of actuation can be illustrated by his drawing reproduced in figure 4.4, belonging to the period when he planned to write his three-part memoir on the "theory of the electrophorus." The immediate antecedent to the simple experimental setup represented here was probably Aepinus's air condenser. In the manuscript draft accompanying the figure, Volta discussed at a qualitative level the phenomena produced by charging one of the discs removed from the other, then bringing it toward the second disc, observing the electroscopic wires of the second disc rising and falling again when the disc was touched by the experimenter, etc. In the same manuscript Volta evoked Aepinus's mathematized discussion of similar phenomena.[88] Apparently, Volta's subsequent effort at quantification was

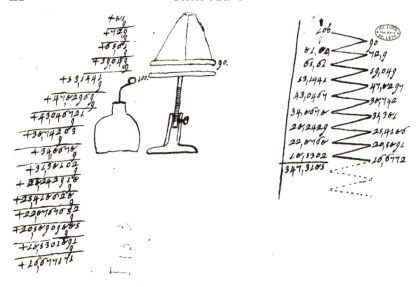

4.5. Measuring setup and calculations on the "actuation" of two metal discs, from VMS, I, 13 (*Istituto Lombardo, Milan*).

prompted by Aepinus's example as well as by Volta's own longstanding commitment to Boscovichian physics. Volta's approach, however, did reveal some distinctive traits which represented a momentous shift in relation to earlier models: in Volta's investigation mathematical calculations and the concrete measurement of real electrical phenomena interacted closely and extensively as they had never done before, other than in Cavendish.

Take for example Volta's treatment of the "oscillation of electricities" in the case of an air condenser like the one represented in figure 4.5. The starting point of Volta's investigation must have been his usual measurements of the tension "actuated" in the suspended, uncharged disc by the lower disc charged to a tension of 100 degrees by a large Leyden jar, also illustrated in the figure. Repeated measurements carried out in a setting similar to that represented in figure 4.5 are documented in many of Volta's laboratory notes.[89] The "oscillation of electricities" procedure, already adopted by Beccaria when carrying out similar experiments, consisted in touching the upper disc after it had been "actuated" by the lower disc. Assuming that the lower disc was positively charged, and a positive tension had been caused by actuation in the upper disc, the latter when touched became charged negatively to the same degree of tension. At this stage the negative charge of the upper disc "actuated" the lower disc, so that it required a certain amount of electricity to be brought back to its original 100-degree tension, which was done by bringing it into contact with the Leyden jar again after each move. The two columns of figures on the right of figure 4.5 contain the estimated quantities of electricity successively added to the lower disc (left column) and subtracted from the upper disc (right column) at each move.

Needless to say, the figures written down by Volta in his notes reproduced in figure 4.5 are *not* the result of direct measurements: no electrometer at the time could possibly allow a four-digit approximation. Here Volta was trying out a method for *calculating* the quantities of electricity involved, as shown on the left margin of the note. He assumed that, if a tension of 100 degrees in one disc causes an observable tension of 90 in the other, when the latter is earthed and attains a 90-degree minus electricity it then actuates the former, bringing its positive tension down to 19 degrees, so that it will draw 81 from the jar when restoring equilibrium and regaining its original 100 degrees. As Volta explains elsewhere, the distance between the two discs was supposed to increase one line after each operation, as illustrated in the notes reproduced in figure 4.6, page 134, where the case of plates $A$ and $B$ is treated.[90]

The same manuscript indicates (by extrapolation from the first column at the left) that Volta calculated that the tension $x$ produced in another disc by a disc having a tension of 100 degrees decreased with distance $d$, so that

$$x_d = \frac{(x_{d/2})^2}{100}.$$

In the same manuscript Volta suggested that the formula giving the total quantity of electricity gained and lost in the process by disc $A$, and indicated by $x$, was:

$$x = \frac{a^2}{a - c},$$

where $a$ is the (constant) charge of a large Leyden jar with which $A$ is touched repeatedly, and $c$ is the decrease in the tension of $A$ subsequent to each operation. According to Volta, the formula for $B$ was:

$$y = \frac{xb}{a},$$

where $b$ is the tension actuated in disc $B$. Attempts made at comparing Volta's conclusions with twentieth-century electrostatics indicate that,

4.6. Volta's calculations of the "actuation" of two discs $A$ and $B$, for distances varying from 1 to 22 lines, using the "oscillation of electricity" procedure. From VMS, I, 13 (Istituto Lombardo, Milan).

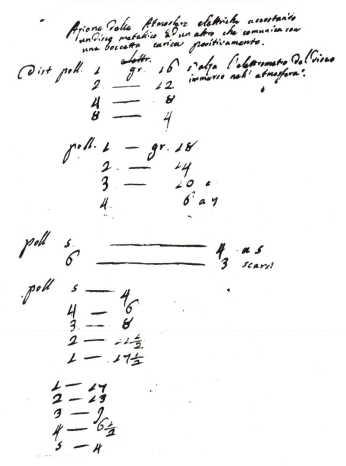

4.7. Record of measurements of the "actuation" of two metal discs, one of which is connected to a positively charged Leyden jar. From VMS, I, 6 (*Istituto Lombardo, Milan*).

to a significant extent, Volta had got it right.[91] Actual measurements, however, were much more rough and laborious, as shown by the notes reproduced in figure 4.7, and by the numbers in the top left window of the manuscript reproduced in figure 4.8, page 136.

Volta's enthusiasm for quantifying, however, did not bend easily. Manuscripts show that he kept comparing actual measurements and calculations, taking note of atmospheric conditions, humidity, etc., and rejoicing when, in a dry climate, measurements seemed closer to the values predicted by calculation.[92] However rough Volta's efforts may seem, he was among the very first electricians to systematically pursue

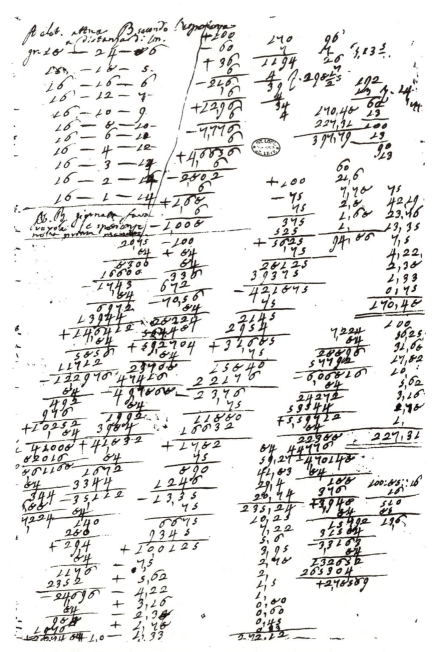

4.8. Measurements (top left window) and calculations of the actuation of two discs at varying distance. From VMS, I, 9 (*Istituto Lombardo, Milan*).

the quantifying strategy that Aepinus had pointed out, but which—if we judge from his published works—he was unable to subtantiate with numerical data.[93] Volta's quantifying strategy, in any case, was deeply affected by his limited mathematical training, that did not go beyond arithmetic, as his works, published and unpublished, reveal. Notwithstanding Volta's appreciation of authors like Boscovich, Aepinus, and Cavendish, in whose works analysis played a crucial role, his position in this important respect remained closer to that of many electricians of a former generation. Needless to say, this difficulty on Volta's part worsened when Coulomb entered on the scene with his series of sophisticated memoirs on electrostatics published in the *Mémoires* of the Académie des Sciences from 1785 to 1789.

### Volta on Coulomb

Volta first knew of Coulomb's first memoir—showing that the electric force varies inversely as the square of the distance—through the brief summary published in Rozier's journal for 1785.[94] Volta's reaction, probably also affected by his earlier frustrations in Paris in connection with Haüy and Coulomb, was harsh, though several more substantial factors explain the reaction. First, Volta was ill prepared to appreciate the kind of analytic mechanics involved in the torsion balance that Coulomb used for his measures. In addition, Coulomb's apparatus must have seemed to Volta, as indeed it seemed to several others,[95] unnecessarily complicated and subject to false readings, as Volta's subsequent adoption of a common beam balance for a similar purpose suggests. Moreover, Coulomb's demonstration of his law was based on measures of *repulsions*, which Volta had long regarded to be apparent, and thus unreliable as a source of information on electrical attraction. Furthermore, Coulomb was soon to proclaim himself to be in favor of the two-electrical-fluids position, while Volta had always supported the single-fluid view.

Among Volta's motives for rejecting Coulomb, the fact that, when Coulomb entered the scene, Volta had been working for several years on the different, successful, and promising line of investigation on what we now call induction, must also have figured prominently. The focus in Volta's investigation had been brought to bear on how to measure and calculate the macroscopic phenomena of "actuation," leaving increasingly aside the controversial issue of the microscopic forces involved, over which he had drawn repeated criticism from his peers. Coulomb forcefully restored microscopic attraction as a legitimate concern of the electricians precisely at the time Volta had at last emancipated himself from his bent for microscopic forces. At this stage Volta was more of an instrumentalist than Coulomb himself.[96]

Volta's instrumentalism, as we know by now, was not so much his choice as the result of the several lessons he had learnt to his own cost during the 1770s and early 1780s. As a consequence, he had come to renounce his old theoretical inclinations of 1769, and to focus on instrumentation as well as theory. It was hard for him to accept Coulomb's resolute reversion to Newtonian-style electrical science in the mid-1780s. The instrumentalist attitude Volta had developed by that stage may well explain what has been described as his difficulty in perceiving "that his [Volta's] macroscopic relations might all be consequences of Coulomb's rules for the particles of electrical fluid(s)."[97]

All this goes some way toward explaining the content, as well as the tone, of Volta's outburst at Coulomb's demonstration of the inverse square law: "But what? [*Ma che*?] if I shall demonstrate that it [the inverse square law] does not apply to the actuation [*pressione*] of the atmospheres . . . , nor to the repulsion of electrified bodies . . . nor even, in general, to [electrical] attraction. . . ." Similar considerations also explain Volta's confidence that he himself had opened "a more direct path, which leads further."[98]

Despite the reaction to Coulomb, Volta showed himself flexible as usual. He soon started measuring the electrical force himself, adapting the means to his abilities (no complex mathematics) and inclinations (simplicity). He first tried out a balance arranged as in figure 4.9, page 139, in which, for a while, he measured repulsion as Coulomb had done.

In Volta's apparatus the repulsion measured was that which took place between two charged discs (on the left in the figure), one suspended from the beam, the other supported by an insulated and adjustable mounting. A ballast was added to the suspended disc so that, when kept in balance by appropriate weights on the other scale, the repulsion required to unsettle the beam could be tested as required.[99] But Volta trusted attraction more than repulsion, and he soon changed his balance accordingly. In an instrument arranged as above, he substituted an earthed conducting plane for the lower, insulated disc supported by the mounting. By charging the disc suspended from the beam, attraction resulted between the disc and the plane.[100] He thus discovered several empirical laws, which led nowhere.

For example, he established that, other conditions being equal, attraction was proportional to the surface of the disc being used. He also found that, if the charge imparted to the disc was increased by $x$, attraction increased by $2x$. He explained the latter finding, which he said he had not expected, by invoking the increased action of electrical atmospheres as the charge increased. As to the correlation between at-

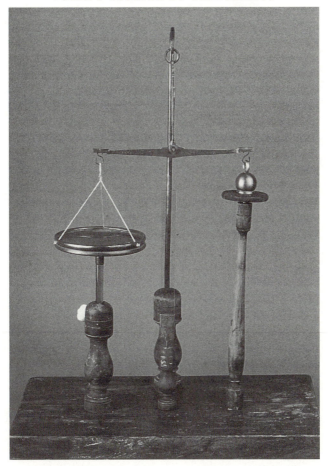

4.9. Volta's electrostatic balance, reconstruction in the Tempio Voltiano, Como (*Musei Civici, Como*).

traction and distance, he found that attraction varied inversely as the *double*, not the square, of the distance, and tested this using his balance for distances varying between six inches and five lines. Below that limit electricity all too easily poured off the disc into the plane.,[101]

Judging from the works Volta published in the late 1780s,[102] the attraction law as such did *not* become a central concern for him. His investigations engaged with a broader range of purposes, linked to the goal of measuring electricity in all its possible manifestations. Within this perspective, the exact measure of the electric force was to be pursued by refining devices and procedures designed to measure the macroscopic phenomena of electricity in general. At the same time, he continued to address what were topics traditional among eighteenth-century electricians, such as atmospheric electricity. The outcome was the development of electrometry as a new, special subfield, though often still perceived as part of the atmospheric electricity business.

These developments of electrometry are reflected in Volta's use of his straw electrometer.[103] The instrument was, per se, a minor variation on similar electrometers introduced by Cavallo and Saussure, using metal leaves instead of straws. Yet with the new instrument Volta pursued a thorough campaign of refinement and standardization of measuring devices and techniques. The goal was twofold. First, he needed to ensure the uniformity of each degree of electricity measured by the same instrument, within a reasonably wide range of the scale. Second, he tried to introduce some kind of "fundamental degree" or unit, so that electricians could compare their instruments and measurements.

Volta himself provided the lead. His straw electrometers, he said, ensured that—between 1 and 20 degrees at least—the progression of the degrees measured paralleled the progression of the electricities applied to the instrument by a standardized procedure. Henley's electrometer, Volta maintained, could achieve that result only for measures above 14 degrees.[104] He then started comparing his own and other instruments. Having found it useful, for different measuring purposes, to have electrometers built using straws of different diameters, he connected them and compared their performances as in figure 4.10, page 141, so that their scales could be made comparable.[105]

With respect to standardization, Volta suggested that recourse should be made to his own electrostatic balance (fig. 4.9, but in the version for attraction measurements), to Henley's electrometer, and to a good dose of the electrician's common sense. Using these tools, the "fundamental degree" could be fixed as that "electric force" required to unsettle (by attraction) the suspended disc (5 inches diameter) of the balance, when the disc is placed at a distance of two inches from the earthed plane and kept in equilibrium by 12 grains on the other scale. Such an electric force, Volta suggested, should be taken as being (arc) degree 35 of Henley's electrometer.[106] By reproducing Volta's measuring setup as accurately as possible, Giovanni Polvani and Giovanni Adorni found that Volta's "fundamental degree" corresponded to 44.5 absolute electrostatic units, or 13,350 volts.[107]

In invoking such standardized procedures, Volta stated emphatically that "instruments produced in Geneva, Paris, London or Berlin would thus at last become *unisonant*."[108] And, he might have added, even the electricians who had long refused to adopt his views of the science of electricity would at last have to agree to take his instruments and procedures into due consideration.

4.10. Volta's two straw electrometers made comparable. From *VO*, 5: plate 1 (*Biblioteca Universitaria, Bologna*).

## Conclusion

During the 1780s and after, instruments and procedures continued to be Volta's best allies. A survey of several textbooks of electricity circulating in those years, and written or edited by people like Lichtenberg, Gren, Adams, Cuthbertson, Sigaud de la Fond, and Cavallo, confirms that the great appeal of Volta's work lay in his impressive machines and technical procedures, more than in his straightforward notions, like those of tension, capacity, and actuation.[109] These notions, if and when they were traced back to Volta's work, followed in the wake of the instruments themselves. Volta's ideas on electricity which, at the beginning of the 1780s, were seen by some to be highly promising, failed in the end to be recognized as a systematic body of knowledge. Volta's machines remained the safest ground for establishing his contributions to the science of electricity, and for assessing his standing in the implicit hierarchy by which the community of electricians ranked its members. As a consequence, Volta was regarded increasingly as a brilliant inventor of electrical apparatuses, rather than as an accomplished natural philosopher.

Besides his instruments and technical procedures, Volta's other effective allies were the many practical electricians he continued to cultivate in his correspondence and travel, and with whom he exchanged thoughtful attestations of recognition and esteem. Indeed, it has been suggested throughout this chapter that these electricians—together with their less amicable colleagues who rejected Volta's claim to be regarded as a genuine natural philosopher—did more than just favor or resist the circulation of Volta's work and his ambitions: *they contributed to the shaping of his investigative style.*

The investigative style Volta came to adopt while interacting with the mixed community of electricians was no longer the style of the traditional natural philosopher he had set out with. In his search for recognition he had had to come to terms with a stage on which a varied troupe of actors—academicians, university professors, independent lecturers, amateurs, inventors, instrument makers, patrons—all played a part. Equally, through extensive correspondence and travel he had come to terms with a variety of different research traditions. For good or for bad the mixed community Volta interacted with refused to grant him the rank and position of natural philosopher to which he aspired. However, by partly accepting the limitations imposed on his theoretical ambitions, Volta made a virtue of necessity. The limitations imposed on him as a natural philosopher contributed to the shaping of qualities that many regard still as his most distinguishing: his laboratory-based theorizing, his accurate measuring techniques, his emphasis on machine as well as theory. In an important sense, Volta's most salient qualities as an investigator were not the result of personal choice only, but rather of an adjustment between his own attitudes and aspirations, and the role the community of electricians came to assign him. Conception, laboratory work, and the quest for public recognition interacted closely in shaping the investigative style he developed during the 1780s.

On the other hand, it should be noted that Volta himself, along with his fellow late-eighteenth-century natural philosophers, was not prepared to fully appreciate the theoretical dignity of the midrange concepts—like capacity, tension, actuation—that stemmed directly from laboratory practice. Hesitation about the theoretical worth of such notions was both the consequence of an old intellectual tradition, and of the implicit hierarchy of ascribed ranks of competence and standings among the physicists. In order to make the concepts stemming from laboratory practice fully acceptable to the domain of natural philosophy, some important changes had to occur within the physicist community.

One might say that it was precisely the work of scientists like Volta, and the impact of instruments like the electric battery—which again (as we shall see in chapter 6, below) stemmed from laboratory practice and for which no convincing theory was available—that slowly

Fig. 4.11. Volta's portrait, after 1805, by an anonymous painter (*Famiglia Volta, Milan*).

changed the eighteenth-century hierarchy of ascribed ranks of competence within the physicist community. By granting more room and more prestige to men like Volta, that change prepared the way for the emergence of the new, nineteenth-century figure of the scientist, and the partial eclipse of the old, eighteenth-century figure of the natural philosopher that Volta in many respects still was. Volta found himself in the middle of this process, and—unaware—he contributed to it above all with his machines.

What was involved in the transition can be captured watching the portrait of old Volta reproduced in figure 4.11, and submitting it to the tools of iconology. The portrait—by an anonymous and modest painter, but very likely taken from real, perhaps following Volta's own recommendations[110]—belongs to a later period in Volta's life. However, it conveys intentions and meanings that were already in place in Volta's private ambitions and public career in the late 1780s, as described in this chapter.

The portrait contains three kinds of symbols. The book alludes to what Volta himself considered his main achievement. He regarded himself as a natural philosopher, and he considered the concepts and theories of electricity expounded in his publications as his major achievement; and note that his publications make up a seven-volume collection. Conceptual contributions to the science of electricity, and to natural philosophy, was what Volta wanted to be remembered for. Appropriately, the book is shown in his hand, and close to his heart. The second kind of symbols contained in the painting are the instruments: a voltaic battery, and an electrophorus, or *condensatore*. They are shown in the foreground. They are not in Volta's hand, and they are a bit farther away from his heart than the book is, but they stand by themselves and occupy a prominent place. Volta's peers and the cultured elites of the late Enlightenment regarded these instruments, rather than Volta's natural philosophy, as his most visible and important achievement. The third symbol is of a rather different kind, but it also has a story to tell. In this portrait Volta is proud of showing a decoration, the ribbon of the Legion d'Honneur on the collar of his jacket. This was a major decoration in Napoleonic France. Bonaparte awarded Volta this and other rewards after Volta had presented the battery in person to the Institut in Paris in 1801, in the presence of Bonaparte himself.

One can say that, just as there are three different symbols in this portrait, there were three different plots going on in the story of Volta's science of electricity. Each plot was characterized by its own, slightly different pace and logic; but each is relevant to our understanding of the interplay between the Enlightenment notions of "useful knowledge," "the quantifying spirit," and instruments like the electrophorus, the *condensatore*, and (as it will be discussed in detail in chapters 6 and 7) the battery. One plot (symbolized by the book) is the story of natural philosophy, that people like Volta still regarded as their main business. The second plot is the story of the instruments that, unlike Volta, many contemporaries came to regard as his main achievement. The third plot (symbolized by the Legion d'Honneur) is the story of the loose, long-term values used to assess achievement and reward both within and outside expert communities.

The three objects symbolically represented in Volta's portrait won him different degrees of recognition in the Republic of Letters of the late Enlightenment. Volta's "books," despite his own frequent travels, his copious correspondence, and the several translations, did not travel easily within the experts' community and across frontiers: they won important but limited support to the thorough "Voltaic" science of electricity that he was circulating in the 1770s and 1780s. Instruments like the electrophorus, the *condensatore*, and the battery, on the other hand, were exceedingly successful travelers, and they were very effective ambassadors of Volta's ambitions and achievements, commanding for him the attention of expert and lay audiences Europe-wide very quickly after they had been introduced.

In order to fully appreciate these instruments' and Volta's own contributions to knowledge, some important changes had to occur in the experts' and in the public perception of what natural philosophy was about.

In any event, the tensions caused by the ongoing shifts in the goals and pursuits of natural philosophers, as well as in the hierarchy of merit and ascribed competence, were felt more acutely within the expert community than outside it. Public opinion and educational institutions could be more flexible. The lack of a specific set of values against which to assess achievements like the electrophorus and the battery did not prevent lay audiences from accommodating, and eventually celebrating, the new figure represented by Volta with his machines just as easily as the traditional figure of the natural philosopher. In the end (as we shall see in chapters 7 and 8, below) the broad, loose set of values associated with achievement, and the Enlightenment's vague but compelling notion of "useful knowledge," allowed Volta to enjoy on the public scene after 1800 the rewards that some of his peers had refused to grant him within the expert community during the 1780s and 1790s.

## CHAPTER FIVE

# The Cosmopolitan Network

### VOLTA AND COMMUNICATION
### AMONG EXPERTS IN LATE
### ENLIGHTENMENT EUROPE

As seen through Volta's biography, the brand of cosmopolitanism shared by natural philosophers in the late eighteenth century had two major roots. One lay in the aristocratic tradition, with its cross-national family ties. Aristocratic networks occasionally intertwined with the ties linking the rich merchant families in several European countries. As a noble man in touch with aristocrats and merchants in Como and Milan, Volta relied on the opportunities offered by both. This brand of cosmopolitanism rested mainly on family allegiance. It had little or no connection, per se, with the intellectual and professional pursuits Volta was engaged in; it presented Volta, however, with effective models, ready to be imitated. The countries included in the aristocratic network Volta was involved in extended to the east and north of Como: they stretched as far as Vienna, the Austrian empire, and the German states. The mercantile network extended to the north and west, offering Volta contacts with France, the Netherlands, and Great Britain.[1] As an example of the opportunities offered by the latter network, consider the merchant family from Como, the Guaita's—members of which Volta met in Amsterdam during his tour of 1781–82, their firm having conspicuous interests in the Netherlands—that helped Volta to keep in touch with scientists in the Low Countries.[2] Or consider the little-known, modest merchant from Como whom Volta chose as a messenger to the Royal Society of London to carry one of the two letters announcing the electric battery, as the merchant was leaving Lombardy to try his fortune beyond the Alps.[3]

The second root of the cosmopolitanism nurtured by Volta had direct links with his science. It stemmed from the notion of a supranational

Republic of Letters, to which scholars were supposed to belong, regardless of allegiances to country or king. This was the cosmopolitan ideal intellectuals had built up since the Renaissance to assert their relative independence from local authorities. As is well known, the idea of a supranational community made up of men and women of letters prospered during the Enlightenment, when the exchange of books and journals across national borders increased dramatically, and writers attained an unprecedented degree of visibility in the eyes of the learned public and governments. This brand of cosmopolitanism was shared to a large degree by natural philosophers, who ascribed special motivations to cosmopolitanism and molded it along a special rhetoric. The cosmopolitan ideal of natural philosophers has been a frequent topic of study on the part of historians.[4] In what follows I shall focus attention on the concrete rather than ideal motives and means that helped Volta build up an impressive network of international relationships, used in support of his science.

By focusing on the concrete motives and means supporting the cosmopolitan network Volta was part of, the present chapter has three main goals. The first is to establish a "map" of the special region of the Republic of Letters to which Volta felt he belonged as a natural philosopher specializing in the science of electricity. The leading "nations," the perceived "capitals," and the assumed "borders" of that part of the Republic of Letters will be reported as Volta saw them. A second goal of this chapter is to ascertain how the map of that ideal, scientific province fitted into the real map of late eighteenth-century Europe, and how its perception was affected by Volta's own political predilections and aversions, and by the allegiances imposed on him as a public servant. A third goal will be to show—by illustrating Volta's conduct as a skilful correspondent, a consummate traveler and a good scientist-diplomat—how natural philosophers and inventors emerged on the European scene as special protagonists of the cultural and social ferments of the late Enlightenment.

Participation in a cosmopolitan network was yet another means by which scientists like Volta asserted their role as emerging figures in the service of the enlightened public administrations of late eighteenth-century Europe. As it would turn out, the role of natural philosophers like Volta on the international scene consisted mainly in acting as informal but effective agents in the imitation-competition game practiced by the European governments to demonstrate their support for cultural and educational initiatives. It was not a new game, of course, but it was booming, thanks also to the spread of Enlightenment culture and ideology. The game had a propaganda side and an intelligence side. On the propaganda side, it involved advertising, by visits abroad of talented and well-supported scholars, the reputation of the governments and individual patrons supporting them at home. On the intelligence side it involved, in the case of Volta, an informal gathering of information, confined mainly to his specialty—higher education in the field of natural philosophy. As evinced from the reports Volta wrote his patrons at home, these expected him to provide information that would allow the public administration to improve educational institutions and the quality and appearance of their scientific collections, clearly regarded as symbols of "enlightenment" and prestige for the government supporting them. In Volta's circle of colleagues, however, the intelligence side of the imitation-competition game could easily merge with real industrial and political intelligence, in which at least two of Volta's acquaintances—Landriani and Magellan—were engaged.

Individual, scientific, and political motivations blended in support of Volta's participation in the cosmopolitan network illustrated in this chapter. Volta's ability to derive personal advantage from participation in the network, however, should not divert attention away from the fact that the main key that gave him access to the game was natural philosophy. Physics and its practitioners were attaining recognition from, and raising expectations in, government officials to an extent that enabled them to play an increasingly important role on the international scene. The semidiplomatic, semipolitical roles natural philosophers played across the borders of the major European countries was yet another sign of the power they were acquiring at home. At the same time, the fact that scientists like Volta could play the role along not too dissimilar lines in Milan as in London, Paris, Amsterdam, Berlin, or Vienna, was also a sign that expanding opportunities for technical experts in the public administrations—from which they drew part of their power—were a cross-national phenomenon. As it will turn out in the case of Volta, the cosmopolitan network of late-eighteenth-century natural philosophers derived its forms and impulse, partly from imitation of already established aristocratic, merchant, and literary networks, partly from the growing need for technical expertise by the public administrations of the major European countries. Imitation and the exploitation of expanding opportunities, however, were far from being passive processes for the experts involved. Physicists like Volta had both the culture and ambition to cooperate actively in forging the figure of the cosmopolitan expert in the field of natural philosophy then in the making.

## Overcoming Isolation

As we know, Volta started his scientific correspondence when he was just eighteen by getting in touch with experts from abroad.[5] The reasons for such a choice are not difficult to find. Despite the popularity of electrical experiments, there was no major figure in the field in Lombardy in the early 1760s. The only scholar to emerge with an interest in electricity comparable to Volta's was Carlo Barletti. He was, however, only ten years older than Volta and had not yet published any significant contribution to the field. Barletti lacked the authority and fame the young, ambitious Volta sought in his scientific advisers.[6] This being the situation in his own country, Volta turned to Giambatista Beccaria from nearby Turin and Nollet from prestigious, fashionable Paris. Scientific prestige and the expectation of reward merged in dictating Volta's choice. From a strictly scientific point of view, Franklin and Philadelphia should have been his first choice: young Volta regarded Franklin as a beacon in his early navigation through the domain of physics. When Volta came on the scene, however, Beccaria already enjoyed the reputation of being Franklin's closest disciple on the continent: he was a good substitute for the admired, but too distant, electrician from Philadelphia. The fact that Beccaria was professor at the University of Turin, some ninety miles from Como, and the fact that Volta knew somebody in Turin who must have been in Beccaria's entourage, added appeal to scientific correspondence with him. In the eyes of Volta all this compensated for what he soon discovered to be Beccaria's poor reliability as a correspondent, and his bad temper.[7]

Beccaria offered Volta the model of a natural philosopher from south of the Alps who was a recognized member of the cosmopolitan network of electricians. He soon gave Volta proof of the position he enjoyed at an international level: among the few signs of consideration he granted Volta was copy of the proofs of an article that he was publishing in the *Philosophical Transactions* of the Royal Society of London. The London connection, that Franklin himself had opened up to Beccaria, was already there for young Volta to imitate. Furthermore, Beccaria's standing as a public figure had been sanctioned by no less a figure than the Austrian, imperial authority of Joseph II, of whom Volta was a loyal subject. The visit Joseph II had paid Beccaria in Turin to attend a demonstration of his electrical experiments had left an indelible impression in the minds of Lombard natural philosophers, who still referred to it publicly years later.[8] The episode honoring Beccaria demonstrated to Lombard physicists that the achievements of Italian natural philosophers, when approved by leading scientific authorities in London, were highly regarded by the court of Vienna as well. All these reasons combined with scientific prestige in making Beccaria a model capable of helping Volta overcome the isolation he suffered in provincial Como.

The choice of Nollet as another, potential rescuer from isolation was rather obvious for Volta in the early 1760s. Since his first stay in Piedmont in 1738,[9] Nollet had maintained frequent contact with Italian electricians. His popularity south of the Alps temporarily waned around 1749 as a consequence of the dispute over medical electricity that saw Nollet acting as a critic of several Italian electricians on the occasion of a second trip to Italy.[10] Nollet's popularity rose again with the Italian translations of his *Lettres sur l'éléctricité*, published in Naples in 1761, and of the *Leçons de physique expérimentale*, published in Venice in 1762.

The well-known controversy over the theory of electricity that pitted Nollet against Franklin and Beccaria added appeal to Volta's contacts with Nollet: by sending letters to both Beccaria *and* Nollet, Volta made sure that his views circulated on both sides of the fence dividing the electricians. Young Volta in fact secured the advantages deriving from the need of the two major parties to win new supporters in the dispute. The little book Volta (then twenty-two) received as a present from Nollet in 1767 was the first tangible, if tiny, reward Volta obtained in exchange for his participation in the cosmopolitan network of electricians.[11]

As seen from the periphery of the community of electricians, around 1770—the year of Nollet's death—a new figure emerged in Europe as an authority and likely dispenser of rewards: Joseph Priestley. His semipopular *History and Present State of Electricity*, published in English in 1767 and in French translation in 1771, was presented as a repertoire of ongoing research, which the author promised to update from time to time with information on the discoveries communicated to him in the meantime.[12] The choice of Priestley in the early 1770s as yet another foreign correspondent shows that, among the motives prompting Volta to expand his network of correspondents, the search for recognition was at least as important as the need to circulate and collect scientific information. When Volta first wrote to Priestley, the latter was not known for any momentous contribution to the science of electricity besides the compilation of the *History*. The prospect of being mentioned in the promised (and never published) additions was a prime mover in Volta's decision to begin scientific correspondence with Priestley sometime before March 1772.

Private and public letters to Priestley remained Volta's preferred means for circulating information on his own work throughout the

1770s. Indeed, after 1777 the choice of Priestley as a mentor[13] led Volta to cultivate Priestley's own scientific specialty, which was (rather than electricity) the chemistry of airs. The tie with Priestley was probably strengthened when the two met in Birmingham in 1782, though Volta's account of his visit there is unfortunately sketchy. The relationship with Priestley had lasting consequences on Volta's work. He once described the network of correspondents that had Priestley as its focus as being in competition with another, French-oriented network of chemists to which Volta's rival in Florence, Felice Fontana, belonged instead.[14] Volta's involvement in the former network strengthened his support of phlogiston chemistry and his resistance to Lavoisier. The tie established with Priestley also played a part in Volta's adhesion to Crawford's theory of heat that Priestley had strongly recommended to his Italian correspondent.[15]

Priestley, unlike Beccaria and Nollet, was generous with Volta. As we know, he was the first among Volta's foreign correspondents to grant him in writing the recognition he had sought.[16] At his third attempt to break the scientific isolation he lived in, Volta at last found in Priestley the kind of authoritative and sympathetic foreign correspondent he was looking for. The success achieved was largely Volta's individual merit: until the mid-1770s he struggled for recognition from abroad supported only by his own entrepreneurship and achievements, culminating in the invention of the electrophorus and the work on airs. After 1775, however, Volta's own drive merged with the interest in international emulation, and the means to support it, provided by the public administration of Austrian Lombardy, such that he joined that year as superintendent to the public schools and professor of physics in Como.[17]

The interest of the public administration of Austrian Lombardy in international emulation was made explicit in a new "Piano degli studj" of 1777.[18] Through the plan, authorities saw fit to encourage their professors to expand contacts with colleagues from other countries, and special funds were provided for travels abroad. The expectation was that the "honor" of the government would benefit from crossnational emulation, especially from the kind of international advertising that "philosophers" engaged in trips of instruction provided for the government supporting them.[19] The travels of the professors and medium- and high-ranking public servants were in fact already a widespread practice in Lombardy in the 1770s. Kaunitz, when inviting Firmian to pay Volta 50 zecchini for his first trip abroad, mentioned the example of Barletti, who had already benefited from a similar scheme.[20] Marsilio Landriani, professor at the Brera schools in Milan and Volta's competitor and correspondent, often traveled at the government's expense in those same years. As is shown in a secret report preserved in Vienna, instruction and industrial espionage sometimes combined as motives supporting Landriani's trips.[21] The state-supported trips devoted to science, technology, and instruction bore obvious resemblances to the old tradition of aristocratic traveling, in which leisure, diplomacy, and business came together. This was the kind of mixed business the travels of Count Carlo Gastone della Torre di Rezzonico, Volta's fellow citizen from Como, were well known for.[22]

Nor was the enthusiasm for scientific and technological trips limited to Lombardy. In Tuscany, then in the hands of a branch of the Hapsburg, the public servant and chemist Giovanni Fabbroni traveled extensively in 1775–80 supported by the government.[23] In that same period Nicolis de Robilant in Piedmont added to the genre of technological reports by compiling a six-volume account of the visits he had paid to mines across Germany and Austria since 1749.[24]

It would be idle to ask whether the provisions supporting travel abroad were the results of governmental or professorial initiative. The close intertwining of the two groups suggests that both played a role. In Volta's case we know for certain that, in the spring of 1777, he himself brought pressure to bear on the government in order to embark on state-supported travel. He sent his recent publications on inflammable air and a gas pistol to Prince Charles of Lorraine in Brussels, whom he knew to be an amateur electrician who had shown interest in the electrophorus, and an earlier boss of Austria's powerful minister in Vienna, Kaunitz.[25] Some time later Volta sent the same evidence of his scientific and possibly military worth (the pistol) to Austria's minister in Milan, Firmian, asking it to be forwarded to Kaunitz. On the planned trip Volta also alerted a dignitary at the Court of Vienna, whom he used both as link and adviser: Baron Sperges. The lobbying maneuver was successful: in a few weeks Kaunitz in person ordered Austria's representative in Milan to pay Volta a sum to be spent on a "literary trip." The trip was presented in Kaunitz's letter as both a reward for Volta's scientific achievements and a means for turning Volta's own reward into "public advantage."[26]

Fifteen years after his first, stunted attempts at overcoming the isolation of provincial Como by establishing a scientific correspondence with leading electricians abroad, Volta could now envision plans enabling him to face his peers in person. He was no longer acting alone. He now presented himself with credentials from both his rising fame as an electrician and a chemist, and his role as a brilliant professor in the service of the Austrian empire.

## Exploring the Republic of Letters: The Neighborhoods

Volta's exploration of the Republic of Letters began with a visit to his closest neighbors. If Priestley and England were his preferred terms of reference, the plans for the first trip in 1777 had to come to terms with the time and money available. With this in mind, Switzerland and Austria were the two possible choices. Switzerland offered itself for the renown of its natural philosophers and scientific institutions, Austria for the opportunity it offered of becoming acquainted with leading figures in the administration of the empire. Volta hesitated a while between the two. Some years before he had been considering the idea of embarking on a trip to Switzerland in the company of Lazzaro Spallanzani, the professor of natural history at Pavia, a move which would have given him an opportunity to campaign for a chair in Pavia. The stated goals of that early plan were figures like Haller and Bonnet, with whom Spallanzani already corresponded.[27] To his adviser Baron Sperges, however, Volta had also mentioned the possibility of a trip to Vienna, which Sperges did not recommend. Himself living in the capital, Sperges thought that the major towns of Switzerland fared better than Vienna for the excellence of natural philosophers and scientific institutions they housed.[28] Volta's final choice was Switzerland.

The first trip abroad—the first of five visits he paid to Switzerland during his lifetime—proved stimulating and fruitful. He crossed the Alps at the beginning of September 1777 traveling in the company, among others, of the brilliant educationalist and philosopher Francesco Venini, who had been a pupil of Condillac in Parma, and who was an admirer of the *philosophes*.[29] As guide they employed the man who had assisted De Saussure on a recent excursion through the region.[30] Both fond of speculations on the history of the earth, Venini and Volta turned the trip into a geological excursion and an opportunity for frequent criticism of the Biblical geologists. Disillusioned with Catholic Lucerne because of the paucity of commerce, population, and scientific life, they moved quickly to Protestant Zürich.[31] Here, on 16 September, Volta performed a number of experiments—including a spectacular explosion of his gas pistol—before a crowded assembly of the local academy to which he had been admitted the year before.[32] Volta's experiments in Zürich were soon reported to Benjamin Franklin by R. Valltravers.[33]

Scientific and mundane meetings kept Volta in Zürich for five days.[34] Subsequent major steps included Schaffausen, Basel—where he met the elderly Daniel Bernoulli—and Strasbourg, where he became acquainted with Barbier de Tinan, who used to impress his guests by inviting them under his big electrical machine, where their hair was raised by electrical effluvia.[35] In Strasbourg Volta also met baron Frederic Dietrich, who was later instrumental in presenting Volta's experiments on inflammable air to the Académie des Sciences in Paris.[36] From Strasbourg Volta moved back to Basel and then to Bern, where he visited aged Haller; then he went to Geneva, where he met Senebier, De Saussure, and Bonnet. Saussure, who became a frequent scientific adviser of Volta's, entertained his guests with experiments in which he melted iron wires and broke pieces of glass by means of electricity.[37]

When in Geneva, a visit to Ferney to see Voltaire was a must for any member of the Republic of Letters. Volta regarded it an honor that Voltaire granted him and his Italian friends a half-an-hour audience. In a letter to his rival Landriani the estimated length of audience grew up to "almost one hour."[38] Volta, apparently, did not report any momentous topic of conversation, nor any brilliant *mot d'esprit*. Volta's friend Giovio was more impressed. He reported Voltaire as saying that talk about God should be "sans politesse." He also reported having seen in Ferney a little church with the inscription: "Deo erexit Voltaire MDCCLI."[39] On the way back from Switzerland Volta called at Turin, where he visited the Academy of Sciences, which he found impressive, and where he met Cigna, who must have argued with him over priority in the discovery of the electrophorus.[40]

It is worth quoting (the allusion to peculiar eighteenth-century hygienic habits excepted) how Volta's phlegmatic travelmate Giovio described his energetic friend during his first trip abroad:

My Volta is always busy. What an industrious scholar he is! When he is not paying visits to museums or learned men, he devotes himself to experiments. He touches, investigates, reflects, takes notes on everything. I regret to say that everywhere, inside the coach as on any desk, I am faced with his handkerchief, which he uses to wipe indifferently his hands, nose and instruments.[41]

Back in Como Volta could weigh the results of his first trip beyond the borders of his native land. Abroad, his scientific reputation had benefited and would continue to prosper from contacts established during the trip. Some of the scholars he had met became the translators of his works into foreign languages. Barbier provided French translations of Volta's letters on inflammable air, and those on conductors for Rozier's journal.[42] Schintz translated the letters on inflammable air into German.[43] Others, like Dietrich in Paris, became instrumental in circulating Volta's experiments and views abroad.[44] The impact of Volta's first trip on his career at home was equally significant. It was pointed out in a warm, flattering letter that his patron Firmian sent him soon after his return. Firmian declared that the contacts Volta had established with foreign scholars contributed to the "communication of enlightenment (*lumi*)," through which "the advancement of useful discoveries *was* unfailingly favored."[45] Above all, Firmian promised Volta the money needed to buy in Geneva and Paris, as Volta had suggested, the machines necessary to equip the physics cabinet in Como.

Volta's plan for a second, longer trip to be made at the government's expense began like the first with a letter to the most prominent aristocratic patron he corresponded with: Prince Charles of Lorraine.[46] The new plan was linked to Volta's efforts to submit an essay to the Royal Society of London. The goal was in keeping with Volta's chosen models (Franklin, Priestley, and Beccaria), and with his new position as professor at the University of Pavia as of 1778. The execution of the plan had to be deferred for some time, however. In the summer of 1780 Volta accepted what was (geographically) a diversion, but one which still enabled him to pursue his goal regarding the Royal Society. The diversion brought him to Tuscany, where Volta longed to see above all Prince Cowper, a Fellow of the Royal Society living in Florence, with whom he had been in contact since 1778 and whom he had alerted about his plans well before leaving Lombardy.[47]

The new trip to Tuscany, endorsed by Firmian and supported by the government with 1,500 lire, was conceived and implemented with the purpose of favoring imitation and competition among the philosophers and the instrument makers of the two countries concerned.[48] Volta set out on the journey in September 1780 accompanied by his instrument maker at the University of Pavia, Abbot Re. They carried with them several machines designed by Volta, which he was eager to show Prince Cowper. Volta and Re, in turn, studied carefully the machines found in private and public cabinets along their way and especially in Florence. On their return they pressed Firmian to obtain similar machines to enrich the collections in Pavia. The desired apparatus was subsequently ordered in Paris and London, while Cowper ordered from Re an eudiometer, built upon Volta's specifications. Re drew several designs of the machines kept in Florence, most of them built in England. Several other designs were supplied to Volta by the director of the cabinet of the grand duke, Felice Fontana.[49]

As a result of the imitation-competition game implicit in Volta's 1780 expedition to Florence, a few weeks after his return home two expensive sets of machines were being built in Paris and London for the cabinet in Pavia. The set commissioned in London was ordered through Jean Hyacinthe de Magellan, a key figure in the European market of scientific instruments, who was Volta's best contact in England after Priestley had put them in touch in 1776.[50] The set—occupying six boxes—included an orrery, an Atwood fall machine, and optical boxes of 'sGravesande.[51] The machines ordered in Paris through Sigaud de la Fond were built under the supervision of Rouland, and took up another ten boxes.[52] Scientific instruments, with their international market, were the most visible objects of the imitation-competition game Volta was joining.

Volta's first two trips abroad thus reinforced the mechanisms of imitation and competition involving natural philosophers, instrument makers, and top public servants like Firmian, fascinated by the "enlightened," expanding business of the physicists and their "useful" discoveries. The trips also established Volta's claim to be regarded by authorities in Milan and Vienna as an important figure in the imitation-competition game of the public administrations throughout Enlightenment Europe. By the summer of 1781, when Volta was still harvesting the seeds sown in the trip to Tuscany—and some of the machines subsequently ordered were still being repaired for damage suffered during transportation—he obtained permission and some of the money needed for his long-desired trip to France and England.[53] The trip was to last more than one year: it offered Volta the unprecedented opportunity of getting in touch in person with leading figures and scientific institutions in Europe.

## Facing the Peers: Paris in 1782

Volta's 1781–82 trip to France and England reveals the working of several cultural models, illuminating for the historian to consider. The journey was conceived, to a large extent, along the old pattern of the Grand Tour tradition that European elites had indulged in for centuries. Volta—who was thirty-six when he left and had recently been appointed to the chair of experimental physics in Pavia, himself having no university training—conceived of the journey as a personal experience, allowing him to refine his training as a physicist and to improve his acquaintance with scientific and aristocratic circles across the continent. Accordingly, about five-sixths of the money needed for the journey was provided by the Volta family, thanks to an agreement Alessandro had arranged with his brother Luigi.[54] Only the remaining sixth came from the government, and that in two installments.[55] In order to share expenses and opportunities, Volta traveled most of the time in the company of a Marchioness Leonora Villani, her young son, and a colonel. The traveling group left Milan toward mid-September 1781. Natural philosophy and Volta's incumbencies as a professor in the service of the Austrian empire impinged, as it were, on the well-established pattern of the Grand Tour rituals: the thirteen months of journey were split roughly between leisure and the learned opportunities offered by traveling and natural philosophy, whenever Volta met experts, amateurs, or instrument makers in the field.

Volta planned the formative side of his trip seriously, and modestly. A good part of the time he spent in Paris from January to April 1782 was devoted to attending, six days a week, lectures by Balthazar-Georges Sage on chemistry and by Jacques-Alexandre-César Charles on physics.[56] Both lecturers were about Volta's age and not particularly distinguished. Sage was indeed a controversial figure: successful as a teacher, he had a dubious reputation as a chemist. He was, however, in contact with scientists like Lavoisier and may have had a role in putting his foreign, temporary pupil in touch with people in the entourage of the Académie.[57] Charles began lecturing in physics the same winter that Volta arrived in Paris. He was later to gain reputation for his studies on the physics of gases and, above all, as an expert on balloons.[58]

While prepared to act as a learner in chemistry and general physics, Volta obviously felt more confident as an electrician. He did not follow any lecture courses on electricity, either in Paris or elsewhere. Indeed, his expertise in the field—unhindered in Paris, unlike London, by language barriers—enabled him to play the lecturer on electricity, which occasionally he did very brilliantly with amateurs like Madame Lenoir de Nanteuil and with De Luc.[59]

The real challenge of the trip for Volta lay in facing in person the peers he had till then cultivated through his correspondence. Volta set out on his journey, as we know, with a long memoir originally intended for the Royal Society.[60] The memoir illustrated—among other things—what he (rightly) regarded as a new, major contribution to the experimental apparatus of the electricians: the *condensatore*. This, plus the fame acquired with the electrophorus several years before and the brilliant memoir on conductors translated into French in 1779, raised his expectations of winning over the attention of the experts in Paris and London. Fulfilling the expectations, however, turned out to be a complex process, especially in Paris. The move from being a foreign, visiting scholar attending courses offered for a fee, to a trustworthy candidate member of the varied community of scholars somehow enjoying the prestige of the Académie des Sciences was an arduous one. Volta explored different paths likely to offer him a way through. Opportunities came as much from scientific circles as from the "salons," to which he was admitted thanks to his double status as member of the nobility and public servant in the service of the Austrian empire.

Volta reported having attended a meeting of the Académie des Sciences a few weeks after his arrival in Paris.[61] A month later he had lunch at Franklin's, whom he reported having met several times, though without giving hints of the topics discussed. Volta also reported that he attended public and private meetings in which Lavoisier, Le Roy, and Sage took part. As a result, he was invited to show his experiments on electrical atmospheres and the *condensatore* at a meeting of the Académie. Weeks later he was expected to submit experiments with his eudiometer, and on inflammable air, to the same Académie.[62] Another, less prestigious society, the Vieux Musée de Paris, had meanwhile admitted him among its members. He reciprocated the honor by reading a paper on the fires of Pietramala, a by-product of his earlier trip to Tuscany.

Judging from the reports Volta wrote home to his brother—who however was more impressed by the rank and names of the aristocrats than by those of natural philosophers—Volta especially enjoyed the private gatherings in Paris in which leading philosophers mixed with amateurs from the nobility and high-ranking public officers. He seems to have particularly liked the evenings offered by the head of Paris's police, Mr. Le Noir, in whose house he met Buffon several times.[63] There he also lectured on electricity and chemistry for Le Noir's daughter, the aforementioned Madame de Nanteuil, who was fond of physics and, like De Luc, developed a high opinion of Volta as lecturer and theorizer of electrical matters.[64]

The assessments—occasionally severe—to which Volta's reputation as an electrician was submitted in Paris in 1782 have already been discussed.[65] Here we can ponder the course of his dealings with the Parisian region of the Republic of natural philosophers, with a view to assessing its attitudes toward a foreigner who aspired to join it.

From this perspective, the balance of Volta's experience in Paris in 1782 was a mixed one. On the positive side, the Académie proved relatively open to Volta's early steps aimed at recognition. It was not too difficult for the foreign scholar to obtain access to the Académie, and he was allowed to address its assembly twice in a few months. Joining leading scientists from circles around the Academy also proved relatively easy. Volta had what every natural philosopher would have regarded as the great honor of being able to team up with Lavoisier and Laplace in making important experiments on the electricity of vapors. At a more modest level, figures like Jean d'Arcet, De Luc, and Madame Le Noir developed a high regard for Volta. The successful side of Volta's dealings with the portion of the Republic of Letters lodged in Paris culminated with his nomination as corresponding member of the Académie, the nomination letter being signed by Condorcet and Lavoisier.[66]

To obtain substantial recognition for the work presented in Paris, however, turned out to be much more difficult for Volta. On this less successful side, it should be noted that Volta's contribution to the experiments on electricity from vapors that he carried out with Laplace and Lavoisier in April 1782—a contribution that he himself regarded, and his colleagues in London agreed to regard, as momentous—failed to obtain approbation. Volta's contribution, which he advertised in a paper published in three different languages and above all in the *Philosophical Transactions*, remained unacknowledged by his Parisian colleagues.[67] In their report on those same experiments, published two years later, Laplace and Lavoisier mentioned Volta's attendance, but on the question of merit limited themselves to declaring that "Volta knew how to make himself useful to us during the experiments."[68] Technical reasons go some way toward explaining prudence on the part of Lavoisier and Laplace. Volta attributed the merit for the successful detecting of electricity from vapors to his *condensatore*. Laplace and Lavoisier, on the other hand, were probably unwilling to base the important result—regarded as able to throw light on the origin of atmospheric electricity—on Volta's newly discovered and still little-known apparatus, rather than on better-known detecting devices, like Cavallo's electrometer, that also proved up to the challenge.[69] It can be speculated, too, that Lavoisier's and Laplace's subsequent coldness toward Volta may have been occasioned precisely by the latter's decision to advertise his merits in London without consulting his French colleagues. In any case, Volta read the episode as showing that natural philosophers in Paris were not easily induced to grant recognition to foreign scholars.[70]

The timing of Volta's exploits and failures in Paris sounds indeed like a counterpoint to the progress of his contacts with the Royal Society in London. The collaboration with Lavoisier and Laplace began soon after Volta's memoir on the *condensatore* was read at the Royal Society; and we know that news of the attention Volta was being paid in London soon circulated in Paris.[71] Volta was admitted to the Académie Royale des Sciences on 21 August 1782.[72] Three weeks before he had left London leaving behind him ready for publication in Italian and English the memoir on the *condensatore* that contained vindication of his role in the experiments on electricity from vapors performed in Paris. In the end, despite the frequent and productive ties established with Lavoisier and Laplace, he failed to open any scientific correspondence with them, despite at least one later attempt on his part made to that end.[73]

Facing the peers in Paris turned out to require complex and uncertain negotiations on the part of the foreign natural philosopher. By dealing with *both* French and British colleagues in 1782, Volta discovered that the philosophers' Republic was far from constituting a single community. It turned out to be, rather, a federation of different constituencies, molded along national lines. Identifying the niches an outsider could occupy to build up his own reputation without interfering too closely with the authorities having control over the established regions was of crucial importance. These were the lessons Volta learned in his search for recognition in Paris and London in 1782, lessons he would again have recourse to at the time of announcing his discovery of the electric battery.

## Anglophilia

Anglophilia was a widespread syndrome on the continent in the 1770s and 1780s. Volta was affected as was his patron Firmian, who procured the additional money allowing Volta to include England in his tour of 1782. Science, technology, and politics—easily merging under the banner of Enlightenment ideals—helped foster Volta's admiration as well as perception that England had a leading place in the cosmopolitan republic of natural philosophers.

One of Volta's letters from England (only bits of his diary unfortunately survive) has been described by Franco Venturi as "the finest page written by an Italian on England as she was coming out of the war with America, France, Holland, and Spain, and entering the industrial revolution."[74] The letter is worth quoting at length:

> Far from finding in England the decadence and weakness some speculate on, one sees in it nerve and vigor to an extent that no other country displays. Commerce seems to be increased, and gold certainly circulates very quickly. The riches of individuals are huge; the class of comfortably-off people is extremely large; the worker is well dressed, better fed, and despite taxes (such as to scare anybody) still earns enough money to throw it away in the taverns. Building, manufacturing and new enterprises flourish everywhere. Public entertainments, that in France and Holland especially suffer like commerce from the distressing effects of the war, and decline and diminish accordingly, seem rather to multiply in England. Here you hear only one-fourth of the complaints about war heard in Holland and France. You can judge from this whether England is a weak and sick body, as those who have not seen it pretend it to be. . . . Those who know it can tell from its appearance and internal motions that the body is

healthy, sturdy, well fed, rich of juice and blood, so that you can still draw a lot from it without enfeebling it too much. . . . It is hardly imaginable how many resources England can rely on, and how much strength she derives from the courage of the nation. She perceives herself as strong, and that itself makes her strong. . . . I shall say no more, except that, having left England, the other countries move me to pity because of the extent to which they seem to be pervaded by misery.[75]

A few weeks earlier Volta had written that he would "willingly spend several years" in England, that England was such an interesting country that he found it "hard to leave it."[76] Every surviving document about the visit conveys the same feeling of sympathetic admiration.

Volta had in London acquaintances (lacking in Paris) who were able to put him in touch quickly with the liveliest circles of natural philosophers. His chief contact was the already mentioned Magellan, with whom Volta had toured Holland, and whom he joined again in Brussels on his way from Paris to England. Magellan was a Fellow of the Royal Society and an active member of the informal clubs of natural philosophers, chemists, instrument makers, and physicians that prospered in London. He had particularly close ties to the club sometimes called the Chapter Coffee House Society, from the place where it met in the early 1780s.[77] Magellan described the society as "notre petit club philosophique."[78] The club met on Friday evenings every second week to discuss a range of topics that fitted perfectly Volta's own interests: experimental, nonmathematical physics (with special attention to electricity, magnetism, and heat), chemistry and scientific instruments (the latter being the business Magellan made his living from), technological innovations, and occasionally medicine. Ordinary members of the club included Richard Kirwan, Benjamin Vaughan, John Nairne, Adair Crawford, John Hunter, Tiberius Cavallo, Josiah Wedgwood, Thomas Percival. The chosen "honorary members" were a group of eminent figures from Birmingham: James Keir, James Watt, Joseph Priestley, Matthew Boulton. It was this sort of people Volta met on several occasions during his nine-week stay in London, and his three-week tour of England that brought him, in the company of Magellan, to Oxford, Birmingham, Manchester, Washington, Liverpool, Norwich, Chester, Shrewsbury, Bridgnorth, Worcester, Gloucester, Bristol, Bath, and Portsmouth.

Volta's access to the circles of English natural philosophers was made easier by the fact that just before his arrival the Royal Society had been presented his long memoir on conjugate conductors in four subsequent meetings, on the initiative of Lord Cowper from Florence.[79] Volta himself, apparently, never attended the meetings of the Society, which stopped during summer, when he was in London. The informal circles of natural philosophers available in London, however, and the presence of emigrés like Magellan and Cavallo prepared to listen to a foreigner like Volta, proved especially conducive to the goals of instruction and recognition pursued by him.

On the instruction side, Volta learned a lot from meeting the group of scholars to which Magellan belonged, as well as from the reports he received—mainly through Cavallo—of the reactions to his views within the Royal Society. If the physicists of the Society proved as impervious as their colleagues in the Parisian Academy to Volta's theorizing on electricity, reworking the long memoir on conductors for publication in the *Philosophical Transactions* was an exceedingly important experience for him.[80] Once published, the memoir became a lasting tie linking Volta to England. The publication sanctioned, after the rewarding correspondence with Priestley, Volta's leaning toward the British province of the republic of philosophers. The leaning led Volta to address again to the Royal Society his main contributions on Galvanism, as well as the first description of his major invention, the electric battery. The Royal Society reciprocated Volta's preference by bestowing further marks of recognition on him. On 18 November 1790 he was proposed for election "on the foreign list." On 15 May 1791 a ballot took place and Volta was elected.[81] On 1 December 1794 he was awarded the Copley Medal. The award was made on the grounds that Volta's memoir on Galvanism had "explained certain experiments published by Professor Galvani of Bologna, which till commented upon by Mr. Volta had too much astonished, and perhaps in some degree perplexed many of the learned, in various parts of Europe." The motivation insisted that Volta had put Galvani's experiments to the test "of sound reasoning and accurate investigation," gainsaying the supposed discovery by Galvani of the "chief arcana of nature," or the "vital principle called by some Physiologists the nervous fluid."[82]

Volta's fascination with England reached its peak in 1782 during his visit to the Midlands. The factories, the machines, and the literary societies he saw there were "cose meravigliose" to him.[83] In Birmingham he met Priestley, Boulton, and Watt.[84] In Manchester he was introduced to Thomas Henry and Thomas Percival, and he probably attended a meeting of the Literary and Philosophical Society.[85] Looms, steam engines, chemical works, canals, and the organization of factories elicited detailed, enthusiastic reports in Volta's diary. The entire region was depicted as "the kingdom of Vulcan."[86] A visit to the fleet of admiral Richard Howe in Portsmouth educed equally enthusiastic comments.[87]

The genuine wonder contained in the following description of the railways between Ketley and Coalbrookdale earns it a place in the history of the impact of iron technology and early industrialization:

> The road is covered with two rails or iron bars on which the wheels of the
> carriages revolve, the wheels having a border that keeps and guides them
> along the said rails. There is not just one such road, but many leading to
> different parts, two, three or more miles away, the road being in several
> places double so that two carriages can pass. Rails are fixed on earth by
> means of wood beams, on which they are nailed down. Can you imagine
> a more lavish profusion of iron? Yet, such an expenditure is compensated
> for by the gain in easing the transport of coal and minerals. This alone
> gives you an idea of how big the enterprise is, and of the amount of iron
> being melted in the furnaces.[88]

Compared to London and the Midlands, flourishing with scientific
and technological innovations, Volta was less impressed by a traditional university town like Oxford. He did not fail to appreciate, however, the astronomical observatory then being built, the rich libraries,
and the fine art collections that he found to his surprise full of paintings by the best Italian artists.[89] He appreciated the English garden,
which he preferred to gardens in the Italian style as he declared in the
report of a visit to Blenheim, because of the better balance between
nature and art.[90] Even the British weather drew positive comments
from Volta because of the absence of extremes.[91]

*Medietas*, indeed, was the key to Volta's fascination with England: in
society the reasonable balance (compared to the continent) between the
enormous riches of a few individuals and the relative welfare of the
lower classes; in culture, the balance between respect for traditional
values and search for new ones, between the humanist tradition and
scientific culture, between the pride for national achievements and appreciation of the contributions of others; in landscape, the equilibrium
between nature and artifact. It would be pointless to separate the reasons underlying Volta's anglophilia—coinciding often with the reasons
for his adhesion to Enlightenment ideals—from those that brought him
to cherish as a natural philosopher the English province of the Republic of Letters: broad cultural, political, and scientific motives merged
in his judgment. Nor did Volta feel alone in his appreciation of British
culture when in Lombardy. Two years after his visit to England, urging
the acquisition by the University of Pavia of the magnificent collection
of books of his late patron Firmian—himself a well-known anglophile,
as his library showed—he remarked that "not only our professors, but
also our young doctors and even our students read English books, and
often earnestly look for them."[92]

British books, like British scientific instruments and British or British
naturalized natural philosophers, were clearly the heralds of a coun-

try—part real, part ideal—that Volta came to regard as playing a leading role in the cosmopolitan network he belonged to.


## Continental Europe and the German-Speaking Countries

Holland was after England the nation that most closely approached what Volta regarded as a civilized, enlightened country. In his appreciation the perspective of the natural philosopher merged again with broad cultural and political motivations. As a philosopher, Volta was pleased to find that "in Holland the amateurs interested in natural history and physics, as well as collectors in these fields, are countless, their collections being often rich and magnificent to a high degree."[93] One such collection Volta was impressed by was of course that of the Teyler Museum in Haarlem, then under the direction of van Marum, which he visited on 28 November 1781.[94] Judging from subsequent publications of van Marum and Paets van Troostwijk, Volta must have elicited a deep interest in the electrophorus in his Dutch colleagues on the occasion of his visit.[95] An important correspondence with van Marum ensued, together with Volta's prompt enrollment in the Dutch Society of the Sciences.[96]

Volta's appreciation of Dutch natural-philosophy experts and collections matched the enthusiastic reports he made as self-appointed judge of the state of civilization in the country, as in the following description of Rotterdam:

> Rotterdam is a large and fine town, crossed by many navigable canals, some of which take big ships carrying more than thirty guns. All the buildings are extremely clean: everything shines like crystal; and indeed most of what is offered to view is polished crystal, because windows occupy a much larger share than walls on the exterior of houses, magazines, and shops. Everything is marked by an appearance of comfort that instills joy. Open roads are large, straight, splendidly paved, and large trees have been planted on both sides, to the right and left of every canal. Bridges, some of which are very fine, are many and the sight of people flooding everywhere—among whom you do not see a single man in rags—provides the best possible form of recreation in the world.[97]

The fact that in the merchant towns of Holland Volta came across members of several Lombard families of distinction he was acquainted with in Milan—the Cravennas (merchants), the Zappas and the Lorlas (bankers), besides the already mentioned Guaitas (merchants)—strengthened his sympathetic attitude toward the country.[98] While in Holland and in England he must have realized that the cosmopolitan network he was building as a natural philosopher imitated—and in parts overlapped with—the merchant and banking networks of which several Lombard families were part.

The planning, financing, and implementation of the grand tour of 1781–82 were to a large extent, as we have seen, the fruit of Volta's own personal initiative. As such, the tour reflected his own individual propensities better than any other foreign trip he undertook. Accordingly, the major goals of the tour, and sources of the best memories in his diary and later recollections, were—in order of importance—Britain, Holland, and France. In the 1780s these were the preferred regions of that part of the Republic of Letters he had explored so far, spurred on by his scientific curiosity and his cultural and political leanings. During the 1781–82 tour Volta also touched on the western part of the German states, though that was mainly because of the engagements of the friends he was traveling with. A thorough exploration of that large portion of the Republic of Letters that spoke the German language was left for another occasion. When the occasion came—very soon—Volta discovered that his personal predilection for Britain, Switzerland, Holland, and France was being gently, but forcibly redirected by the policy of the Austrian government he served as professor at the public University of Pavia.

Volta's first patron, Firmian, died in July 1782 while Volta was in London.[99] Two years earlier Empress Maria Theresa had also died, leaving her son Joseph II sole arbiter of the empire. In the new circumstances, Volta seems to have relied for patronage mainly on his contacts in Vienna, and above all on Joseph II's prime minister Prince Kaunitz rather than on Firmian's successor in Milan, Count Wilzeck.[100] Kaunitz liked to tinker personally in the affairs of the University of Pavia. The tour of Austria, the Czech regions, and the German states that Volta undertook in 1784 in the company of anatomist Antonio Scarpa, also professor at the University of Pavia, was planned and executed under the direct auspices of the court of Vienna. That had both advantages for and demands on the professors benefiting. On the one side, money: in contrast to the 1781–82 tour, Volta's 1784 trip was generously supported with government funding.[101] On the other side, expectations: Kaunitz let his protegé know, when they met in Vienna, that he expected from Volta an increased acquaintance with "German literature," and more frequent contacts with German-speaking scholars in return.[102]

Volta was eager to meet his new patron's expectations. The death of Firmian had left him with the task of redrawing his patronage connections; this he did partly by applying to Kaunitz for additional facilities for physics in Pavia, partly by circulating abroad the news that he was contemplating resigning from his chair, and partly by accepting his new patron's invitation to strengthen ties with the German-speaking province of the Republic of Letters.[103] For a physicist like him, however, it turned out to be easier and more rewarding to establish contacts with scholars from the German states than with those from the Austrian empire. From this perspective, the interests of the patron and the scientist did not coincide, and compromise—toward which Volta inclined—was more necessary than ever.

Volta's four-month journey through the German-speaking province of the Republic of Letters in 1784 was, by contemporary standards, a tour de force. Giovio, Volta's literary friend, all too obviously compared the trip to an eagle's flight:

> I say flight because to spend one month in Vienna, to reach Berlin, to make a run to Hannover, to cross all of Germany, and then to be back after having visited natural history cabinets, academies, courts, gardens, mansions and universities within the span of four months is something easier for eagle's feathers than for post-horses.[104]

Giovio expressed admiration for his friend: "You know the art of traveling as well as that of studying, and your movements are equally rapid when running across countries as when discovering new provinces of true knowledge."[105]

The high spots of the 1784 trip, as far as science was concerned, were Göttingen and Berlin. It was the individual scholars—especially Georg Christoph Lichtenberg and Giuseppe Luigi Lagrange—rather than scientific circles such as those found in Paris and London, that made the German part of the trip valuable for Volta.

Lichtenberg's report portrays Volta as the archetypical traveling natural philosopher of the eighteenth century, at home in Göttingen as in Pavia, performing experiments with his portable set of physical apparatus:

> He opened his traveling bags in front of me. They were packed with instruments that remained scattered in my house during his stay. The instruments were smith work, but he could perform everything with them. He stopped for five days, two-thirds of which were spent in my apartments. . . . The next day he arrived at dawn and set up by himself the experiments he subsequently performed.[106]
>
> He then wanted to show me the experiment proving that vapors carry positive electricity with them. He isolated a brazier with a few coal pieces lightly burning in it; he then wetted a linen rag and threw it on the coals. An iron wire connected the brazier to a very sensitive electrometer. Nothing happened and he cursed in French and Italian. . . .[107]

Volta often popped in at seven-thirty in the morning and stayed until noon. We performed the experiments on atmospheric electricity with balloons twice. . . . The experiments delighted him so much that he truly did not hear or see anything. I introduced some gentlemen to him, but he did not take his hat off, nor did he pay attention to what I intended to do; he rather ran to the balcony repeatedly. . . .[108]

Through his guest's experiments and conversation, Lichtenberg came to cherish Volta's science and instruments, which he had not particularly appreciated before. As John Heilbron has put it, via Lichtenberg "in 1784 Volta invaded Germany," and his German colleague became "prepared to grant him the Newtonship of electricity."[109] Volta, in turn, learned to appreciate Göttingen and German science via Lichtenberg. He was highly impressed by the library, lavishly funded, which he found housed in the largest building in town.[110] He was impressed by physics textbooks and journals as well, and soon decided to use, when preparing his lectures at Pavia, the textbook of physics by J.C.P. Erxleben, Lichtenberg's predecessor in the chair at Göttingen, a textbook the successor kept updating over the years and that Volta for a while considered translating into Italian.[111]

Via Lichtenberg and Göttingen, German scientific instruments, as well as German books and science, found their way to Pavia. A stopwatch, several eudiometers, and glasswork for various purposes were ordered by Volta. Kaunitz's recommendation to visit Göttingen, the additional money he provided for Volta and Scarpa to include that university town in their tour, and the invitation to order scientific instruments with government's funds during the trip, were all put to good use.[112]

Similar, though less pervasive, effects followed Volta's visits to Halle, Brunswick, Augsburg, Munich. Berlin, where he spent sixteen days, was a case apart. Of the Berlin natural philosophers, Franz Karl Achard probably had interests closest to Volta's. Together they performed experiments on the *condensatore* and on airs. It was Lagrange, however, who attracted Volta's attention and admiration most. Volta was surprised to find that the great mathematician was well versed in physics and chemistry too. They spent several evenings performing physical and chemical experiments together.[113] Admission to the Berlin Academy, in any case, had to be postponed by two years: as Frederic of Prussia informed Volta, the high number of scholars already admitted as foreign associates meant he could not be added immediately to the list.[114]

If the German states outdid Austria and Prague in fulfilling Volta's expectations as an expert physicist, as a man of the Enlightenment and citizen of the Austrian empire he learned some important lessons from his stay in Vienna too. He was impressed by the huge program of hospital and educational facilities being developed by the state under the supervision of a powerful group of medical men and engineers.[115] In Volta's report of Vienna, the new hospitals and the facilities for medical and engineering education under construction equaled in importance the displays of military power he was invited to attend as part of the ritual visit to the capital. As he saw them at work in Vienna, medical, educational and construction experts were acquiring a social standing on a par with traditional administrative and military experts in the fabric of the empire.

Among natural philosophers, one of the figures that impressed Volta most was Baron Ignaz Edler Born, an expert mineralogist with broad cultural and political interests.[116] Born was an active propagandist of the secularized philosophy of the state endorsed by Joseph II. Side by side with seminal mineralogical treatises, he was then publishing a spirited satire of the church, in which he liked to describe religious orders as many entomological species. Volta liked the satire and circulated it among his friends in Italy.[117] If Vienna and Austria could not match Berlin or Göttingen for the number and quality of the physicists and scientific collections, the reality Volta was exposed to in Vienna gave him another, no-less-important message: it showed how closely natural-philosophy experts interacted with the administration of the empire, both in the running of key educational and medical institutions, and in the propagation of ideas that endorsed the policy of the emperor.

Like all previous trips abroad, the 1784 journey bore fruits at home as well. The increased use of textbooks in the German language and of physical instruments made in German-speaking countries in the teaching of physics at Pavia, were just two of the consequences. Another was the renewed credit Volta earned as a key figure in the imitation-competition game that educational authorities felt they were engaged in. Ably exploiting the signs of consideration shown him in Vienna and the German states, Volta managed to convince authorities in Milan that a new lecture hall specifically devoted to physics should be built in Pavia. Plans to that end had been under way for years, but only after 1784 did things start moving.[118] Volta's active presence in the cosmopolitan network of natural philosophers and public administrators continued to pay dividends at home.

## After 1789

Events following 1789 had dramatic effects on Volta's international contacts. French troops led by General Bonaparte invaded Lombardy in the spring of 1796, putting a halt to Austria's ninety-year rule over the region. On 14 May 1796 Volta—who had served Austria for more than twenty years—and his friend Giovio as representatives of the town of Como paid homage to Bonaparte on the occasion of his entering Milan.[119] Political and military events brought Paris and the French, who had till then frustrated Volta's quest for recognition, to the center of his professional and public life.

The new circumstances, as we shall see in chapter 7, below, did not bring Volta's science any closer to that practiced by his French colleagues. In that sense, Volta's ideal map of the philosophers' republic— a republic that did not have its capital in Paris—was not significantly altered. However, Volta knew how to adjust his personal ambitions to the changed if troubled map of power in Lombardy and on the European scene. The successes he harvested in the new situation were due partly to his own realistic attitude and readiness to compromise, partly to the fact that he found in Bonaparte and the administrators in his entourage a set of assumptions, concerning the role of experts and natural philosophers, that had substantial continuity with those shared by the administrators Volta had dealt with during the previous regime.

Volta's battery (chapters 6 and 7, below) was discovered and announced to the public during the brief interval when the Austrians regained temporary control over Lombardy after the French invasion three years before. This, plus Volta's well-known fondness for Britain and the recent award of the Copley medal, explain why in March 1800 he chose the Royal Society of London to announce his discovery. In fact, the temporary restoration of Austria, allied to the British, had dramatic consequences for him as a professor. In retaliation for supposed or actual disloyalty of the professors, the Austrian authorities shut down the University of Pavia, and Volta—whose stipend had already been suspended by the French in 1797—lost his job and stipend accordingly. On their return to power, the French declared the university open again in July 1800 and confirmed Volta in his position. However, another incident befell him. As part of the policy to help sustain the military effort, the French imposed huge fines on a number of citizens from Como and elsewhere. Volta and his brother were among them. The sum amounted to almost twice Volta's yearly stipend. He appealed against the measure but had to pay two-thirds of the sum before being finally exempted and refunded.[120] The political and military turmoil were creating a degree of hardship for Volta. Already in 1795 his teaching at Pavia had been subjected to public criticism, and the following year he was accused of maneuvering for the transfer of the university from Pavia to Milan. With this in mind—plus the fact that he was then fifty-five—it is understandable why in the autumn of 1800, with the discovery of his new brilliant apparatus in his favor, Volta began to search for a kind of rec-

ognition that would place him above the uncertainties of the times, or
at least offer him shelter of some kind. A trip to Paris to present the
battery there seemed the right kind of initiative.

On 9 February 1801 the Lunéville peace treaty was signed. Only Britain remained at war with France, and that too soon changed when
peace negotiations, started in London, were concluded in October that
same year. In May Volta resumed contacts with Gaspard Monge and
Claude-Louis Berthollet in Paris, both of whom had attended Volta's
experiments on galvanism in Como at the time of their mission for the
French government in Italy in 1797.[121] Volta also sent a description of
the battery to chemist and minister Jean Antoine Chaptal in Paris.[122]
Early in July an illustrated description of a fine pocket battery was sent
to Dieudonné Dolomieu, whom Volta also solicited to press his case
with the French government.[123] Meanwhile, pro-French administrators
in Milan realized that Volta's trip to Paris might benefit them too, and
accorded money and facilities allowing him to go.

When Volta arrived in Paris, Bonaparte was close to achieving in
Europe the kind of peace, favorable to the French, that he had pursued
as First Consul of the French Republic. Volta's three subsequent presentations of his work and instruments at the Institut National (more
on this in chapter 8, below)—all three attended by Bonaparte—took
place during the same weeks when Paris was celebrating with great
pomp the peace agreed to by the European powers after a series of
French military victories. Bonaparte's unusual request to the Institut
National that a gold medal be awarded to Volta for his discovery was
presented at a meeting held two days before the key ceremony of the
peace celebrations, scheduled for 9 November 1801.[124] Rewarding
Volta—whose name was by then known in cultured circles throughout
Europe thanks to the controversy over galvanism and his recent discovery of the battery—carried with it several symbolic implications
under the circumstances. The man being rewarded had served Austria
for twenty-five years; he had announced his discovery in London
(using indeed the French language), but was now a citizen of the Cisalpine Republic within the French sphere of influence, and was presently
conferring with natural philosophers in Paris. By rewarding Volta Bonaparte's aim was to convince intellectual circles throughout Europe
that Paris was the ideal "capital" for the pursuits of distinguished savants from any country or allegiance.

Other political reasons for rewarding Volta can be found in Bonaparte's policy of favoring savants as a social group. An example in this
regard will suffice. While Volta was in Paris, a huge Italian Constitu-
tional Assembly was being organized to meet shortly in Lyon. One of
the purposes of the meeting was to draw up lists of citizens from the

various regions who were to be allowed full political participation under the new, pro-French regime. Following instructions coming from Bonaparte himself, the lists grouped citizens into three classes: owners, savants (*dotti*, a third of whom were to be priests), and merchants.[125] Volta himself was invited to take part in the Lyon assembly. Bonaparte apparently realized that, in the aftermath of the Enlightenment and the French revolution, by rewarding savants he was doing more than pleasing an upwardly mobile social group. By celebrating especially the natural philosophers—who were so popular in cultured circles in and outside France, as Volta was finding out to his advantage on his trips—Bonaparte pleased the growing sectors of the bourgeoisie and aristocracy that were fascinated by their achievements. Judging from the many further rewards Bonaparte bestowed on Volta,[126] he must have found the diplomatic and political benefits deriving from his celebration of Volta rewarding.

Genuine, if vague, intellectual curiosity was also among the reasons inclining Bonaparte toward Volta. This intellectual curiosity stemmed in part from suggestions nurtured by galvanism and similar topics fashionable in those days; in part it was also associated with expectations of "utility" that had long surrounded the work of physicists in the mind of the learned public and administrators.

Allusions to fashionable galvanism are discernible in Brugnatelli's report of the first brief conversation Bonaparte had with Volta in Paris. On that occasion the First Consul was keen to know whether Volta's science implied that the entire "animal machine" was "ruled by the electric fluid."[127] This philosophical issue, as the controversy over galvanism clearly showed, had broad if ambiguous implications for the most general views concerning nature and the human species. Bonaparte, like many contemporaries, took a lively interest in such issues, which the Institut National itself echoed when adopting Bonaparte's resolution to establish an annual prize for discoveries in electricity and galvanism.[128]

The expectations of usefulness nurturing Bonaparte's curiosity about Volta were related to the technological, and possibly military, applications of the battery.[129] Bonaparte alluded to such applications when he first invited the Institut National to reward Volta. He mentioned them again when attending Volta's experiments with a big battery of eighty-eight metallic pairs, capable of producing impressive shocks and sparks, and of smelting iron wires with considerable noise.[130] To satisfy the expectations of the general, on the occasion Volta also exploded his electric pistol,[131] a device he had conceived some twenty-five years earlier and circulated since to arouse the curiosity of his patrons and audiences in military matters. By sharing in such curiosity Bonaparte fell in line with expectations that—though recently magnified by the French revolutionary and Napoleonic wars—had already been cultivated by several previous generations of philosophers and public administrators Volta had been in touch with.

Thus Bonaparte found a number of reasons that encouraged him to meet Volta's request for recognition, while using Volta as a symbolic figure for his own political purposes. The Parisian scientific elite, which had accepted Bonaparte as a member of the Institut National, must have shared at least some of Bonaparte's reasons.[132] However, they seem to have encountered more difficulty than Bonaparte in making Volta and the battery suit their own ends.

The pace of Volta's scientific work declined rapidly after the invention of the battery, a fact which makes it difficult to judge the impact his exposure to Parisian electricians in 1801 might have had on his work in the long run. What is certain is that, despite the further honors bestowed on him by Napoleon and the presence of French-dependent governments in Lombardy until 1814, Volta's contacts with French scientists remained as scanty as ever after 1801.

Volta's 1801 mixed success with the philosophers from the Institut (more on this in chapter 7, below), his memories of the frustrations suffered in 1782 (discussed in chapter 4, above), and his own views on the place of natural philosophers in society at large, must all be borne in mind in accounting for the fact that in Paris in 1801 he liked to mix—alongside with people of top civil and scientific distinction—with amateur philosophers and even a well-known charlatan: Eugène Guillaume Robertson. The connection with Robertson (as we shall see in chapter 7) shows that Volta, who for several reasons was still an outsider in Paris despite the success sanctioned by the Institut, continued to address and try to win over to his science a range of audiences, including some of dubious scientific reputation.

Volta's broad, not-strictly-professional, or not-national view of the experts worth addressing was apparently rooted in his background and experience. This was the experience of a self-trained natural philosopher in the age of the Enlightenment who had worked for decades far away from well-structured research centers, who was in frequent contact with amateurs of many sorts, and who had learned to strive for recognition by dealing with scholars from several cultural, social, and national backgrounds.

## Conclusion

Several figures and institutions participating in the cosmopolitan network of natural philosophers in late Enlightenment Europe have been dealt with in the present chapter. The biographical approach adopted throughout does not allow generalizations going beyond the individual case considered. That approach, however, opens up some in-depth views of the motives and means that, at an individual level, sustained the exchange of information and recognition among natural philosophers across national borders. Our concluding remarks are focused on these motives and means, and on the ways cosmopolitan networks of natural philosophers managed to survive amid the clash of national interests after 1789.

Volta's early steps beyond his native land were inspired by motivations that for centuries had nourished the idea of a cross-national Republic of Letters. The basic motive was the search for peers, not available at home, able to inspire, judge, and eventually reward the intellectual pursuits of the isolated scholar. Among the electricians there were some very good examples for young Volta to imitate. The works of Franklin, Nollet, Priestley, and Beccaria, through which he had been introduced to the science of electricity, gave ample proof of how fascinating and rewarding participation in a cosmopolitan network of experts could be. These were not, however, the only models available to Volta. Evidence discussed in this chapter shows that the entrepreneurial spirit of an ambitious cadet from the lesser nobility exposed Volta to other models as well. The international networks that several Lombard aristocratic, merchant, banking, and manufacturing families were part of, and which Volta was in touch with, offered him important stimuli for developing his contacts beyond national borders. Until 1775, when he was still an independent scholar, these models worked on him at the cultural level. Joining the public administration of Austrian Lombardy offered Volta new, concrete opportunities and means to pursue his scholarly and entrepreneurial activities across national borders while acting in the name of his patrons.

The opportunities presented to Volta as a professor and a public servant grew out of the imitation-competition game in which public administrations of Enlightenment Europe were engaged. Volta ably put the talents and international contacts he was developing as an independent scholar to the service of his patrons' ambitions. These ambitions implied that educational and scientific institutions within the Austrian empire should be regarded as on a par with the ideals of Enlightenment culture and ideology that were shared across national borders by European elites. To encourage and finance the trips abroad of brilliant, gentlemanly scholars like Volta was, on the part of Austrian public administrators, a way both of advertising their achievements at home and of acquiring new information that might improve their future chances in the game. To join in the international competition to establish the most prestigious scientific collections was another side to the same game, and one which Volta cleverly encouraged in the administrators of Austrian Lombardy to his own advantage. Scientific instruments and collections, with their international market, were the most visible of the objects involved in the imitation-competition game. An examination of Volta's experience has shown us how tightly interwoven the market of such instruments was with the international pursuits of natural philosophers and their patrons.

Volta's participation, as a professor in the service of the Austrian empire, in the imitation-competition game outlined above clearly improved his chances of recognition as a natural philosopher on a European scale. However, the complex rules governing the assessment of scientific excellence and reward within expert circles ensured a degree of independence between his quest for recognition as a natural philosopher and his activity as a public servant working and traveling at his government's expense.

To illustrate the situation more clearly, the metaphor of the different "maps" of the European scene available to Volta is helpful. We can regard Volta in his international pursuits as an explorer approaching a little-known region, and having recourse to several different maps of the region depending on need and situation. Three such maps are discernible. While the basic features of all of them naturally overlapped, the "borders," the "capitals," and the hierarchy of countries drawn on each one did not necessarily coincide.

The first map was the expert's map. This was the map Volta relied on as an electrician: it was the map of the region of the Republic of Letters to which he felt he belonged as a natural philosopher specializing in that field. A number of countries were painted on the map in brighter colors, e.g. Britain, Switzerland, Holland, the German states, and France. Scientific excellence and opportunities for reward blended in guiding Volta's choice of which countries were more prominent in the republic of electricians. Their order of merit changed during Volta's career and experience. There was no single capital for the multinational confederation represented on this map. However, should Volta have been asked before 1796, he would very probably have pointed to London as the most obvious choice. The lack of a single capital in the republic, or indeed the simultaneous presence of several—as seen from Volta's position at the periphery—had notable advantages for him in his search for recognition. He could negotiate rewards with several capitals at a time—as indeed he learned to do in Paris and London in 1782. He could identify niches an outsider could fill to establish his own proper competence, without interfering too closely with authorities already established. The competition among established research centers could also eventually be exploited to the same end.

Volta's second map was the enlightened lay person's map. The countries highlighted here were to a large extent the same as on the expert's map, with the notable addition of Maria Theresa and Joseph II's Austria. Broad cultural and political views combined to shape this second map. We have already seen at work the criteria that inspired Volta as self-appointed judge of the civilization and culture of the countries he visited. A well-run public administration, lively commerce and cultural pursuits, substantial secularization of the state and society, a reasonable balance between individual riches and the welfare of the lower classes—the *medietas* he liked so much in Britain—these were his chief criteria, and they together molded his own map of Enlightenment Europe. The many overlappings of the expert's map with this second map show just how closely the natural philosopher in the age of the Enlightenment saw his own pursuits as being intertwined with those of lay citizens and public administrators.

Volta's third map was the civil servant's map—the map of allegiance and power. Its features changed drastically in 1796, with the passage of Lombardy from Austrian to French rule. The new contours took on a relatively permanent shape only around 1800, with the advent to power of Bonaparte. Both before and after 1796 the major features, borders and capitals marked on this third map did not coincide with those of the maps Volta relied on as a natural philosopher and a citizen of Enlightenment Europe. Such discrepancies did not cause Volta any major troubles throughout his career; but the pressures on him resulting from his position as public servant grew steadily during the 1790s and after.

Until 1784 he was relatively free to choose the experts he wanted to deal with independent of the obvious predilection for Austrian and German culture encouraged by his patrons. When, after 1784, he was pressed to reinforce his acquaintance with "German literature," he was able to comply by strengthening ties with natural philosophers from the German states rather than from Austria. We may conjecture (though direct evidence is lacking) that Volta's continued and reciprocated predilection for the Royal Society of London between 1789 and 1800 was linked to his aversion for revolutionary Paris, and to his desire to preserve a degree of independence from the powers confronting each other in Lombardy. What is certain is that when at last, in the autumn of 1801, Volta chose Bonaparte as his new patron—and Paris took the place of Vienna on his public servant's map—he could still claim that the new patron, in the running of public administration and his dealings with natural philosophers, pursued values and goals that did not clash with those he had pursued in previous decades under the enlightened government of Austria. If this seemed a convenient excuse for compromise, there was nonetheless a good deal of truth in it. As to the objection that he was in this way swearing loyalty to a foreign power, that was no novelty for an Italian who had served Austria for more than twenty years and lived in a country that had not known independence for centuries.

Thus, despite the dramatic changes on Volta's map of allegiance and power and the growing pressures from political turmoil after 1796, a good degree of continuity was and is detectable in Volta's actions as a natural philosopher and civil servant before and after that date. That continuity, apparently, was rooted in values and goals laid down by the two other "maps" Volta continued to rely on.

If continuity of Enlightenment ideals and goals before and after the advent of Bonaparte was the important thing for people like Volta, the historian can nonetheless detect something new and symptomatic in the emphatic way Bonaparte celebrated Volta's merits on the public scene. Neither Volta nor Bonaparte probably grasped the significance of this entirely. Volta perhaps perceived and expressed it confusedly in his reaction of genuine wonder, recorded in letters home from Paris, following the conferment of the magnificent rewards on him in 1801.[133] A further strengthening of ties between natural philosophers and public power was implicit in the rewards Bonaparte bestowed on Volta and the emphasis that went into advertising them on the European scene. Expectations, rituals, and rewards, which for decades lay audiences and public administrators of Enlightenment Europe had associated with natural philosophers, their "useful truths," and their ingenious pieces of apparatus, had reached a new climax.

It was implicit in Bonaparte's gesture that the greater role accorded natural philosophers in public culture carried with it increased public responsibilities and closer ties with political power. Bonaparte's celebration of Volta meant, of course, that the map of allegiance and power, helping natural philosophers navigate through public culture and institutions, was gaining importance over the expert's and the lay person's maps. To judge from Volta's case, however, that did not mean that the other maps lost their efficacy. Volta's science did not become more "French"—either in conception or style—after 1801. That could not be part of the agreement between Volta and Bonaparte, whatever both might have wished in subscribing to it: the redrawing of the expert's map required far more than an agreement between two individuals. Even the pervasive use in Europe during the eighteenth century of the French language as the lingua franca of natural philosophers (as well as the aristocracy and the merchants) had not brought undisputed dominance of French natural philosophy. Long-established practices for excellence and reward assessment, which had not yielded to French supremacy throughout the period when French was the lingua franca, could well resist Bonaparte's propaganda. After 1801 Volta naturally continued to ponder what was being done by his colleagues in London, Geneva, Göttingen, Haarlem, as well as Paris. The simultaneous presence of several capitals on the expert's map continued to function and bear fruit. Habits of thought and behavior that for decades, indeed for centuries, had forged the transnational exchanges of expert communities were unlikely to bow to political pressure easily.

CHAPTER SIX

# The Battery

## INVENTION, INSTRUMENTALISM, AND COMPETITIVE IMITATION

This chapter offers a reconstruction of some three years of Alessandro Volta's investigation that culminated in his epoch-making invention of the electric battery late in 1799. Among the materials used are laboratory notebooks and unpublished writings, including drafts of Volta's letters to the President of the Royal Society announcing the battery, not previously discussed by historians.[1]

A central point of the chapter is that until spring 1799, and probably even later, Volta was *not* in any sense engaged in a search for a device like the battery. The *goal* of building such a device and some important *hints at its construction* were suggested to Volta by a paper written by William Nicholson, which Volta read sometime in the autumn of 1799. Nicholson's article, so far neglected by historians, dealt with the best way of imitating the shocks *and* anatomy of the electric fish, particularly the torpedo.[2] In this respect, Volta's invention followed a fruitful interaction between physics and the life sciences. Contrary to some current views, during the late-eighteenth century their interaction was intense and creative, and not confined to the rhetorical, sometime inconclusive, arguments adopted in the controversy over galvanism.[3] The chapter concludes with a few remarks on the modes of invention within what has been called the instrumentalist bent of late eighteenth-century science.

The circumstance that the battery, later depicted as a momentous turning point in the history of physics, was the outcome of a program that included the goal of "imitating the electric fish," is used in this chapter as a cautionary tale against the sort of history of science that prevailed when the disciplinary boundaries imposed by twentieth-century developments oriented the work of historians. In fact, the reconstruction of the invention of the battery developed in the present chapter invites a reassessment of the forces that, around 1800, were driving the science of electricity toward unprecedented achievements.

One such driving force was no doubt the "quantifying spirit," that historians of science have pointed out as a crucial trait of eighteenth-century science, and we have seen at work in Volta's laboratory practice discussed in previous chapters. The "spirit" consisted in the brand of "sound reasoning and accurate investigation" (and, above all, exact measuring techniques) that the Royal Society of London commended in 1794 when conferring upon Volta the Copley Medal for having tested Galvani's experiments on animal electricity, which had "too much astonished," and seduced, experts and amateurs alike in several parts of Europe.

Another, momentous driving force behind the invention of the battery, however, was of a quite different kind. It was curiosity: a curiosity unrestricted by the usually severe rules of exclusion enforced by strong research programs and expert groups. It was a curiosity nurtured, in an ambitious natural philosopher like Volta belonging to the periphery of the Republic of Letters, by exposure to the diversity of the agendas pursued within the mixed community of electricians, scattered in several European countries, that he had been in touch with in his search for recognition over several decades. Thanks to this curiosity and experience, Volta was often able and willing to compare his notions and results with other investigators, whether they were the independent lecturers and instrument makers he liked to mix with in London in the early 1780s, or well-established Parisian authorities like Lavoisier or Laplace, or—in the 1790s—physicians suspected of being on the verge of scientific unorthodoxy like Galvani, or, occasionally, even quacks, as we shall see in chapter 7, below.

Toward the close of the eighteenth century, Volta's unrestricted curiosity merged with his particular brand of the quantifying spirit (simple, but stern, measuring techniques)—and with a number of unpredictable circumstances generated by imitation and competition within the mixed community of electricians to which he belonged—to bring about the superb apparatus that Michael Faraday still depicted as a "magnificent instrument of philosophic research."[4]

## Galvanism, Electrometer in Hand

Late in his life,[5] Volta regarded 1796 as a watershed in the path that led him to the discovery of the electric battery. There is much to be said for the date, even though Volta's involvement with one of the main threads of the story, the search for weak electricity, went back to his works on the electrophorus in 1775, the *condensatore* in 1780, and electrified vapors in 1782; and his involvement with another major thread, the controversy over animal electricity and galvanism, dated back to the spring of 1792, a few months after the publication of Luigi Galvani's *De viribus electricitatis in motu musculari*. In 1796, however, Volta's approach to the study of weak electricity and of "animal electricity" changed in one important respect. The change may be illustrated by comparing the means he employed before and after 1796 to detect the small quantities of electricity in play in experiments on galvanism.

Until July 1796, Volta measured weak electricity via frogs and other physiological apparatus, such as his own tongue and sense organs. This was consistent with Galvani's own assumption that in such experiments frogs acted as "most exquisite electrometers." According to Galvani and his followers, however, frogs did more than that: they provided the electric fluid, supposedly proper to all animals, that caused the contraction of their legs. Against this conception, Volta had maintained consistently—after a short-lived adhesion to Galvani's views—that the frog played only a *passive* role, proper to an instrument for measuring electricity. According to Volta, the motion of the electric fluid evidenced by the contractions of the frog's legs arose from the contact of different metals and wet substances employed in experiments on galvanism.

During 1794 and 1795, the twofold role of frogs and sense organs as stimulators *and* detectors of electricity became critical as a consequence of a momentous turn in the controversy between Galvani and Volta. Galvani and his followers had succeeded in obtaining the contractions without metals, by just putting certain parts of the frog's body in contact with other parts.[6] Volta reacted to the challenge in two ways. For one, he corrected his earlier assumptions, admitting that, where no metal stimulated the electric fluid, the contact between different humid conductors, like animal substances, could do so. Second, he decided to find a way out of the uncomfortable position that compelled him to use animals as detectors of electricity in experiments conceived to show that animals do not have a special sort of electricity.

Volta estimated that the stripped frog commonly used in galvanic experiments (see fig. 6.4, p. 189) could reveal a tension corresponding to 0.0002 degrees of his straw electrometer. No electrometer then available to Volta could match the frog's sensitivity. In June 1796, however, Volta decided to make use of a new instrument. He asked Giuseppe Re, the instrument maker who assisted him at the University of Pavia, to build for him a special version of a "doubler" of electricity that William Nicholson had described in the *Philosophical Transactions* for 1788.[7]

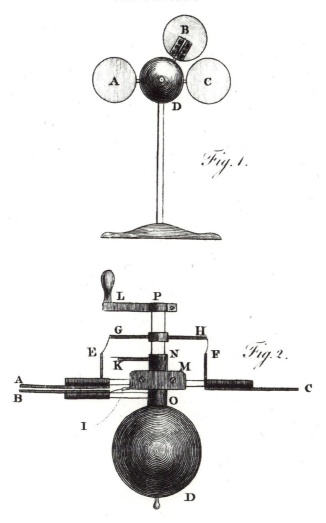

6.1. Nicholson's doubler. From W. Nicholson, "A Description of an Instrument," *PT*, 38 (1788) (*Biblioteca Universitaria, Bologna*).

The instrument was ready in July 1796.[8] Soon afterward Volta started a long and impressive series of experiments with it.

Nicholson's revolving doubler of electricity (fig. 6.1) consisted of three brass discs, two fixed (*A*, *C*), and one mobile (*B*), and a rotating brass ball (*D*) that revolved on the same axis as *B*. Disc *B* faced the other discs without touching them; brushes acting as switches in *E*, *F*, *K*, and *I* connected and disconnected the different parts of the machine during operation. As Volta viewed it, the doubler was based on the principles of his own *condensatore*, which had been used since 1780 to make weak atmospheric electricity detectable by the electrometer (see chapter 4, p. 113). To operate the *condensatore*, one raised the weakly charged plate, which had induced an equal charge on a fixed plate, touched the moving plate to ground, recharged it weakly, reset it, and so forth, until the fixed plate acquired a detectable electricity. In Nicholson's doubler, the trick of the consecutive approaching and separating of plates and the appropriate grounding (provided by the brass ball *D*) was performed easily and effectively by a crank and gears.[9]

In a few weeks, Volta managed to detect with the doubler the electricity deriving from the contact of two different metals, something only frogs' legs and men's tongues had achieved before. This brought with it a significant shift in Volta's own perception of the place his work on metals and galvanism held in contemporary research. In October 1796, resuming correspondence with Martinus van Marum after a four-year interruption, Volta stressed his own recent success in making "metallic" electricity instrument-detectable. He presented it as a step for bringing his earlier work on galvanism back into mainstream physics research. He observed that there was an important parallel between van Marum's work on large quantities of electricity with the powerful Teyler electrostatic machine and his own work on extremely weak electricity; a parallel that appeared more firmly established now that, thanks to Nicholson's instrument, the weakest electricities could be treated by the ordinary means and criteria known to electricians.[10]

The fact that the first extremely weak electricity that Volta detected by physical methods arose from the contact of metals alone had further consequence for his work. Until 1796 he had maintained that in experiments on galvanism, the contact between metals ("first-class conductors"), and humid matters ("second-class conductors") stimulated the electric fluid. After summer 1796, he inclined to believe that the contact between two different metals was a much more effective "motor" of electricity than a mixed junction of first- and second-class conductors.[11]

There was a final, and subtler, influence of Nicholson's doubler on Volta's investigation: it contributed to a change in the shape, handling, and combination of the metal pieces on his worktable. Before the summer of 1796, Volta used metals arranged in the ways common in experiments on galvanism, that is, shaped either as "arcs" (monometallic, bimetallic, or partly composed of insulating material), or as rods and bars, in the form usually available to craftsmen. Occasionally, for experiments on the tongue's sensitivity to electricity, Volta had used coins, which, however, were usually absent from his bench before summer 1796. With the introduction of Nicholson's doubler, metal discs became common again on Volta's worktable when experimenting on galvanism. After a while, he found it useful also to shape humid conductors as discs. Volta's manuscript notebooks from summer 1796 onward bear overwhelming evidence of the important role the shape, handling, and combination of metals and humid conductors held in his day-to-day experimental work.

Volta spent many months, from summer 1796 to early spring 1797 experimenting on weak electricity with the new machine.[12] At first, he used Nicholson's doubler as a substitute for frogs and sense organs; we find him connecting Nicholson's device in circuits made of various metals and/or humid conductors just as he had connected frogs in circuits to detect electricity.[13] We know that Volta's main purpose in substituting the doubler for the frog was precisely to eliminate the animal that Galvani and his followers assumed had an active role in exciting electricity. Soon, however, to Volta's disappointment, the doubler turned out to be not so passive either.

Nicholson had warned that his doubler was dependable only when the electricity to be detected and communicated to its plates, was "strong enough to destroy and predominate over any other electricity the plates may possess."[14] In fact, as Volta soon discovered, by turning the handle 50, 60, or 100 times, the instrument *always* gave distinct signs of electricity,[15] even when none had been communicated to it. To protect against such false results, Volta tried many different ways of manipulating the instrument, but to no effect. He calculated that the machine had to retain no more electricity than corresponded to 0.004 degree of his straw electrometer to give a positive reading.[16] This increased Volta's admiration for Nicholson's device, and had other important consequences.

Volta's estimate of the apparatus's "residual" electricity made it an extremely weak source similar in strength to those he had dealt with in his experiments on atmospheric electricity, on vapors, and on contact electricity. Thus, it was quite reasonable for Volta to assume that at least part of the doubler's "residual" electricity came from the atmosphere, or from *the metals forming the instrument itself*.[17] Following up this last possibility, Volta started a new series of experiments, substituting discs of various metals for the brass ones of Nicholson's machine. This step, dictated by the theory of "metallic electricity" that Volta had opposed to Galvani's "animal electricity," would not have occurred to Nicholson.

For Volta, attributing an *active* role to Nicholson's doubler in setting up electricity in motion meant renouncing the purpose for which he had taken up the instrument. Now, like the frog in galvanist theory, the doubler contributed to setting electricity in motion. No more than the frog could it be assigned the role of a neutral observer. That by no means made it useless to Volta. On the contrary: by substituting discs of different metals for Nicholson's brass ones, he could find new ways for testing his theory that electricity can be set in motion by the contact of different conductors. Instead of a neutral observer, Nicholson's machine could become an active supporter of Volta's views.

Here a short excursus into Volta's theories and concepts concerning electricity in motion is necessary. The tests to which he submitted Nicholson's machine were suggested by concepts developed during the early arguments over galvanism. In 1792, Volta had assumed that the electricity manifested by the frog's contractions derived from an "imbalance" of the electric fluid caused by the metallic coats or arcs used in Galvani's experiments on frogs.[18] This imbalance caused the motion of the electric fluid detected as soon as the metal coats were connected "in a circle" through the animal. Volta also assumed that in these circumstances the motion of the fluid would have quite different characteristics from the almost instantaneous discharge of the Leyden jar. He conceived of the electricity observed in galvanic experiments as a *continuous current*, kept in motion as long as the circle formed by the different conductors was uninterrupted.[19] As Volta admitted, this continuous motion of the electric fluid, maintained independently of any motion or friction from the outside, did not fit the notions then adopted by electricians.[20] Yet, as he liked to say, it was so real it could be touched, as it were, through the prolonged contractions of a frog's legs and the continuous taste delivered to the tongue by the contact with two different metals.

Volta attributed this "spontaneous" electrical motion to a power, intrinsic to metals, of "pushing" the fluid into humid conductors.[21] He thought of this "pushing" and "pulling" in analogy to similar explanations in mechanics and hydraulics. Thus, dealing with the combination of many different metals and humid conductors in the same circle, he argued that, under certain circumstances, the powers of pushing and pulling the fluid could counteract one another, so that no signs of electricity were observed.[22]

While experimenting with Nicholson's machine late in 1796 and early 1797, Volta wrote of "mutual forces of attraction and repulsion" acting between metals in contact with one another, and considered them responsible for the motion of the electric fluid.[23] This was reminiscent of his juvenile speculations on attractive forces, discussed in chapter 3, above. He then contrasted this electricity in motion with the "local," "stagnating" electricity previously known to electricians.[24] In April 1798, however, Volta displayed considerable reticence in committing himself publicly to any conclusive explanation of the motion of the fluid. He then labeled the causes, which apparently disturbed the spontaneous "equilibrium" of the electric fluid, as "arcane."[25]

Whatever his uncertainties and reticence concerning the true causes, Volta—as we know—availed himself of a number of definite, midrange concepts, which oriented his investigation very effectively. Some of them had the form of empirical laws, summarizing the results of a long series of experiments. Others were concepts hinting at some general, and possibly quantitative, feature assumed to be proper to all electrified bodies, or to the electric fluid itself.

One of Volta's empirical laws, most relevant in our context, attained full maturity in the autumn of 1795.[26] It established a series or scale of metals, arranged according to their decreasing power of "pushing" the electric fluid into humid conductors. The scale ranged from zinc, in which the power expressed itself in the highest degree, to charcoal, which Volta included among metal conductors. The more distant in the scale were the metals put in contact with each other, the more vigorous was the motion they impressed on the electric fluid.[27]

Another example of Volta's midrange concepts was resistance. Contact of different metals could set electric fluid in motion inside them because conductors opposed very little "resistance" to its motion. In insulating bodies, the fluid experienced a high "resistance," hence the need for moving the fluid. Toward the end of 1797, this conceptualization led to the enunciation of an empirical law that ranked the different actions responsible for the motion of the electric fluid—from mere contact through percussion to rubbing—in a continuous scale according to their increasing efficacy.[28] Supplementary to the concept of resistance were those of "tension" and "electric atmosphere," which we already know (see chapter 4, p. 116ff).

The following example shows the close interaction between experimental setup and conceptualization that characterized Volta's work on galvanism during the 1790s. It concerns work on the doubler and dates to the end of 1796 and early 1797.

In figure 6.2, L (figs. a-d) and O (fig. e) designate brass discs, A designates silver, S tin, and aa a humid conductor. Keep in mind the notion of a scale of the different powers attributed to metals according to their efficacy in "pushing" the electric fluid:

> The discs of the doubler are separated either by a very short gap or, more effectively, by a thin, humid cardboard, which, if not too wet, is quite a bad conductor. The electric fluid excited and put into motion by the usual actions will accumulate in the brass disc (L, figs. a, b), which is touched by the silver rod (A). It will also accumulate in the tin disc, touched by the

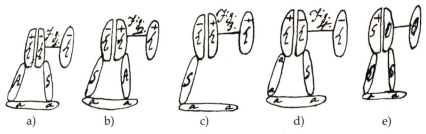

a)          b)          c)          d)          e)

6.2. Volta's representation of his experiments with Nicholson's doubler, where he substituted different metals for Nicholson's brass discs. From *VO*, 1:469–73 (*Biblioteca Universitaria, Bologna*).

brass rod (*O*)(fig. e). In contrast, the electric fluid will be diminished in the brass disc, which is touched by the tin rod (figs. a, b, c, d). And this will occur the more easily since these two opposed electricities—balancing each other in the facing discs—will also sustain each other, and will somehow compensate each other. And this because the mutual influence of their atmospheres, which are so close, will relax and diminish their *electric tensions.*[29]

The concept of tension, as distinct from that of quantity of electricity, helped to explain an amazing circumstance in experiments on galvanism: the fact that frogs reacted as spectacularly to the weak current put in motion by metals as to the fluid drawn from a powerful electrostatic machine. This happened, Volta explained, because the physiological reaction depended on tension and quantity together. Whereas the tension of electricity set in motion by metals was very weak, its *quantity* was large enough to give it the effect of a high-tension, low-quantity source like electrostatic machines.[30]

### The Hunt for Weak Electricity

Volta's enthusiasm for Nicholson's revolving doubler was not shared by other leading figures in the small international community of physicists engaged in the hunt for weak electricity. These were, besides Nicholson and Volta, Abraham Bennet and Tiberius Cavallo. Cavallo had added caution to Nicholson's own caution in presenting the instrument that Volta had used so effectively in the course of 1796.[31] The uncertainty deriving from what Nicholson and Cavallo explained as the instrument's own "residual" electricity was common to all the devices then employed in these investigations. Half a dozen such instruments were available to electricians in the 1790s. A brief survey of

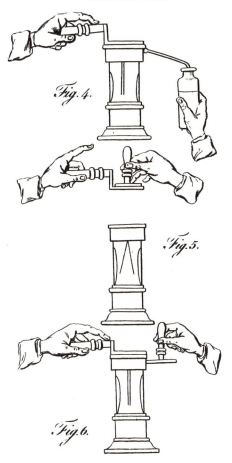

6.3. Bennet's doubler. From A. Bennet, "An Account of a Doubler of Electricity," *PT*, 77 (1787): 288–96 (*Biblioteca Universitaria, Bologna*).

these instruments will help us to understand why Volta, after obtaining impressive results with Nicholson's doubler, abandoned it in favor of his own older, and simpler, *condensatore*.

As contemporaries viewed it, the main function of the *condensatore* was to collect an expanded quantity of electricity into a small space.[32] To use Volta's terminology, the *condensatore* augmented the tension of a small quantity of electricity until it affected the electrometer. This was different, as Cavallo noted, from augmenting electricity. Cavallo regarded Bennet's "doubler," first described in 1787, as the first one able to multiply charges. It consisted of three brass discs mounted on an electrometer, and combined as shown in figure 6.3.[33]

A small quantity of electricity, communicated to one of the discs, was augmented through a delicate series of operations until it became detectable by the electrometer. As electricians knew, Bennet's doubler had the same source of uncertainty to which the instrument later introduced by Nicholson was liable; when manipulated 20 or 30 times, it always gave signs of electricity even when none had been purposely communicated to it. According to Cavallo, the friction on the insulating parts of the instrument caused the superfluous charge. Hence Cavallo had introduced in 1788 a new instrument, with plates insulated from one another by air. In that same year, as we know, Nicholson had described his own revolving doubler, which combined Cavallo's artifice of the air as insulating medium, and the practical handle, whereby the plates could be turned facing each other, reducing the risk of accidental friction.

In 1795 Cavallo introduced a new device, which he called a "multiplier,"[34] a sophisticated version of the electrophorus introduced by Volta in 1775. Volta's original instrument is illustrated in chapter 3, figure 3.1, page 74.

In Cavallo's version (fig. 6.4, p. 189) "the plate A may represent the excited resinous plate, B may represent the metal plate of the electrophorus, and C is a kind of reservoir, into which the successive charges of the plate B [moved by the lever KL] are collected." According to Cavallo, his multiplier was less subject than earlier doublers to adventitious charges since the metal plate A was likely to retain little residual or accidental electricity.[35]

Nicholson doubted this claim and offered a new "spinning" device, which attracted less attention than his doubler.[36] Despite his own new contribution to the field, Nicholson expressed scepticism concerning all the machines for multiplying electricity built since Bennet's doubler. As Nicholson wrote in December 1797, Bennet's old, simple instrument was still the most reliable device for detecting weak electricity, despite the awkward manipulating techniques it required. Volta preferred to return to the *condensatore*.

The experiments that resulted are described extensively in Volta's third letter to Friedrich A. C. Gren, dated March 1797. He began them much earlier, however, since he alluded to them already in his second letter to Gren, written the previous summer; yet, as late as February 1797, a friend saw Volta demonstrating metallic electricity with Nicholson's doubler.[37] Apparently for some time after he had decided to substitute the *condensatore* for the doubler, Volta still used Nicholson's device to demonstrate metallic electricity. The two experimental settings—the doubler and the *condensatore*—probably coexisted for some time, as did the conceptualizations associated with them.

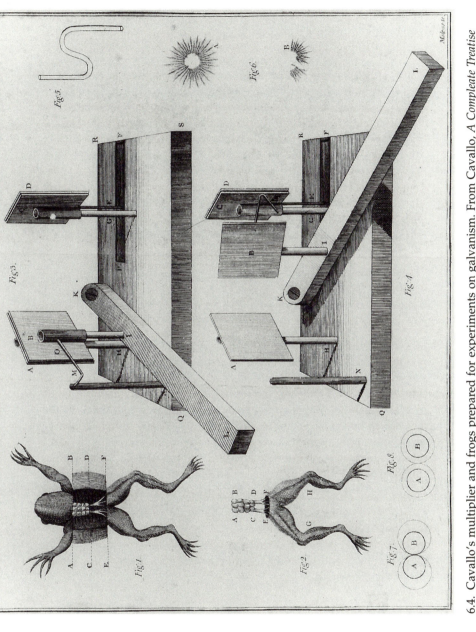

6.4. Cavallo's multiplier and frogs prepared for experiments on galvanism. From Cavallo, *A Compleate Treatise* (London, 1795), plate 6 (*Bodleian Library, Oxford*).

In the experimental setup using the *condensatore*, Volta required only metal plates, a small Leyden jar, and a waxed cloth. He described his operations as follows:

> I charge the small Leyden jar in the same way I would an electrophorus; that is, I put it into contact 20 or 30 times with the insulated metal disc, whose electricity I want to test, and which is brought each time from contact with another metal disc, communicating with the earth. Assuming that the plate's electricity is 1/4 of a degree, after 20 or 30 such contacts the Leyden jar will be charged to 1/20 or 1/30 of a degree. If the Leyden jar is now discharged in the way used with a *condensatore*, that is, by putting it into contact with another metal plate lying on the waxed cloth, and if this last metal plate, instantly lifted, is put into contact with an electrometer, it will develop 2 or 3 degrees. That is, the leaves of my straw electrometer will diverge of 1 line, corresponding to 4 or 6 lines of the golden leaves of Bennet's electrometer.[38]

This sort of experiment, or demonstration, brought a clearer though not entirely new conceptualization of the condensing power that Volta ascribed to metals in addition to their powers of moving, conducting, and resisting the electric fluid. He wrote:

> . . . each metal has a certain power, which is different from metal to metal, of setting the electric fluid in motion, and has also the power of conducting it, this power being about the same in all metals. When metals are put into contact, those powers combine so that electricity attains a certain power or tension, proper to that combination of metals, and such that it can preserve itself given the metals' conductivity. Their conductivity, however, will be reduced as a result of the contrast [between differently charged metals], so that they will acquire a degree of insulating power higher than when no motive powers affect them.[39]

By this last remark, Volta meant that under such circumstances metal plates acted as *condensatori*, and so could retain an amount of electricity they could not hold when alone. As in the case of the revolving doubler, adoption of a new instrument—the *condensatore*—strongly influenced Volta's thinking about the mechanisms responsible for the motion of the electric fluid.

## The Electricity of Animals

A prominent event in the controversy over galvanism in 1797 was the publication of Galvani's *Memorie sull'elettricità animale*.[40] As they had done three years before with the anonymous treatise on the *Arco conduttore*, Galvani and his followers attacked Volta's views in a book of considerable scope and ambition. No more did Alexander von Humboldt's big book published in the same year,[41] or the then recent third volume (1795) of Cavallo's *Complete Treatise on Electricity*, which did not adhere to Volta's views on metallic electricity, give him any support. The delicacy of his position was the greater because he had so far defended his views only in unsystematic articles scattered in many journals. He therefore took Galvani's latest challenge very seriously. To Volta's relief, four of the five challenging memoirs presented arguments and experiments that he had already dealt with in his letters to Gren. Volta disposed of these memoirs by restating his previous arguments, in a more systematic way, in two anonymous letters addressed in April 1798 to Giovanni Aldini, Galvani's nephew and supporter.[42]

The fifth memoir was more challenging. It described observations and experiments on the torpedo fish. Galvani offered the fish as living proof that a fluid similar to or identical with the electric existed in the animal world, and had nothing to do with metals. The argument from electric fish to animal electricity was an old one.[43] Galvani brought in new experiments aimed at establishing a close link between the brain, nerves, and electric organs of the torpedo.

Volta was not equipped to confront his adversary over electric fish. The experiments required the abilities of an anatomist. Probably for this reason, Volta delayed answering; he did not mention the torpedo in replying to Aldini about Galvani's other memoirs. There is evidence that he worked on an answer during 1798. That autumn he told a correspondent that he almost had an essay ready "on the power of the will to move the electric fluid in the brain of animals."[44] He did not finish the essay and changed his mind about the nature of torpedo action in 1799. The fish remained prominent in his investigations until it received full treatment in a long essay published in 1814.[45]

Volta had entertained unusual ideas about the nature of the mind since very early in his career. At the age of sixteen, as we know, he is said to have written notebooks "on the soul of animals," maintaining to the dismay of his Jesuit teachers that animals also had a spiritual power.[46] The issue was much debated in eighteenth-century philosophy and science. It related directly to the idea that nerves acted via a fluid very likely identical with ordinary electric fluid. In 1771 Volta refuted alleged experimental proof of this identity adduced by Comus (Nicholas P. Ledru), who had built an electrostatic machine with a disc made of nerves dried in an oven. The machine produced electric signs, whence, according to Comus, nerves can generate electrical fluid. Volta had built a similar machine with a wooden disc, we know (chapter 3, p. 92). He judged Comus's demonstrations inconclusive.[47]

Newton had hinted at the identity of the nervous and electric fluids in Query 24 of his *Opticks*.[48] As Volta knew, one of the leading authorities in the life sciences in the eighteenth century, Albrecht von Haller, had opposed the idea.[49] Many arguments and experiments had been fielded in support of one or the other champion. Galvani knew this literature, which played an important role in directing the work culminating in *De viribus*.[50]

In 1782, in London, Volta met the leading authority on the electric fish, John Walsh, who had first demonstrated conclusively the electrical nature of the shock given by the torpedo. From Volta's report, which contained information Walsh had not yet published, it appears that they discussed the electrical character of the fish at length.[51] During the visit, Volta formed the idea that the torpedo presented the only certain case of electricity in the animal world and became acquainted with Henry Cavendish's seminal paper describing an artificial torpedo charged with a large set of Leyden jars.[52] And with Walsh he probably discussed John Hunter's view that the special extension of the nerves going from the brain to the lateral "muscles" of the torpedo accounted for their electric power (fig. 6.5, p. 193).

Although he regularly discussed the torpedo in his lectures in the 1780s,[53] Volta does not seem to have devoted special attention to it before 1797. He had not lacked opportunities to discuss the subject. Lazzaro Spallanzani, Volta's colleague and professor of natural history at Pavia, used to show his students large drawings of the anatomy of the torpedo, drawings that a competent visitor, Anton Maria Vassalli, found impressive in 1790. Later on, Spallanzani's successor in the chair of natural history, Giuseppe Mangili, used to show his students preserved specimens of the torpedo's electric organs.[54] The torpedo had an impressive following in Pavia.

The necessity of using frogs to repeat Galvani's experiments returned Volta to his earlier, occasional interest in the nervous fluid. He soon made some interesting observations on the connection between electricity and the supposed nervous fluid. He observed that four to eight times less electricity was needed to excite leg contractions when the electricity communicated to the frog ran from the nerve to the muscle than when it run from muscle to nerve. He concluded that in the first case the electric fluid followed the natural stream of the nervous fluid, and in the second ascended against it.[55] This reinforced Volta's "rather well-established conjecture," as he called it, that the soul and will exerted their power on the body through the nerves by means of the electric fluid.[56] The conjecture seemed further reinforced by Volta's

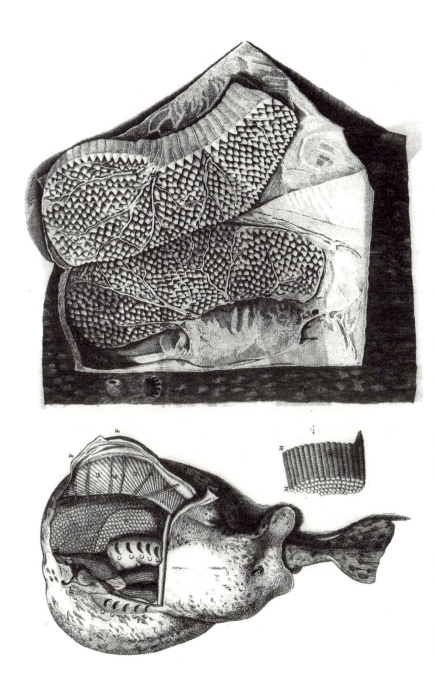

6.5. The torpedo fish as represented by Réaumur (left), and details of the organs giving the shock as represented by Réaumur and by John Hunter (center and right). From R.A.F. de Réaumur, "Des effets que produit le poisson appellé en François Torpille," Académie Royale des Sciences, *Histoire* (1714): 344–60, and from J. Hunter, "Anatomical Observations on the Torpedo," *PT*, 63 (1773): 481–89 (*Biblioteca Universitaria, Bologna*).

observation that the organs controlled by will are easily excited by electricity, while other organs like the heart, whose action does not depend on the will, react less to electric than to mechanical or chemical stimulations.[57] In that same year, 1792, Felice Fontana in Florence destroyed this symmetry by showing that the heart too reacted to electricity. Nonetheless, celebrated anatomists like Antonio Scarpa adopted Volta's new electric technique for distinguishing between voluntary and involuntary muscles.[58] Giuseppe Mangili—who had the mixed pedigree of a pupil of Volta's, Spallanzani's, and Scarpa's at Pavia, and of Fontana's in Florence—contributed a study of the electric stimulation of the nervous system of "worms."[59]

During 1794 and 1795, Volta gave his speculations on the nervous and the electric fluids the form of a definite conjecture, combining his notions on physiology and on the power of conductors to set electricity in motion. The clearest formulation appears in a letter of 13 April 1795. In it Volta invoked the old idea that will acts on the body through the nerves by means of the electric fluid and expressed the view that his theory of electricity arising from the contact of different conductors applied very well to animal organs like the brain and the nerves, composed of a variety of humid and liquid substances.[60] Volta supposed that the brain produced an electricity similar to that arising "from the simple contact of two different metals."[61]

With this conjecture, Volta updated speculations on the ways in which electricity might be produced or collected in animals. An earlier tradition had compared the brain to the sulphur or glass sphere of an electrostatic machine, and the nerves to the conductors or wires through which electricity and volition were distributed to the body.[62] Alternative speculations assumed that electricity derived in animals from the friction blood caused in its circulation, or from the food ingested and assimilated in the body.[63] Volta substituted his idea of mere contact for the old frictional view.

Volta further exploited his theory of electricification-by-contact to account for some new, more sophisticated conceptions of the anatomy of the nerves and the transmission of volition. The old electrical models were still dominated by the hydraulic imagery that viewed the brain as a collector of the electric fluid, and the nerves as pipes distributing it. In contrast, Volta's model was compatible with new evidence showing that the nerves were not hollow inside. Also, his model offered a new way of conceiving how electricity could act from the brain through the nerves, on the muscles and organs without transportation of the fluid. Perhaps the will, a spiritual entity, imparted a minimum impulse to the electric fluid, a material entity, at the root of the nerve, thus modifying the electric state of the whole nerve. Thus Volta's account of the interaction between spirit and matter; in the same vein, he described the electric agent of volition as a vapor rather than a fluid.[64]

Volta's view that the will, through electricity, acted only at the root of the nerve seemed to account for a phenomenon he himself had noticed. Electric impulse stimulated motion *or* the different sensation proper to a particular sensory organ according to the nerve to which the electricity was communicated. Volta deduced that the electric impulse played the limited role of activating an "energy" specific to each muscle or organ and not yet known in its nature and way of operation.[65]

From this rapid survey, it is apparent that Volta's views on the physiology of the brain and the nerves were very well developed in the mid-1790s. They allowed ample room for the traditional view that conceived of the brain and the nerves as special, privileged organs where the possible action of the electric fluid in animals could be located. Yet, having introduced here too the notion that electricity was set in motion by the mere contact of different conductors, Volta had implicitly weakened the border that traditionally kept the brain and the nerves apart from other humid and liquid substances making up organic bodies.[66] Volta oscillated between the traditional view and an inclination to apply his contact theory of electricity to organs other than the brain and the nerves. Galvani's memoir of 1797 on the torpedo intruded in this undecided setting.

Galvani aimed at demonstrating that in the torpedo, and arguably in all animals, a fluid similar or identical to the electric fluid was collected in the brain and distributed to the body through the nerves. Galvani also aimed to counter Volta's objection that this electricity, being associated with life and will, could act only in *living* animals, not in the dead frogs used in galvanic experiments.

Galvani's experiments on the torpedo were conceptually simple and straightforward. A first series consisted in extracting one of the fish's two electric organs, (see fig. 6.5, p. 193) leaving the other in its place. Galvani noted that the organ extracted produced no signs of electricity, while the other attached to the animal still gave them. A second series removed the torpedo's heart without damaging the brain and the nerves, to see whether the animal gave signs of electricity. Galvani found that it gave shocks until the brain too was cut out, when all signs of electricity disappeared. This second experiment was particularly delicate. Galvani apparently tried it twice and, still uncertain about the outcome, asked a colleague to do it again. After this repetition confirmed the pattern of events, he reported them in his memoir as a demonstration that the torpedo and probably all other animals were endowed with electricity, that this electricity stemmed from the brain, and that it persisted for some time after life and arguably also will had been suppressed by removing the heart.

We have two reports of Volta's reaction to Galvani's memoir on the torpedo. One of them, dated October or November 1798,[67] probably contains abstracts of the essay, "on the power of will to move the electric fluid in the brain of animals," which Volta had promised a correspondent on 19 October. Volta substantially adopted the conclusions Galvani had drawn from his experiments. According to Volta, they agreed with his own view that will acted on the body through the electric fluid. Volta also agreed with Galvani that the brain and nerves had special functions to perform in the special electric machine that is the living animal. Volta also seemed perfectly convinced that the source of electricity in the torpedo fish had its place in the brain and the nerves, not in the animal's special electrical organs that anatomists and electricians had compared to a set of Leyden jars or Franklin squares.[68] Rather, Volta argued that the experiments on the torpedo demonstrated convincingly that all animals with brain and nerves have some sort of animal electricity: the presence of so-called electric organs in the torpedo did not negate this basic truth. The only fact that mattered was the presence of a brain, and, above all, the presence of life: whenever Galvani and his followers had recourse to "animal electricity" in experiments on *dead* animals, they violated their own assumptions. But Volta could apply his contact theory of electricity to both the quick and the dead, as he had been doing since 1792.

On all accounts, by the end of 1798 Galvani's memoir on the torpedo had reinforced that part of Volta's views on physiology that he shared with the anatomists' tradition and that accorded the brain and nerves a special role in the "electric" living machine. Apparently, Volta had also swung away from an inclination he had occasionally indulged in before, for example in 1795,[69] to view the *whole* living organism as a combination of different humid and liquid conductors, all potentially able to put electricity into motion.

Volta's anatomical opinions soon shifted again. Something happened in the course of 1799 that changed his mind about the source of electricity in animals, and particularly in the torpedo. It was a new model suggested to explain the electric fish. In an unexpected and surprising way, it led Volta to the construction of the electric battery.

## Nicholson's Contribution to Volta's Discovery

Speculations on the will, the brain, and the nerves did not keep Volta away from his bench work for long. Laboratory notebooks kept be-

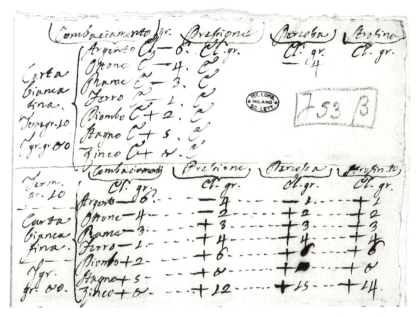

6.6. Detail of a page from Volta's laboratory notes, 1799(?). VMS, J, 53 ß
(*Istituto Lombardo, Milan*).

tween December 1797 and March 1799 show him engaged in extending
his scale of conductors' electromotive powers with a variety of chemical substances. He was also looking for a law able to combine electromotive powers and the different kinds of mechanical interaction that
conductors might exercise on one another.

Notes taken from December 1797 to March 1799 record measurements of the electricity produced either by mere contact, or by pressing, hitting, or rubbing discs of different metals against discs made of
dry or wet wood, paper, and cloth, and against his own body.[70] While
measuring, he kept note of atmospheric conditions, especially heat and
humidity (see fig 6.6, above). When electricity was weak, he used a
Leyden jar and a *condensatore* to detect it. Meanwhile, as he announced
in his second letter to Aldini in April 1798, he had succeeded in making
the electricity arising from the contact of two metals (silver and tin)
detectable by the electrometer without the artifice of the *condensatore*.[71]
With the same instrument he had also obtained some extremely weak
signs of electricity arising from the contact of two wooden discs moistened with different liquids.[72]

In another set of laboratory notes, we find measurements of the electricity arising from the contact between nonmetal discs moistened with

6.7. Volta's *condensatore*-electrometer, as sketched in Mar. 1799. From *VE*, 3, 440 (*Biblioteca Universitaria, Bologna*).

acid substances, soda, or potash, and other discs dampened by natural atmospheric moisture.[73] Here Volta used a new *condensatore*-electrometer (fig. 6.7, above), and Cavallo's multiplier.[74]

In March 1799, Volta arranged the results of his measurements in a new empirical law that ranked the different motors of electricity in four classes according to their decreasing power.[75] First came those consisting of two different metals; second, a metal and a wet conductor; third, two different wet conductors; and fourth, two similar wet conductors. This classification and the contact theory behind it apparently applied to galvanic experiments with *dead* animals. In the case of *living* animals, however, in accordance with his reactions to Galvani's experiments on the torpedo, he inclined to preserve a special role for the will, and for the brain and nerves as able to convey the electric fluid. The classification, the contact theory, and the special role of the will were no doubt the ingredients Volta had in mind when he announced to his correspondents in March that during the following summer he expected to complete a long promised essay on galvanism and his own theory of electricity.

He did not write such an essay. During the late summer or, more probably, the autumn of 1799, something changed his plans. This something reoriented Volta's investigation and his publication plans as well, giving him *a new goal, a new model of the electric fish to work with, and new ideas that led him in a few weeks or months to the construction of the electric battery.*

From time to time Volta ordered recent books and journals from Ambrosius Barth, a bookseller in Leipzig. On 10 April 1798, he ordered a long list of works, including William Nicholson's *Journal of Natural Philosophy, Chemistry and the Arts.*[76] Given that Nicholson was, like

Volta, a member of the small international group of electricians visibly engaged in the hunt for weak electricity, Volta's interest in Nicholson's *Journal* does not require special explanation. We do not know exactly when Volta received the *Journal*. On 20 September 1799,[77] Barth sent Volta a bill for books he had probably sent him on the same date or sometime earlier. Volta had the issue of the *Journal* for November 1797 when he wrote the first drafts of his letter of 20 March 1800 to Joseph Banks, announcing the discovery of the electric battery. It appears that an article of Nicholson's in this issue dramatically reoriented Volta's own investigation.

Nicholson's paper dealt with detection of weak electricity, the electric fish, and the construction of ingenious electrical devices. The title of the article refers to two of them and, indirectly, to Volta himself. It read: "Observations on the Electrophorus, Tending to Explain the Means by Which the Torpedo and Other Fish Communicate the Electric Shock."[78]

The observations on the electrophorus contained nothing new: it is a Leyden jar whose capacity diminishes as its shield is raised. Nicholson then recalled that the torpedo's electric organs (see fig. 6.5, p. 193) contained from 500 to more 1000 columns, about 1 inch long and 0.2 inch in diameter filled with many transverse films, 0.0033 inch thick and distant from one another by 0.0017 inch. His own experiments showed that the laminae or plates of Muscovy talc (common mica) electrified oppositely when separated. He compared the capacities of mica and glass and deduced that the two organs of the torpedo must have a capacity equivalent to that of a jar of 162,000 square inches. How did the torpedo jar actually work? Nicholson submitted a detailed conjecture, based on the further assumption that *a mechanical and electric machine could be built to give a shock like the torpedo's.*

Nicholson's account deserves to be quoted at length:

> It has appeared to me, from the observation of the high electric state which talc naturally possesses, and from the innumerable shocks the electrophore is capable of giving by mere change of arrangement, that a machine might be constructed also capable of giving numberless shocks at pleasure, and of retaining its power for months, years, or to an extent of time of which the limits can be determined only by experiment. I will not here describe the mechanical combinations which have occurred to me in meditating on this subject, but shall simply shew that the dimensions of the organs of the torpedo are such as by certain very possible motions, and the allowable supposition of conducting and non-conducting powers, may produce the effects we observe. How far it may be probable must undoubtedly be left to future experimental research.

In new talc, which had never been excited nor electrified, and exhibited no signs of electricity when applied to Bennet's electrometer, I found that the laminae were naturally in strong, opposite, electric states, counterbalancing each other. When they were torn asunder in the dark, they gave flashes at least 1/10th of an inch long to each other. This is 1,875 times the intensity of the torpedinal electricity, as before deduced. If, therefore, one or more columns of talc, or other thin electric plates 1/300th of an inch thick, and making up the surface of the electric organ of the torpedo, were so constructed as that the plates might touch each other by pairs only, naturally in opposite states, and coated on the outside; if moreover, there were one common conductor communicating with the upper plate of every pair, and another in the same manner with the lower;—then a separation of all the pairs at the distance of only 1/18,750 inch would produce the torpedinal intensity; the equilibrium would be restored by the two conductors if made to communicate, and whatever living creature was in the circuit would receive a shock: and on restoring the original situation, a shock might also be given. The force of these shocks would differ according to the quantity of the apparatus made use of at the time, or the distance to which the plates were drawn asunder. If different columns were exploded in rapid succession, the quick repetition of small shocks would produce the tremulous sensation.

If we were to conjecture that the torpedo actually operates like a machine of this kind, we should find our supposition to include the following subordinate parts:—1. The membranes may be non-conductors, and the fluid between them a conductor. 2. They may act as electrophores. 3. The white reticular matter between the columns may consist of conductors separately leading to the two opposite surfaces. 6. [sic] These separate conductors, in all their subdivisions, may be well kept asunder by a covering of non-electric matter. If this be of the same kind as the membranes, and 1/10,000th of an inch thick, it would be sufficient to the purpose, because the intensity of the electric state is deduced from its power of breaking through a much more permeable electric, namely, air, at nearly twice that interval. 7. [sic] The effects may also be produced by the motion of conducting plates in a non-conducting fluid.[79]

As the long discussions he devoted to Nicholson's model in the early drafts of his letter to Banks indicates,[80] Volta both deeply admired the model and dissented from it on basic issues. Among the points he accepted, at least in part, were

• An apparatus modeled in detail on the anatomy of torpedo's electric organs might be constructed from known electric devices. Volta had known somehow of the electric organs of the torpedo since his visit in London of 1782, when he became acquainted with Walsh's, Hunter's, and Cavendish's works on the subject. He had not followed their lead, however, and as late as November 1798 directed his attention to the brain and nerves of the fish, as Galvani himself did in his memoir of 1797, rather than on its electric organs. After reading Nicholson's article and building the battery, Volta never again accorded a special role to the brain and nerves in accounting for the torpedo's electricity.

- The basic units of Nicholson's model were pairs of discs in opposite electric states. This was probably his most important hint to Volta. It required considerable reworking, however, to fit Volta's ideas about the powers able to move electricity.
- Nicholson suggested that the pairs of discs making up the apparatus should be connected by two common conductors and that the columns so composed should be joined to give shocks of adequate intensity. This too required emendation on the part of Volta.[81]

Volta's qualified assent to Nicholson's model stopped here. The following points he could not accept:

- Nicholson assumed that the torpedo had internal insulating parts, like the resin cake in the electrophorus.
- Nicholson tended to assume that bodies were either perfect insulators or perfect conductors. For Volta, organic bodies consisted of second-class or imperfect conductors. The torpedo therefore could not contain a resinous cake or an electrophorus.

Side by side with Nicholson's positive hints, this realm of disagreement with him provided conceptual room for Volta's great invention. He had only to substitute his pairs of self-acting conductors, especially metallic couples that he knew to be most effective in setting electricity in motion, for Nicholson's electrophoruses. By replacing Nicholson's electrophoruses with metallic couples, Volta arrived at a strictly electrical apparatus: he had no need for a mechanical device simulating the lifting and lowering of the electrophorus's shield. Nicholson had not elaborated the mechanical side of his model; probably he had in mind something similar to the multiplier presented by Cavallo as a mechanical electrophorus (fig. 6.4, p. 189).

One challenging problem remained. Nicholson had suggested that each pair of plates making up the model torpedo should be connected to other pairs *in parallel*, in analogy with the arrangement adopted for connecting many Leyden jars together. This suggestion did not fit Volta's ideas about circuits including couples of conductors. His earlier experiments on circuits containing identical couples, each made of two different metals, had shown that under certain conditions the impulse given to the electric fluid by one couple was counterbalanced by an-

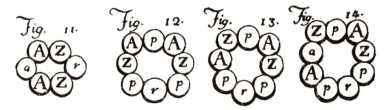

6.8. Volta's representation of some of his experiments with arcs or circuits made of metals (A, Z), frogs (r), and people (p). From VO, 1: 398 (*Biblioteca Universitaria, Bologna*).

other.[82] With a proper arrangement, however, the counterbalancing effect could be avoided, and the motion of the fluid restored (see Volta's fig. 13 in our fig. 6.8, above).

Volta thought that because of the balancing, the total impulse given to the electric fluid in a circuit composed of many different metals was equal to the impulse imparted by the two conductors put at the circuit's ends.[83] Thus, in Volta's view, connecting the couples in parallel through two main conductors, as suggested by Nicholson, would have introduced contact points able to counterbalance the motion imparted to the fluid by the couples of plates.

The net result of this second, complex interaction with Nicholson (the first being over the doubler) was to set Volta to search for a way of connecting many couples together to boost the motion of the electric fluid. A succinct autobiographical fragment—one of the very few we have from this period—confirms that this was a crucial point in Volta's process of discovery. He wrote that he discovered the battery "towards the end of 1799, when I conceived the idea of connecting many [couples of different metals] together."[84]

### Building the Battery

Volta's laboratory notebooks do not allow a further step-by-step reconstruction of the process leading to the voltaic battery. The following conjectural suggestions are based on evidence offered by Volta's correspondence, by his writings immediately after the discovery, and by a few laboratory notes kept between 1800 and 1804, which may be assumed to display a pattern of investigation similar to the one followed up to December 1799. Although no new, major conceptualization took place before the battery was built in its two basic forms, the pile and the "crown of cups" (fig. 6.9, p. 203), more was required than mere manipu-

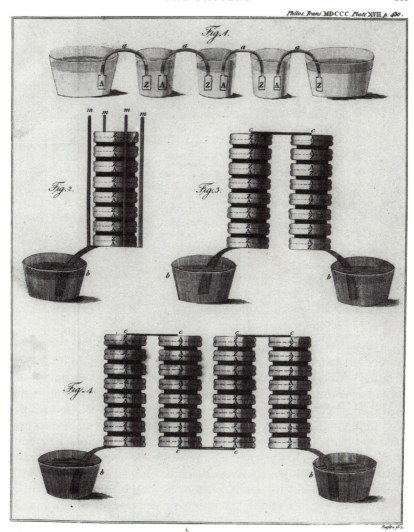

6.9. Volta's crown of cups and column batteries, spring 1800. From *PT*, 90 (1800): part 2 (*Biblioteca Universitaria, Bologna*).

lation by a man who, to use Derek de Solla Price's striking phrase, had his "brain in his fingertips."[85] Assuming that Volta adopted a "trial and error" strategy at this stage of his investigation is useful in order to describe, though perhaps not to explain entirely, his next steps.

For example, within the framework of galvanic experiments, to which Volta substantially adhered, it was natural to put metals and

6.10. Frog legs, arcs, and cups from the original drawings for L. Galvani, *De viribus electricitatis in motu musculari* (Bologna, 1791), table 4 (*Bologna Academy of the Sciences*).

conductors side by side to form arcs or circuits, but not to stack them in piles.[86] Volta might have been further encouraged to imitate the torpedo's columns or piles by Alexander von Humboldt's then recent observation that the frog could be stimulated by two different metals put in contact with it side by side, *without forming a circle*. In May 1798, the *Journal de physique* had reported as important news that arcs were not essential to the success of galvanic experiments.[87] The crown of cups was probably a compromise between the two mechanisms to excite the electricity of animals—the arc and the pile. The crown showed a lineage ascending back to experiments by Galvani himself (figs. 6.9, 6.10, pp. 203 and 204).[88]

Perhaps reflecting this tie to a galvanic investigation, Volta considered the crown of cups to be a more "instructive" form of the battery than the pile, and hence better suited for demonstrations.[89]

Evidence from close to the time of discovery makes it likely that Volta tried to build an all-metal battery by interposing a third metal between the usual metal pairs.[90] He thought that an all-metal battery would have been cleaner and more reliable, requiring no maintenance. Trials showed him that it did not work. Even the failure, however, had consequences at the level of conceptualization. If, as Volta maintained, conductors formed a single series, albeit divided into the two classes

6.11. Volta's sketch of a battery that combines the column and the cup, drawn before 13 Nov. 1800. From VMS, H, 48 (*Istituto Lombardo, Milan*).

of metals and wet conductors, why did a wet conductor work where a third metal did not? He came to recognize that, after all, a greater discontinuity existed between metals and wet conductors than he had formerly allowed.

The pile of metallic pairs and wet cardboards symbolized what Volta liked to depict as the victory of metallic over animal electricity. Yet, any battery containing metals was imperfect as an imitation of the electric fish. Thus Volta went on trying to build a nonmetal battery to complete his program of imitating the torpedo. Figure 6.11, above, probably represents an attempt at a nonmetal battery (Volta used small letters to designate his second-class conductors), which combines the column and the cup.

Figure 6.11 was drawn before 13 November 1800.[91] Laboratory notes dated 10 May 1804[92] show Volta trying a nonmetal battery, made with discs of bone soaked with fresh water, sulphuric acid, and potash. This battery, containing eight pairs of discs, gave an electricity equal to 2 or 2.5 degrees with the help of the *condensatore*.

The unexciting results of 1804 suggest that Volta did not publish his previous attempts when announcing his discovery of the battery

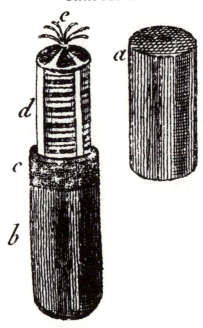

6.12. A pocket battery that Volta brought to Paris in the autumn of 1801. From J. N. Hallé, "Exposition," *Bulletin des sciences, par la Société Philomatique,* 3, Nivose an X (Dec. 1801): 74–80 (*Biblioteca Comunale dell'Archiginnasio, Bologna*).

because they had been indecisive. He did not doubt that a nonmetal battery could work; the electric fish guaranteed as much.[93] Some ideas he had formed while building his batteries (both successful and unsuccessful) helped him to locate the problem with nonmetal batteries. In an all-wet battery, what special body could act as an appropriate gap in conductivity (or increase in resistance) between the couples? If a certain discontinuity was needed between the couples, as ordinary batteries showed, an unknown subclass of wet conductors might be supposed to exist that could act with conductors as wet conductors did with metals in ordinary batteries. This assumption, and Volta's partial success in imitating the fish, stimulated further research on the fine anatomical structure and chemical composition of the torpedo's organs.[94]

Although it could not be taken for a fish, the classic, mostly metal voltaic battery had the great advantage of convenience, robustness, and power over the delicate, easily dried, wet batteries. It lent itself to portable types like the pocket pile Volta brought to Paris in the autumn of 1801, to show his fellow scientists and Bonaparte (fig. 6.12, above).[95] He found, however, that the First Consul, as we know, was moved especially by powerful and menacing apparatuses—batteries that could break metal bars. Bonaparte asked Volta to superintend the construction of such a battery in Paris.

The voltaic battery's success—to be discussed in the next chapter—and the almost immediate realization that important chemical phenomena were associated with it (a discovery in which William Nicholson again had a part) probably helped to obscure the role played by the electric fish in the discovery of the new apparatus. Volta himself contributed to this amnesia. Only his unpublished writings convey an adequate idea of the prominence the episode had in the process of discovery. With one late and partial exception,[96] his published writings after 1799 gave spare and decreasing space to the torpedo. Volta had two reasons for this de-emphasis. The battery's debt to the electric fish was Volta's own debt toward Nicholson, which he found embarrassing on two accounts. It might have challenged his originality and, more importantly, it might have mixed up the vague principles behind Nicholson's *ideal* model with the effective ideas behind Volta's *real* battery. From this perspective, Volta could maintain that he had contracted no debt with Nicholson.[97] Nonetheless, Nicholson's paper was a precipitating event, perhaps *the* precipitating event, in Volta's invention of the pile.

### Conclusion: Invention, Instrumentalism, and Competitive Imitation

Although further study of Volta's laboratory notes and unpublished writings will no doubt improve the reconstruction of Volta's path to the battery given here, we are unlikely ever to have a full, step-by-step account. The limits arise not only because of the incomplete records in this particular case, but also because of the general reasons Frederic L. Holmes has convincingly called attention to in his book on Lavoisier and scientific creativity: ". . . for human thought does not proceed in the step-by-step fashion that its formalized reconstruction exhibits. It circles endlessly around on itself, views the same ideas over and over from slightly different perspectives, repeats itself many times, and occasionally in the midst of these iterative processes thrusts forward."[98] The schema of figure 6.13, page 208, may, however, be expected to stand as a reasonable representation of the main factors involved in Volta's invention.

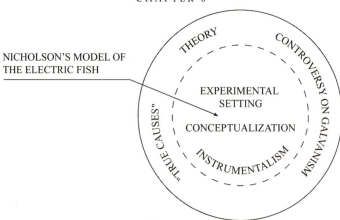

NICHOLSON'S MODEL OF
THE ELECTRIC FISH

6.13. Inventing the battery. A schema representing Volta's invention of
the battery.


The circle represents Volta's investigation: it helps distinguish between the parts likely to have been more easily involved in his interaction with other scientists and their research programs, and those closer
to his own personal laboratory work and day-to-day thinking. The external belt of the circle, accordingly, represents the controversy over
animal electricity and galvanism. As the literature on the controversy
abundantly shows, the most visible, if not the main ingredients of the
controversy were highly general discussions about the nature of the
electric fluid and of life itself—discussions Volta knew how to contribute to, and also how to resist.

Volta had a definite bent for what John L. Heilbron has called the
instrumentalist orientation of late-eighteenth-century physics.[99] Mature Volta's declared indifference to the search for "true causes," and
the continuous, at times drastic, revisions to which he submitted his
own contact theory of electricity show that he kept a flexible, sometimes agnostic attitude toward highly theoretical issues. At the "core"
of his investigation stood no theory as such, but conceptions that
helped to save the phenomena better and that kept close to the experimental setup with its natural objects and artifacts, instruments, and
manipulating techniques. Hence "experimental setting" and "conceptualization" belong near the center of our representation and very close
to each other. The discontinuous line, separating the core from the belt,
alludes to the fact that the border between these areas of Volta's endeavor, sanctioned by the instrumentalist attitude, occasionally could
be crossed: instrumentalism made trespassing an undramatic event.

In our sketch, Nicholson's contribution comes from outside. His model for the electric fish was conceived according to scientific concepts significantly different from those Volta had independently developed in quite a different context. Nonetheless (to keep to our visualization of the process), Nicholson's model penetrated the outward belt of Volta's investigation and reached its core, reorienting Volta's goals, experimental setting, and conceptualization. Volta's instrumentalism, his interest in the torpedo stimulated by Galvani's publication of 1797, and his earlier admiration for Nicholson's revolving doubler, all assisted the penetration of Nicholson's model of the electric fish into Volta's investigation, despite their dissent over theory. Within a few months—perhaps even weeks—of this penetration, Volta managed to translate Nicholson's work into his own terms and to replace Nicholson's ideal machine with a working apparatus. To account for the rapidity and depth of this assimilation and its results, we must refer to Volta's "genius for inexpensive, effective apparatus," and to his qualities as an investigator "with brain in his fingertips."[100]

As to the broader cultural and social pursuits involved in Volta's path to the battery, they conform to a pattern that can be described as competitive imitation and appropriation; something we have already seen at work in Volta's career on previous occasions and will be seeing again in the work of those who built on Volta's battery (chapter 7, p. 228ff).

When inventing the electrophorus (as shown in chapter 3, above) Volta had imitated and improved on Aepinus's sulphur and cup apparatus, urged on by the controversy he was involved with Beccaria over "vindicating electricity," and seconded by his own ideas about electrical atmospheres, which in the early 1770s had helped him to define his place within the expert community. In the case of the battery, he imitated and improved on Nicholson's model for the torpedo fish, this time urged on by the controversy over galvanism, and seconded by his own contact theory of electricity, which had identified his stance within the controversy throughout the 1790s.

In both cases the products of Volta's pursuits—his inventions—embodied something of the models being imitated, and something else deriving from the ongoing competition with other expert electricians. The inventions thus contained much that had been known for some time, and something else that was unquestionably new, and the result of Volta's endeavor. The latter, however, was itself the outcome of competition with other experts, and thus part of a pursuit that Volta's competitors had themselves contributed to shape.

The novelty that contemporaries perceived (and we can still see) in Volta's instruments was the result of this process of competitive imitation and appropriation, in which the goals of the inventor interacted continuously with those pursued by the other actors involved in the game. Novelty was as much the product of this interaction (unpredictable in its detailed outcome[101]) within the expert community and with scientific objects, as of Volta's own goals, ideas, and creativity.

It is more than a passing coincidence that a similar pattern of competitive imitation and appropriation has been detected (in chapter 5, above) in the game that the public administrators and the wealthy amateurs supporting fashionable science played in several countries in late Enlightenment Europe; a game at which, we know, Volta himself was an exceedingly good player.

## CHAPTER SEVEN

# Appropriating Invention

## THE RECEPTION OF THE
## VOLTAIC BATTERY IN EUROPE

$A$part from a few, quick mentions in the works of Thomas S. Kuhn, the assessment of the voltaic battery in the history of science rests entirely on the concluding pages of John L. Heilbron's magisterial *Electricity in the Seventeenth and Eighteenth Centuries*.[1] Here the pile is described in a few, powerful brush strokes as "the last great discovery made with the instruments, concepts, and methods of the eighteenth-century electricians." By making available for the first time a steady electric current, Heilbron noted, the battery "opened up a limitless field," which included electrochemistry and above all electromagnetism, the study of which "transformed our civilization."[2]

The new, limitless field and the powerful transformations involved, however, fell beyond the chronological limits of Heilbron's great book. The impact, meanings, and implications of the battery for nineteenth-century science remained unexplored, and still are, with few exceptions.[3]

The present chapter explores the early reactions to the voltaic battery in several European countries. Amateur as well as expert opinions will be discussed, the former playing an important role in those days, as we shall see. The attitudes of various local communities will be compared. The success of the battery with expert and lay audiences in the early nineteenth century makes it an interesting and rarely pursued case study allowing us to assess the status of the physical sciences, chemistry, and their practitioners around 1800 in society at large, and to verify to what extent similar developments were taking place across Europe's cultural frontiers. The early circulation of the voltaic battery also lends itself to a discussion of a number of historiographic issues. These include the interaction among instrument, theory, and interpretations at a time that was crucial for the transition from eighteenth-century natural philosophy to nineteenth-century science. The early impact of the voltaic battery offers grounds for testing the comparative merits of historical approaches based either on realist or social constructivist views of scientific instruments and practice, or a combination of both. The apparently easy replication of the voltaic battery, on the one hand, and the circumstance that widely different meanings were attached to the same basic instrument by different practitioners, on the other, will be discussed in detail. As will be seen, the fact that, compared with other scientific instruments, the battery traveled comparatively easily across cultural frontiers, did not save it from getting involved in a maze of interpretations and unintended developments. These circumstances contributed to the notion, frequently expressed by Volta's contemporaries, that the battery had opened up a new continent for the endeavors of scientific investigators. The present chapter is a contribution to the geography of this new continent as it emerges from a survey of the battery's early travels throughout Europe; a geography in which the instruments, their interpreters, and the differences as well as the similarities of the cultural contexts all receive due attention.

## Spreading the News

Volta's battery happened to be invented in Como during the thirteen months—from 28 April 1799 to early June 1800—when the Austrians, allied to the British, regained temporary control over Lombardy after the French invasion of the Italian peninsula three years before.[4] This, plus Alessandro Volta's well-known anglophilia and the award of the Copley medal in 1794, explain why in March 1800 he chose the Royal Society to announce his invention. On scientific grounds, too, he must have thought that the magnitude of his latest invention was such as to make it worth announcing in London, which as we know (chapter 5, above) he regarded as the capital city of the ideal experts' community to which he belonged. As to the language for the announcement, Volta chose French, a language he had mastered completely, and that at the time made communication across Europe easier.

The troubled political and military situation of the continent in spring 1800 played a role in Volta's decision to split the paper announcing the battery into two parts. The first part was a four-page, self-contained, one-thousand-word account of the column or pile apparatus, with no diagrams and no discussion of its effects. It was dated Como, 20 March 1800, and sent to the president of the Royal Society, Joseph Banks, by mail.[5] The second part, announced in the first, was a five-thousand-word account accompanied with an elegant diagram, describing both the pile and the crown of cups apparatus. It discussed

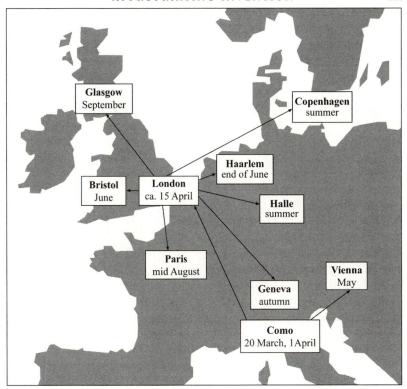

7.1. The main information flows announcing the voltaic battery, from 20 March to autumn 1800. Details in the text.

the effects of the new machine on sense organs in detail and assessed the merits (but above all the shortcomings) of William Nicholson's suggestion on how best to imitate the organs and effects of the torpedo fish by means of an electrical device, similar in some respects to the battery (chapter 6, above). In Volta's manuscript the text dealing with Nicholson was kept separate from the rest by a dotted line, with the implication that the author left to Joseph Banks the decision whether to include the discussion of Nicholson in the published version or not. Volta dated the second part of his paper 1 April 1800, and put it in the hands of a spice and drug merchant from Como, Pasquale Garovaglio, who was leaving Lombardy to try his fortunes in England.[6] The first, short description of Volta's battery arrived in London around the middle of April;[7] the second part weeks later, but apparently before 3 June.[8]

Figure 7.1 visualizes the early dissemination process of the news concerning the battery, as well as the major directions and timing of

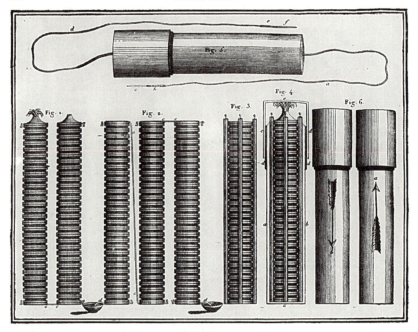

7.2. Voltaic batteries for the travelling natural philosopher. From A. Volta, "Description de la pile électrique communiquée par Brugnatelli," *Journal de chimie et de physique* (Van Mons, Bruxelles), 6 Nov. 1801, in *VO*, 2:127–35, 132 (*Biblioteca Universitaria, Bologna*).

the information flows. Figure 7.2 represents column voltaic batteries, contained in boxes, of a type Volta described in 1801 and recommended especially for the traveling natural philosopher, like himself.

As we shall see, relying only on Volta's first account that Banks circulated through acquaintances, by the end of April and early May 1800 several "philosophers" were making experiments with voltaic batteries in London.

Meanwhile, Volta began spreading the news of the battery closer home. During April, several "cognoscenti" in Milan saw a voltaic battery in operation in the house of the chemist Valentino Brugnatelli, Volta's colleague and friend. On 26 April Brugnatelli reported to Volta that he had observed a number of chemical effects in the battery though not the decomposition of water performed at the same time by Anthony Carlisle and William Nicholson in London.[9]

Sometime during April, Volta sent another description of the battery (including an account of the crown of cups apparatus ) to his main contact at the court of Vienna, who in those days was the physicist, chemist, and official Marsilio Landriani. By early May Landriani and his Austrian colleagues had built in Vienna a sort of "trough" voltaic apparatus, of the kind that became common later in England and the German states. Despite his success in operating with the battery and improving on its design, Landriani declared to Volta that he had problems in understanding "the theory of these new, surprising phenomena."[10] This, as we shall see, was to be a common reaction among those who engaged in experiments with the battery.

Landriani and his friends in Vienna became an important center in the informal network of acquaintances that kept Volta informed of the early career of the battery on the European scene. It was from Vienna and via Landriani that Volta received a detailed account of the decomposition-of-water experiments performed with the battery in London, soon to be repeated in Vienna.[11]

It took some more time for news of the battery to reach Paris. For all we know, news arrived there via London's newspapers. Before considering this side of the story, however, the early circulation of the voltaic battery in London is worth examining in detail. Here news of the battery became immediately tied up in an intricate story, involving also the French, that brought the announcement of Volta's achievement to an unexpectedly wide audience, while exposing the tensions internal and external to the expert community of natural philosophers in the London daily press.

Upon Banks receiving Volta's first account of the battery, as already said, he passed it on to some of his acquaintances, including Carlisle, a distinguished physician at Westminster Hospital and a former pupil of William Hunter.[12] Carlisle shared the information in turn with his friend Nicholson, and on 30 April they together began a series of experiments with a voltaic pile "consisting of 17 half crowns [containing silver], with a like number of pieces of zinc, and of plasteboard, soaked in salt water." Using this as well as another, more powerful battery made with thirty-six half crowns on 2 May—impinging on an observation of Carlisle and adopting an experimental arrangement devised by Nicholson—they observed a series of chemical phenomena that they interpreted as the decomposition of water.[13]

Nicholson continued the experiments on his own, built a one-hundred-half-crown battery and obtained some very impressive phenomena with it. They included loud detonations, clouds of bubbles, gleams of light, shocks felt by up to nine people holding each other by the hand, and a ramified, metallic vegetation, nine or ten times the bulk of the wire, when the wire was kept in the circuit of the battery for four hours.

Nicholson realized the importance of the new experiments quickly. He refrained, however, from immediately publishing the results in the journal he edited because he found himself in a delicate position in connection with the Royal Society, and with Volta. Having had access to Volta's first message prior to its being submitted to a session of the Royal Society, Nicholson felt he should not publish on the new apparatus until formal submission had taken place. Furthermore, he could not rule out that the impressive chemical phenomena he and Carlisle had observed with the battery were mentioned in Volta's own announced, but delayed, second part of his paper to Banks. Nor could Nicholson anticipate that Volta had quoted him and his intended artificial torpedo in the second part of his communication to Banks. All these uncertainties kept Nicholson from publishing his results during May; but he must have talked to many people, if we are to judge from the events that took place in London toward the end of that same month.

Acquaintance with the new apparatus and the decomposition of water was such in London during May 1800—that is, within weeks after the arrival of Volta's first description—that the recently appointed lecturer of the newly established Royal Institution, Thomas Garnett, decided to present Volta's apparatus and the decomposition-of-water experiment to the Institution's fashionable public. Garnett chose for the purpose his Wednesday evening lecture, 28 May 1800. He used a voltaic apparatus lent by Edward Howard. The lecture was a great success as far as the battery and the public were concerned; but it was also a source of trouble for the lecturer and his patron Rumford.[14]

Prompted by the success of the Wednesday lecture at the Royal Institution, and clearly inspired by Garnett himself, the London *Morning Chronicle* published a detailed account of the battery and the decomposition-of-water experiment performed at the Royal Institution on the following Friday, 30 May 1800.[15] The account—the first ever published of the voltaic battery—duly mentioned Volta, the president of the Royal Society, Nicholson, and Carlisle, but it conveyed several misunderstandings about their respective merits, as well as about some of the scientific issues at stake.

Concerning the merits, the newspaper's report led readers to believe that Volta had conceived the new apparatus as well as the decomposition-of-water experiment, and that Nicholson and Carlisle had merely repeated the experiment, rather than devised it in London. This was of course unjust to Nicholson, who by then must have read the second part of Volta's message to Banks, where no mention was made of the decomposition of water, and Nicholson's earlier suggestion on how to imitate the electric fish was discussed. The newspaper's inaccuracy, plus Nicholson's already mentioned delicate position in connection with the Royal Society and Volta—and now also qua editor of a journal that found itself unexpectedly in competition with the daily press, as well as with the venerable but slow *Philosophical Transactions*— prompted Nicholson to a spirited reply.

Published on the *Morning Chronicle* the following Tuesday, Nicholson's letter to the editor revealed the mixed feelings generated among experts by the unexpected enthusiasm the daily press was displaying for scientific matters:

> As I have never made any secret of my experiments, theories, or intentions, and observe with pleasure that philosophical men are at present much busied with the new exhibition of electricity [associated with the battery], I will not attempt to discuss how far it was perfectly regular in the Doctor [i.e., Garnett] to anticipate the publication of those who had worked at the discovery, and waited only to ensure precision, and to see the claims of Volta first laid before the Royal Society, before they themselves addressed the scientific world: but I must make so free as to express a wish, that if the Royal Institution is to become an office for collecting the conversations of philosophers, and immediately publishing them in the Daily Papers, due care may at least be taken not to lend its respectable name as a sanction of misinformation.[16]

Nicholson perceived that changes were under way in the informal rules controlling the circulation of scientific news, as well as merit assessment, because of the role played by the daily press, and by new social arenas like those provided by the Royal Institution. These were not, however, the only tensions that mingled with the early news of Volta's achievement in London. Other tensions were caused by the clash of national feelings exacerbated by the wars.

On introducing the battery to the public attending his 28 May lecture, Garnett—who very likely knew that the first account of the battery circulating in London was written in French, and maybe also knew that Volta, the author, was from Lombardy, and that Lombardy had been under the French since 1796; but perhaps did not know that the Austrians had temporarily regained control over the region, and certainly could not predict that they would lose it again to the French within days—attributed the discovery of the battery "to the French."[17] News of Garnett's statement quickly reached the president of the Royal Society, who immediately wrote to Rumford, patron of the Royal Institution, pressing for public emendation. The statement of rectification agreed to by Banks and Rumford, and read by Garnett at his next public lecture, retracted the attribution of the battery to the French. It emphasized, instead, that the first news of the discovery had been addressed to the Royal Society and acknowledged the merits of the discoverer, "professor Volta of Milan," as well as those of Nicholson,

Carlisle, and Howard.[18] The issue of nationality was clearly a sensitive one, and national rivalries, embittered by war, added to the usual individual and corporate jealousies.

The impression made by the battery and its chemical effects on London's "philosophical men" (to use Nicholson's phrase) and their public was remarkable. Two months later, another newspaper thought it appropriate to report on Garnett's public lecture again. The *Courier de Londres*, a journal printed in London in French for a continental audience, published a translation of the *Morning Chronicle*'s piece on the battery and the decomposition of water on 8 August 1800.[19] The French translation was seized upon in Paris by the *Moniteur universel*, and it was published on 17 August 1800.[20]

The *Moniteur* was no ordinary newspaper: it was general Bonaparte's mouthpiece and styled itself as France's "only official journal"; a definition that, in the wake of the battle of Marengo, had widespread implications for the rest of Europe. It was apparently through the *Moniteur* that news of the battery first arrived in Paris via London, and it was through the French newspaper that Volta first learned about the reception of his new device in London.[21]

It was not unusual for the *Moniteur* to report on scientific matters. Spring 1800 issues of the journal contained frequent reports on the metrical system, emphasizing the involvement in the project of scientists belonging to countries recently included in France's sphere of influence: General Bonaparte and his men knew how to use the popularity of science and its practitioners among the ruling classes to convey a reassuring image of Paris's ambitions. Volta, who in the summer of 1800 found himself again in a region within French control, was soon to be involved in the French campaign aimed at the European intelligentsia. Meanwhile, in Paris the battery won the attention of electricians from a variety of backgrounds as it had in London.

To judge from his promptness in publishing a report on Volta's new device, the person who was most quickly won over to Volta's invention in Paris was Étienne Gaspard Robertson.[22] He was then a popular figure in Paris, because of a peculiar show he ran each night near the Opera. It was called the "Fantasmagorie de Robertson," and was a fashionable entertainment Parisians flocked to see. The key attraction was apparitions: Robertson produced ghosts using specially adapted magic lanterns mounted on wheels.[23] In addition, Robertson included electrical experiments that gave his public very special sensations: a big crank friction machine used in the electrical part of his show was probably the most widely known electrical device in Paris before the advent of the voltaic battery.[24]

Relying only on the information published by the *Courier de Londres*, Robertson quickly had some voltaic batteries built for him and ran a series of experiments in which he declared he had involved "more than fifty people."[25] He read a report of his experiments to the Institut National de France on 30 August 1800; a mere three weeks after the first arrival of news of the battery in Paris. A long-time adept of galvanism, Robertson was intrigued by the new sensations that the steady current of the battery produced on the human body; sensations that could be turned into interesting additions to the repertoire of his "Fantasmagorie." Robertson's amateur results were later published in the experts' *Annales de chimie*, a not unusual step in those days. The paper he had read at the Institut National thus became the first article-length report on the voltaic battery to be published in a French scientific journal. By September, experiments had also been carried out at the École de Médecine, and soon afterwards several groups of experts and amateurs engaged in experiments with the battery in Paris.[26]

Through the daily press, in Paris as in London, the battery and its chemical effects were made known to a public that went well beyond that of expert electricians. An analysis of the press reports, however, helps clarify what exactly was regarded as news in connection with the battery among lay audiences in the early months following Volta's announcement.

What made the news in spring and summer 1800 in several European countries was not so much the battery per se, as we would now expect from our own, technological perspective, as much as the decomposition of water *plus* the instrument that had achieved it. What was presented as big news was the notion that water, still regarded in many quarters—despite decisive experiments to the contrary by Lavoisier and his followers—as a simple element, had been shown to be compound, and this surprising result had been achieved by means of an intriguingly simple instrument, of a kind familiar in popular experiments on so-called animal electricity; the notions of animal electricity and galvanism being linked in turn with speculations on the nature of life, matter, and the nervous fluid. The voltaic battery, in other words, was being presented to lay audiences as closely, if mysteriously, linked with very broad issues debated within the domain of natural philosophy; a circumstance that enabled every cultivated person to form an idea, however vague, of the nature of the novelty announced, and of Volta's own achievement in building the new instrument.

If this was the message conveyed to lay audiences by the daily press in London and Paris, the experts had their own networks for spreading the news: personal contacts, correspondence, and the scientific journals. In the early stages of the case under examination these channels seem to have supplemented, rather than led, the information circulated by the press.

Joseph Banks, for example, described Volta's new device to Martinus Van Marum, of Haarlem, Holland, in a letter dated 14 June 1800.[27] Van Marum, however, joined the frontline of research on the battery only one year later, when Volta sent him Christoph Heinrich Pfaff, then at Kiel, whom Volta had met in Paris, inviting the Dutch colleague to carry out experiments aimed at reinforcing Volta's own interpretation of the battery, against the galvanic interpretations.[28] Compared to the kind of correspondence exchanged by Banks as president of the Royal Society with colleagues on the continent, more effective in spreading an interest in the battery was Nicholson's *Journal*, that appeared monthly.[29] Following the experiments carried out in London on the decomposition of water, Nicholson published a long series of articles on the subject, written both by him and a number of British authors who had joined the rush to experiment, speculate, and publish on Volta's new device and its performance. Several of the articles published in Nicholson's journal were quickly translated and reprinted on the continent.

The *Bibliothèque Britannique, Sciences et arts*, of Geneva, having direct connections with circles in London, and especially the Royal Institution, first announced Volta's new "electrical or galvanic apparatus" in an autumn issue of 1800.[30] It also announced that a Dr. Marcet, who had witnessed the experiments carried out in London with the new apparatus, had repeated them in Geneva, where the Société de Physique et d'Histoire Naturelle was to devote considerable attention to the matter. The next issue of the *Bibliothèque*, for spring and summer 1801, opened with no less than four papers devoted to the voltaic battery and its chemical effects, all taken from Nicholson's journal, the authors being Nicholson himself, William Cruickshank, and William Henry. A number of additional contributions on Volta's apparatus by British authors, including Humphry Davy, were published in subsequent issues.

The *Bibliothèque* made the philosophy supporting its interest in the battery explicit: "The experiments in question need but a little apparatus to be carried out; they can be performed easily by any of them [the readers], and they are fascinating because of their intimate connection with the animal organization."[31] The experiments with Volta's battery fitted nicely within the philosophy stated in the "Prospectus of the Royal Institution," that the *Bibliothèque* had reviewed enthusiastically in the same issue that announced Volta's battery, a philosophy proclaiming that "the sciences are our veritable riches."[32]

From autumn 1800, the same papers taken from Nicholson's journal, and published by the *Bibliothèque* in French translation in Geneva, were also published in German in the *Annalen der Physik*, edited by Ludwig Wilhelm Gilbert in Halle. During the following three years the "Voltaische Säule" attracted the attention of the majority of the authors that contributed to Gilbert's journal, a few issues being devoted almost exclusively to the topic. The journal's interest in the voltaic apparatus culminated in the fall of 1801, when it published reports and a letter written in Paris by Pfaff—who had meanwhile been won over to Volta's interpretation of the battery—describing the experiments carried out by the Italian there.[33]

If we accept the testimony of young Hans Christian Ørsted, news of the voltaic battery arrived in Copenhagen from London.[34] Thus, by the late summer and early autumn of 1800, through the daily press, the specialist journals, and the experts' informal networks, information on the voltaic apparatus had reached a large number of both expert and amateur electricians from a variety of backgrounds across Europe.

From London, news was communicated quickly to the British provinces as well. Within weeks after the first battery had been built in London, voltaic batteries were built in Bristol and Glasgow. In the summer of 1800 Thomas Beddoes, the patron of the celebrated Pneumatic Institute in Bristol—the institutional birthplace of Humphry Davy—added a voltaic battery of 110 metallic pairs to the paraphernalia of the place.[35] Early in the autumn John Robison had a battery of 72 pairs ready for experiments in Glasgow, and expected to learn more about its functioning from James Watt and an unnamed, young chemistry student who had promised to assist him.[36] Volta's choice of London when first announcing the battery proved the right one: the mature, well-traveled natural philosopher knew how best to ensure an appropriate circulation to his latest invention.

Given the simplicity of Volta's apparatus—quickly remarked upon by amateurs and semiamateurs like Robertson and the editors of the *Bibliothèque britannique*—the wide European circulation of accounts describing the voltaic battery implied that a correspondingly high number of batteries were built in the first few months after Volta's announcement. However, the passage from the availability of accounts describing the battery to the actual building of the instrument itself deserves to be analyzed with care.

### Replicating the Instrument

The case of the battery seems especially revealing in connection with the issue of replication, which has often been debated by historians and philosophers of science, but never (to my knowledge) with reference to the voltaic battery.[37]

The anonymous journalist of the *Morning Chronicle*, who first gave the news of the battery to the public, managed to describe Volta's astounding device in barely 131 words. These are as good now as they were then for anybody wanting to build a working voltaic battery:

> A number of pieces of zinc, each of the size of a half crown, were prepared, and an equal number of pieces of card cut in the same form; a piece of zinc was then laid upon the table, and upon it a half crown; upon this was placed a piece of card moistened with water; upon the card was laid another piece of zinc, upon that another half crown, then a wet card, and so alternately till more than forty pieces of each had been placed upon each other; a person then, having his hands well wetted, touched the piece of zinc at the bottom with one hand, and the half crown at the top with the other: he felt a strong shock, which was repeated as often as the contact was renewed.[38]

This was not, however, the first ever description of the battery to circulate. The first such description, of course, was Volta's own manuscript, being the first part of his now famous paper addressed to the Royal Society. In that manuscript Volta himself described the battery in just 360 words[39]; and note that Volta was a prolix writer. As with Nicholson, his first published account of the battery was succint as well as practical, and placed appropriate emphasis on the easy availability of the ingredients needed to build the device:

> Most of our philosophers have used half crowns for the silver plates. The zinc may be bought at 8d. per lb. at the White Lion in Foster Lane, and cast in moulds of stone or chalk. A pound makes twenty thick pieces of the diameter of half a crown, or 1.3 inches diameter.

As to the wet discs to be placed between the metallic pairs, Nicholson recommended cloth, rather than cardboard, as "woollen [sic] or linen cloth appear to be more durable, and more speedily soaked." Nicholson's first description of the voltaic battery rivaled in brevity that of the *Morning Chronicle*: 144 words.[40]

Such straightforward descriptions of the battery, as we have seen, were easily translated from French into English, and vice versa. Nicholson and his colleagues in London had to rely on Volta's original manuscripts, written in French. Volta himself found the *Moniteur*'s description of his device—taken from the *Courier de Londres*, which had translated it from the English of the *Morning Chronicle*—"passable."[41] Judging from the early success of the battery in the German states and the Low Countries, further translations into German, Dutch, and Danish caused no problems.

Such exercises in word counting and translation seem to show that the voltaic battery *was* a simple apparatus, at least from the point of view of the instructions needed to describe it. What about building the instrument itself and making experiments with it?

Another easy exercise, this time in chronology, shows that building the battery and replicating some of its effects was also a relatively simple matter. As we already know, Volta's first letter describing the battery was in the hands of Joseph Banks around the middle of April 1800. From that moment, it took only two weeks to have several voltaic batteries built in London, and another two weeks to have the electrical decomposition of water, or electrolysis, observed for the first time. In another three weeks, the decomposition of water and the battery were being displayed to fashionable audiences in London: within six weeks of the arrival in London of Volta's first letter, the battery, the decomposition of water, and their popular representations in the lecture hall of the Royal Institution were making headline news in London newspapers.

Building the battery, at this early stage of its history, was not a business reserved to a well-defined circle of experts. In London Volta's first account of the battery passed quickly from the hands of Joseph Banks to those of a brilliant surgeon, Carlisle. Carlisle had many friends in scientific London and an occasional interest in galvanism: these are the only reasonable explanations we have for the fact that he was shown Volta's letter first. In trying Volta's apparatus Carlisle the surgeon joined forces with a chemist, technologist, and science writer: Nicholson. Nicholson was less well connected in London's gentlemen circles than Carlisle, but he had an excellent record as an electrician. Neither Carlisle nor Nicholson, however, can be regarded as members of a closely knit circle of London expert electricians, of which people like Henry Cavendish or Tiberius Cavallo would be more prominent and visible members.

Similar considerations apply to many experimenters who were soon building batteries in London and elsewhere in England. They received building instructions either from Nicholson, or from reports published in newspaper accounts. Nicholson was apparently the link that led Edward Howard to the battery. A chemist and a meteorologist, Howard, we know, passed the news to Garnett. A physician by training, Garnett was not himself a recognized expert electrician. Also lured to the battery in those early weeks of the instrument's public career were a dozen London amateurs, who soon contributed lengthy (and often inconclusive) reports of their experiments to Nicholson's *Journal*. Easy construction and replication were no doubt responsible for the special appeal the new instrument exerted over amateurs, as well as expert electricians.

Difficult though it is to estimate the number of voltaic batteries actually built in different European countries in the first few months after the announcement, there is enough indirect evidence to maintain that they must have been several dozen, as there were dozens of published reports describing new experiments performed with the instrument. Easy replication, and the common availability of the ingredients needed to build a battery, also meant easy disposability: that explains why only a few voltaic batteries from the early nineteenth century are preserved in museums today.[42]

The fact that the voltaic battery was easy to replicate, and that even some of its effects were comparatively easy to reproduce, did not impede a great variety of interpretations of the functioning and meaning of the instrument being put forward. In the case of the battery, easy replication did not carry uniformity of interpretation with it. Even assuming (as I am inclined to) that different experimenters in several European countries around 1800 were all facing the same, basic apparatus, the multiplicity of interpretations vis-à-vis the new machine *is* a challenge for historians.

## Appropriating the Battery

Indeed, the early history of the battery is just as challenging in a realistic perspective as it is in a social constructivist perspective, which would probably resolve the diversity of interpretations into the obvious diversity of the expert communities.[43] In analyzing issues of this kind in the following paragraphs my aim is to explore how instruments, interpreters, and the differences (as well as the similarities) in the cultural contexts interacted in the assessment of a device that was almost instantly recognized as bringing about a momentous breakthrough. In order to cover a significant spectrum of interpretations and contexts, six major interpreters belonging to five different milieus will be dealt with: Alessandro Volta himself, in Lombardy; William Nicholson in London; Étienne Gaspard Robertson and Jean-Baptiste Biot in Paris; Johann Wilhelm Ritter in Jena; and Hans Christian Ørsted in Copenhagen.

### Volta's Battery, 1

Volta's paper announcing the battery, finally published in the *Philosophical Transactions* late in 1800,[44] was the result of a combination concocted in London of the two letters already mentioned dated Como 20 March and 1 April. For the sake of convenience we will call the presentation of the battery contained in these documents "Volta's Battery, 1."

Volta's presentation strategy embodied all the knowledge he had acquired in dealing with electricity *and* a variety of expert communities of electricians around Europe in the previous thirty-seven years of his career.[45] Volta's 1800 presentation focused on the following issues in order of decreasing importance: first, the instrument itself; second, its performance; third, its relevance to the controversy over galvanism; fourth, and last, Volta's own explanation of what happened inside the instrument.

In adopting these priorities in such an order Volta was taking advantage of lessons he had learned on previous occasions within the international community of electricians he belonged to. When presenting his other major achievements—like the electrophorus and the *condensatore*—Volta had learned that his instruments tended to be more easily accepted and adopted by fellow electricians than his reasons explaining how the instruments worked, or how he had come to build them.[46] The battery was no exception, and Volta, now fifty-five, was even less willing to risk what he had experienced on previous occasions: by 1800 Volta was well aware that electrical instruments enjoyed (up to a point) a life of their own within the mixed community he belonged to.

Accordingly, the paper announcing the battery to the Royal Society of London began with a short, straightforward description of the instrument itself, offering a hint of its marvelous performance.[47] The performance of the instrument was conveyed easily by saying that it behaved like a battery of Leyden jars that did *not* need to be charged and charged again to produce its effects. Volta's allusion to a "perpetual impulse," "action," or "motion" of the electric fluid achieved by the battery added a sensational touch to his description of the device[48]; while the fact that the instrument, in the construction he recommended, could be built using just a few dozen discs made of three easily available materials added the appeal of simplicity.

The instrument, its performance, and the description of a repertoire of experiments to be carried out with it occupied more than two-thirds of Volta's first presentation of the battery. The remaining third, to the puzzlement of subsequent generations of physicists until recently, was devoted to an analysis of the effects of the battery on sense organs, and to a comparison between the battery and the electric organ of the torpedo fish.[49]

The two latter issues were of course connected with the controversy over galvanism Volta had been involved in over the previous eight years.[50] Yet, when first introducing the battery, Volta did not think it appropriate to present the new instrument as just another step in the controversy, nor as a winning move in his fight against galvanism. Popular as these interpretations became later, they were not encouraged by Volta when he first presented the battery.

The instrument itself, in Volta's presentation, came in two different formats: the column or pile battery, and the crown of cups (see fig. 6.9, chapter 6, p. 203). In his presentation Volta claimed to regard the two formats as perfectly equivalent, though crediting each one with special strengths and limitations. According to Volta, the pile was more handy and compact, and it could be easily built in pocket or traveling versions: an important feature for a natural philosopher who liked traveling and showing his expertise in electrical matters to varied audiences throughout Europe.[51] The crown of cups battery, on the other hand, was more suitable for demonstrations, as the cups and the bimetallic arcs could be added or subtracted easily in a way that, Volta emphasized, "spoke directly to the eyes."[52] Volta also mentioned that he had developed the same basic device in other formats as well, but he thought this was not an issue.[53]

Thus, already in its author's first presentation the battery was indeed a *family* of instruments rather than a single device. There is no evidence, however, that either Volta or the other early experimenters with the battery wished to emphasize the multiplicity of formats. Volta, for sure, was calling his colleagues' attention to the instrument regarded as the same basic device, and to its extraordinary performance, rather than to its different formats.

The fact that, in his first presentation of the battery, he did not emphasize theoretical and interpretive issues was probably part of the same strategy. This is quite clear from the contents and line of argument developed in Volta's texts addressed to the Royal Society, though the title added in London to Volta's 1800 paper—Volta's original letters bore no title[54]—has misled generations of interpreters, conveying the false impression that the contact theory was Volta's *chief* concern on the occasion.

Here and there, in his 1800 presentation, Volta did evoke the contact theory of electricity, which he had developed during the controversy over galvanism, and presented his latest invention as a consequence and application of that theory. As a perceptive contemporary reader, Nicholson, summed it up (adding a note of prudence of his own), Volta's theoretical claims in spring 1800 amounted to the following:

The theory of the learned inventor, if I rightly apprehend him, is, that it is a property of such bodies as differ in their power of conducting electricity, that when they are brought into contact they will occasion a stream of the electric matter. So that if zinc and silver be made to communicate immediately by contact, there will be a place of good conducting energy; and if they be made to communicate mediately by means of water, there will be a place of inferior conducting energy; and wherever this happens there will be a stream or current produced in the general stock of electricity. This is not deduced as the consequence of other more simple facts; but it is laid down as a general or simple principle grounded on the phenomena.[55]

This—with the proviso that the emphasis on "energy" was Nicholson's, not Volta's—is a fine description of the *limited* efforts Volta made in order to explain the functioning of his battery in 1800. Volta in fact alluded only in passing to a *theory* of the battery, without entering into any detailed discussion of it; when he mentioned it, his theory sounded as if it was something he had already proved true long before in his writings on galvanism.[56]

It must also be noted that, in his first presentation of the battery, Volta did not develop any rigorous argument aimed at showing that the power of his batteries, made of many metallic pairs, was the sum of the power of the electricity set in motion by each metallic pair; a point that, as we shall see, he insisted was a crucial one in his later presentations of the battery, especially the one addressed to the Institut National in Paris one year later.[57]

All these circumstances are worth taking note. Volta's strategy when first presenting the battery was in keeping with the steps that had led him to the battery a few months earlier. According to my reconstruction,[58] no new, decisive conceptual turn had taken place in his investigation in the three years preceding the building of the battery. The event that precipitated the building of the new instrument—Volta's reading of an article by William Nicholson that proposed building a device imitating the electric organ of the torpedo fish—took place at a quite different level of expertise. This was the kind of expertise De Solla Price described as requiring, on the part of the experimenter, the ability to operate with one's "brain in the fingertips": a level of expertise in which measuring techniques, ingenuity, and manipulating ability were just as important as theory and natural philosophy. Volta's chosen strategy when he first presented the battery, focusing on the instrument itself, reflected this background, as well as his awareness that instruments tend to be less controversial than the theories explaining them.

Instruments also tend to have a life of their own, however, and (after the electrophorus) Volta was well aware of that too: when dealing with the battery soon after spring 1800, he had to adjust his strategy to the needs deriving from the interpretations that other experts were suggesting for his own creature. This leads us back to William Nicholson again.

### Nicholson's Battery

Nicholson was one of only two authors (the other was Tiberius Cavallo) whom Volta mentioned in his first description of the new instrument. Although Volta did not acknowledge the role Nicholson's article on the torpedo fish had played in leading him to the battery, he did discuss Nicholson's paper at length.[59] Thus Nicholson had several reasons for taking note of the battery, including some personal ones. As already mentioned, while operating in the company of Carlisle with one of the first voltaic batteries built in London, he soon noticed something chemical going on inside the instrument; something Volta had *not* mentioned in his account of spring 1800.

Having devoted about two months to experiments with the battery, Nicholson was ready to acknowledge in his *Journal* that Volta's latest invention was a "most curious and important combination."[60] He also acknowledged that Volta's discovery "must for ever remove the doubt whether galvanism be an electrical phenomenon."[61] Nicholson, however, showed amazement at finding that Volta had entirely neglected the chemical phenomena associated with the battery. As he put it in July 1800: "But I cannot here look back without some surprise, and observe that the chemical phenomena of galvanism, which had been much so insisted on by Fabroni [*sic*], more especially the rapid oxidation of the zinc, should constitute no part of his [Volta's] numerous observations."[62] The allusion was to a paper by Giovanni Fabbroni published in Nicholson's journal some time earlier.[63]

Nicholson's emphasis on the chemical phenomena of the battery, that had been ignored by Volta, may have been prompted by his desire to adopt an original stance vis-à-vis Volta's contact explanation of the battery, which he found less than perspicuous. Some jealousy generated in Nicholson by Volta's critical remarks about his earlier suggestions on how to imitate the torpedo, voiced in the same paper in which Volta announced his new, brilliant apparatus, should also be taken into account. But Nicholson's enthusiasm over the powerful, chemical performance of the battery was genuine. Having combined Carlisle's battery and his own to make an apparatus totaling sixty-eight metallic pairs, Nicholson reported the following impressive observations:

A cloud of gas arose from each wire, but most from the silver, or minus side. Bubbles were extricated from all parts of the water, and adhered to the whole internal surface of the vessels. The process was continued for thirteen hours, after which the wires were disengaged, and the gasses decanted into separate bottles. On measuring the quantities, which was done by weighing the bottles, it was found, that the quantities of water displaced by the gases, were respectively, 72 grains by the gas from the zinc side, and 142 grains by the gas from the silver side; so that the whole volume of gas was 1.17 cubic inches, or near an inch and a quarter. These are nearly the proportions in bulk, of what are stated to be the component parts of water. . . . From the smallness of the quantity no attempt was made to detonate the air from the zinc side, but a portion of that from the silver side, being mixed with one third of atmospheric air, gave a loud detonation.[64]

Reflecting on the power of the battery to set the electric fluid in motion, professor Volta in Lombardy had evoked the old, mechanical notion of a "perpetuum mobile." Viewing the same instrument from the standpoint of an independent philosopher, fully immersed in the scientific and technological ferment of Britain's first industrial revolution, Nicholson perceived the battery instead as a powerful chemical machine; a machine to be compared with, and tested like, other chemical apparatus or, perhaps, like the steam engines of his times. As he put it:

We are in want of a measure of the intensity of the action of these machines. Will this be derived from the quantities of water decomposed, or of gas extricated under like circumstances in given times? Or from any change of temperature? Or what other commensurate incident?—Mr. Carlisle has not found that the water in the tube, while under this agency, did produce the slightest effect on a very small and delicate thermometer.[65]

The impressive chemical phenomena Carlisle and Nicholson had observed with the battery made their apparatus a quite different kind of instrument from the one presented by Volta a few weeks earlier in his letter to Banks. The apparatuses built in Como and in London were indeed basically the same, but the uses described by Nicholson and Carlisle had several distinct features. In a manner unanticipated by Volta, in London (and, via London, in several other places in Europe) the battery was being perceived as closely linked to long-debated chemical issues, and especially to discussions on the nature of water. Volta's work on the battery subsequent to spring 1800 was devoted to coming to terms with this unexpected turn of events: he had to rescue the battery, and his own contact theory, from the swamping effect produced by the chemical interpretations, and by reiterated, public assertions of faith in galvanism on the part of many philosophers interested in his new invention.

## Robertson's Battery

Galvanism was quite popular in Paris in the summer of 1800, when news of the battery first arrived there. To an amateur physicist and showman like Robertson—the first, we know, to publish on the battery in Paris—the new experiments made possible by Volta's device appeared "vraiment étonnans." The most obvious framework within which to interpret the new phenomena was, according to Robertson, the one offered by galvanism. Robertson thus entitled his report on the battery to the Institut National "Sur le fluide galvanique."[66] To him, the effects produced by Volta's battery were distinctly "galvanic," and different from those generated by the ordinary, physical electric fluid. Reading Robertson's report one feels that, precisely because building a voltaic battery was easy, experimenters could be led to focus on its effects rather than the instrument itself. For Robertson the most intriguing effects were those produced on the human body.

He tested the "colonne métallique" by touching the top with his nose, chin, limbs, teeth, tongue, eyes, and "other parts of the body where the skin is especially delicate and sensible," while he held the bottom of the pile with a hand. He also tested the column by touching its extremities with his thumb and little toe previously grazed with a knife, experiencing unbearable pain as a result.[67] Robertson satisfied himself that, contrary to what happened with previously known electrical devices, the battery's effects lasted only while the contact was preserved. The circumstance, again, qualified the effects as "galvanic" rather than electric. He also found that women and babies, having more delicate skins, were more sensitive to the effects of the fluid than men.

As Robertson summarized his tentative speculations on the battery: "Couldn't this extraordinary fluid be the first of the acids available in nature? Couldn't it be the first agent of the living movement, that the ancients called *nervous fluid*? Couldn't it be a veritable poison?"[68]

Experiments with the voltaic battery on the human body naturally revived speculation as to the possible features, powers, and distribution in nature of the "galvanic fluid." Similar speculations were indulged in, as we shall see, by another prominent amateur, also associated with the success of the voltaic battery in Paris: General Bonaparte.

Robertson reported experiments also on the decomposition of water and described an apparatus for measuring such effects that he called *galvanomètre*. This resembled the apparatus described in those same months by Nicholson, Carlisle, and Ritter, who however were not mentioned. Robertson announced that he was planning to carry out additional experiments on the subject in public, on "the first and the fifth day of each *decade* during the evening sessions in Cour des Capucines," as part of his *Fantasmagorie* show. Volta himself attended Robertson's show repeatedly one year later. On the occasion, as we shall see, Volta managed to convert him to his own explanation of the battery.[69]

### Ritter's Battery

News of the battery arrived in Jena probably via Gilbert's *Annalen*. By the fall of 1800 Johann Wilhelm Ritter, then twenty-three, had already worked and published extensively on galvanism within the framework of the Romantic philosophy of nature, earning the approbation of some major representatives of that philosophy.[70] Viewing Volta's new instrument from the perspective of an adept of Humboldtian science, Ritter too perceived the battery as firmly embedded in a tradition of galvanic research. Judging from Ritter's enthusiasm for the battery, the circumstance could add to the fascination exerted by Volta's new apparatus.

In 1798 Ritter had dedicated a treatise on galvanism jointly to von Humboldt and Volta, and had established a correspondence with the Italian.[71] In a letter to Volta, and in a subsequent paper published in Gilbert's *Annalen* for 1799, he had reported experiments pointing to some important connections between galvanism and chemistry. He had shown, for instance, that metallic pairs immersed in water produced oxides. Ritter had interpreted the result as showing that galvanism permeated *inorganic* as well as organic nature. Volta's battery and its chemical effects could thus be made to fit into Ritter's previous line of research, with the important qualification that Ritter regarded water as a simple element rather than a compound.

Accordingly, Ritter's early experiments with the battery—described in a paper published already in the course of 1800—were aimed at showing that the chemical phenomena displayed by the new instrument "were by no means due to a decomposition of water."[72] To this purpose Ritter further elaborated on a circumstance that had seemed problematic already to Nicholson and Carlisle, who had first called attention to it:

> We [Nicholson and Carlisle] had been led by our reasoning on the first appearance of hydrogen to expect a decomposition of water; but it was with no little surprise that we found the hydrogen extricated at the con-

tact with one wire [connected to one end of the column battery], while the oxygen fixed itself in combination with the other wire at the distance of almost two inches. This new fact still remains to be explained, and seems to point at some general law of the agency of electricity in chemical operations.[73]

Being convinced that water was a single element, Ritter tried to disentangle the issue by keeping the water close to each one of the two wires of a voltaic circuit separate by means of concentrated sulfuric acid, placed at the bottom of a V-shaped tube. Further convinced that the concentrated acid did not get involved in the chemical and electrical process, and observing that hydrogen and oxygen were produced close to the two wires also in his new arrangement, Ritter deduced that "the decomposition of water is not the basis of all other events."[74] The topic, that Humphry Davy also dealt with from September 1800, remained controversial. Seven years later Volta still regarded it as crucial, and unsettled.[75]

The revival of galvanic and chemical speculations of a very general kind, together with the issue of what happened in the water subjected to the action of the battery, contributed to taking attention away from what happened *inside* Volta's new instrument. Already in the autumn and winter of 1800 several experimenters in Europe were using the voltaic battery as a source of an electric current, independently of any assessment as to how the current was produced. The "black-boxing" of the battery, that Jan Golinski has rightly described as an important part of Humphry Davy's research strategy,[76] was already in place in the investigative efforts of experimenters like Nicholson, Carlisle, Robertson and Ritter a few months after the first announcement of Volta's new instrument.

In the case of Ritter, the black-boxing of the instrument was favored also by his intense fascination for experiments carried out with the battery on the human body. Already present in the work of Robertson in Paris, in Ritter the fascination reached the form of a self-destroying paroxysm. As Stuart Walker Strickland has rendered it:

> Throughout the first decade of the nineteenth century, Ritter systematically applied the voltaic column to all his sense organs. He used the battery to produce tones in his ears, colors in his eyes, heat and cold in his fingertips, and to induce fits of sneezing. Nor did Ritter neglect the pains and pleasures the battery brought to the lower extremities of his body. He published experiments on the influence of galvanism on the "organs of reproduction," on his pulse, and on "the organs of evacuation" as well as "other choice parts of the body."[77]

The sort of symbiosis Ritter realized with the battery was in keeping with a central tenet of the Romantic philosophy of nature that he endorsed, and expressed with the following words: "For everything in nature appears to be joined through one chain [Kette], of which the voltaic column is only a fortunate member."[78]

Volta, whom Ritter visited in Como in 1803 when they spent hours together discussing research topics of common interest, found Ritter's ideas "too transcendental"; a judgment he had made since reading Ritter's first letter to him in 1798.[79] By 1801 Volta had chosen Pfaff instead of Ritter as his informal ambassador to secure what he regarded as an appropriate understanding of the battery in the German-speaking countries.

### Ørsted's Battery

If we accept the testimony of young Hans Christian Ørsted, news of the voltaic battery arrived in Copenhagen from London.[80] To judge from the same source, chemistry was given an important role in the interpretation of the new device in Copenhagen too. Again according to Ørsted, German philosophy, in the form of the important variant represented by Immanuel Kant, had a conspicuous following there, and intermingled with the interpretation of the battery.[81] Accordingly, Ørsted's early experiments with the battery were inspired by both the "English chemists" *and* German philosophy. Given the context, Ritter's experiments on the battery and the (controversial) decomposition-of-water experiments were given due consideration.

Early in 1801 Ørsted published a paper in Danish illustrating his first experiments with the voltaic apparatus. Prompted by the chemical interpretations of the battery and by his own acquaintance with chemistry, he convinced himself that "it is possible to produce a galvanic effect not only by means of two metals of different oxidizability, but even with oxidizable metals and graphite, which is not a metal."[82] Behind Ørsted's experiments with batteries of new types was a notion he expressed as follows: "it seems very likely that the generation of a galvanic effect is merely a matter of bringing bodies of different oxidizability in contact with water in an appropriate manner."[83] Thus, Ørsted's "voltaic apparatus" (his phrase) contained graphite plates (in fact, graphite mixed with clay) instead of the usual silver plates. With sixty graphite plates Ørsted's battery "generated both shock and spark, though the latter was not always perceptible."[84]

After performing several experiments with a variety of batteries, Ørsted's conclusions turned out to be at odds with Volta's theory of contact electricity. Ørsted credited the new apparatus with showing that "we now have two kinds of electricity, the one generated hitherto by friction and the galvanic." From the circumstance that "each end of the battery produced a different gas from water" he further concluded that "there are two different galvanic electricities."[85]

Needless to say, Ørsted's conclusions placed him worlds apart from the stance that Volta had adopted in his battle against galvanism over the past eight years; a stance inspired by the old maxim according to which "We are to admit no more causes of natural things than such as are both true and sufficient to explain their appearances."[86]

### Volta's Battery, 2

As already hinted at, news of the reception of the battery and the decomposition-of-water experiments carried out in several European countries reached Volta in Como after delays and in a patchy manner. We know that he first learned of the experiments of Nicholson and Carlisle of May 1800 in London from the Parisian *Moniteur,* and again via his friends in Vienna, late in August.[87] We also know that, as a subscriber to Nicholson's *Journal*, he must have seen sometime in the autumn of 1800 or early in 1801 the long series of papers devoted to experiments with the battery printed in that journal.

Volta was of course very pleased by the big impact his latest invention was making on expert and lay audiences alike. Nevertheless, he was definitely worried seeing the limited attention the battery and his contact theory were receiving, compared to the emphasis given to the chemical phenomena that he had not mentioned when announcing the battery. As a consequence, in his correspondence and few publications of 1800–1801 Volta adopted a guarded stance toward the chemical phenomena of the battery; a stance he would preserve for the rest of his life. Indeed, when reading declarations like the one published by Davy in Nicholson's *Journal,* in the fall of 1800, to the effect that "the oxydation of the zinc in the pile, and the chemical changes connected with it are *somehow* the cause of the electrical effects it produces,"[88] Volta must have realized that a prudent stance on the chemical effects of the battery was not enough if he wanted to keep a grip on his invention, to assert his own interpretation, and obtain the rewards he expected from it.

The need to adopt a new strategy in dealing with the battery (compared to the one adopted in his letters to the Royal Society, described in "Volta's Battery, 1") merged with the new needs urged on Volta by the political and personal situation in which he found himself during the summer of 1800. As already mentioned, after the battle of Marengo in June that year the French had regained control over Lombardy, and the Austrians, who had closed the University of Pavia and suspended Volta's stipend as professor of physics, left. In the autumn of 1800, with the recent discovery of the new brilliant apparatus in his favor and the University of Pavia reopened by the French, Volta began to search for a kind of recognition that would offer him protection in the new situation. Under the circumstances, Volta realized, the protection was to be sought in Paris, not in London.

On 28 September 1800 Volta wrote to general Guillaume Brune, head of the French troops in Italy. He had already been introduced to the general and informed him about the invention of the battery. A few weeks earlier, we know, Bonaparte's newspaper had announced the discovery of the battery in appreciative terms. Volta asked General Brune two things: first, leave of absence to go to Paris, formally to thank Bonaparte for the restoration of the University of Pavia; second, protection for his relatives and a list of friends during his absence.[89]

Volta still doubted, however, whether the times were ripe for a mission to Paris: the sudden changes in fortunes on the European military and political scene advised caution.[90] But on 9 February 1801 the Lunéville peace treaty was signed. Only Britain remained at war with France, and that too changed when peace negotiations, begun in London, were concluded in October that same year. In May 1801 Volta resumed contacts with Gaspard Monge and Claude-Louis Berthollet in Paris.[91] Volta also sent a description of the battery to chemist and minister Jean Antoine Chaptal in Paris.[92]

When Volta—still personally uncertain about the advisability of the trip—asked the pro-French administrators in Milan for a substantial sum to cover travel expenses, it was granted. The letter written on the occasion by the Minister of the Interior of the Cisalpine Republic in Milan to finally arrange for Volta's trip to Paris, spoke of two aims of the mission. One was "cementing an alliance of talents and knowledge" between the Cisalpine and the French Republics. The other was to ensure "the progress of scientific pursuits," that "benefits immensely from the fast communication of enlightenment."[93] The rhetoric of the new minister was more emphatic than that used before the French revolution, but the substance of the message was the same as Volta's old Austrian patrons had had recourse to on similar occasions during the twenty-five years in which Volta had worked for them.

Volta left for Paris on 1 September 1801, and arrived there three weeks later, having stopped in Geneva to meet his several friends there and to win them over to his interpretation of the battery. Volta brought with him some fine "pocket" batteries and drafts of a long paper he had written over the previous twelve months, often in counterpoint to the interpretations of the battery being put forward by others, with the

purpose of establishing his own explanation of the instrument, oppos-
ing the galvanic and the chemical interpretations, and reappropriating
his own invention.[94]

As on previous, similar occasions, Volta's arguments were the product of close interaction with other researchers. The more so this time
because he had decided to involve the Paris natural philosophers in
his new campaign for reward, with whom (we know) he had had some
problematic relationships since his visit there in 1782.[95] As on previous
occasions, when he had also acted to defend his claims to the invention
of a new instrument, he developed an elaborate set of theoretical arguments and empirical proofs to secure his rights.[96]

Volta's key step in the new argumentative strategy developed in
1801 with Paris (no longer London) in mind, consisted in rooting the
"demonstration" of the battery on a fundamental experiment, in which
a single metallic pair without any humid conductor gave tangible, if
extremely weak signs of electricity. The experiment—that he had already used many times since 1796 in the controversy over galvanism—
was now conceived to exclude any role for chemical (as well as galvanic) phenomena. However, while the instrument by then popularly
known as Volta's battery, combining many metallic pairs and humid
bullets together, was easy to replicate even by amateurs, and produced
surprising effects to the satisfaction of lay and expert audiences alike,
Volta's "fundamental experiment" using a single metallic pair was extremely difficult to replicate, and its effects controversial.[97]

In order to obtain signs of electricity from a single metallic pair, and
to show that the "force électromotrice" generating the current of the
battery resided in the mere contact of two different metals, Volta
needed his *condensatore*, a device that, like other "doublers" or "multi-
pliers" of electricity could be suspected (see chapter 6, above) of generating by unintended friction the electricity it was supposed to detect.
The elaborate set of manipulating rules that Volta kept repeating in
written and oral presentations of his "fundamental experiment" of
1801—ostensibly with the purpose also of reasserting the merits of his
earlier inventions like the *condensatore*, and the straw electrometer—
was not such as to dispel all doubts. In fact, Volta admitted that this
experimental setting made only the electricity of the *silver* disc detectable. In order to detect the electricity of the zinc disc, a piece of wet
cardboard had to be interposed between the metal and the plate of the
*condensatore*.[98] Doubts about the possible chemical or galvanic role of
wet substances were also nurtured by the fact that, in the experiment,
the discs came in contact with the fingers of the operator, hardly perfectly dry.

A *quantitative* estimate of the "force" (that Volta also called "tension," "déplacement" or "impulsion") generated in the electric fluid by the contact of the two metals further depended on an estimate of the condensing power of his *condensatore*.[99] In order to assess the latter, delicate manipulations were needed, operating with electricity at a much higher degree of tension than the one involved in experiments with a single metallic pair. From such a separate set of tests, Volta calculated that his *condensatore* had an average condensing power of 120 times.[100] In assessing this, Volta further relied on his notion of electrical atmospheres; a notion that the French philosophers he now addressed had refused to take seriously on the occasion of Volta's earlier visit of 1782, and Volta took the present opportunity to reassert.[101]

In the new experimental and conceptual setting, that we call "Volta's Battery, 2," when Volta's straw electrometer measured a tension of two degrees on the plate of a *condensatore* to which the electricity generated by a disc formerly part of a metallic pair had been communicated, Volta calculated that the real tension of the disc was 1/60th of a degree of his electrometer.[102] Needless to say, the complex intertwining of conceptual assumptions, manipulating techniques, and instruments used to find this measure was distinctly "voltaic" from beginning to end. In order to share the results, and the conclusions, anybody willing to repeat Volta's "fundamental experiment" had to adopt Volta's own concepts and tools to a degree which few of his colleagues were prepared to do.[103] There could be no sharper contrast between the easy replication of the voltaic battery and its popular effects, and the extremely difficult replication of the experiment Volta now wanted to use to demonstrate the principle on which the battery worked.

There was one more difficulty with Volta's "fundamental experiment" in Paris. Volta's 1801 demonstration of the principle of the battery rested on the *condensatore* and on Volta's own straw electrometer, not on Coulomb's torsion balance. Parisian commentators repeatedly complained about that, and remarked that Coulomb's principles and the torsion balance should have been brought in in order to place the battery on firm scientific grounds.[104]

As a consequence of all these difficulties, it took repeated demonstrations and all the personal qualities of Volta—including his well-known proselytizing ability—to convince the *commissaires* of the Institut National in November 1801. He managed it, however, through an impressive tour de force.

Several contemporary reports of Volta's two-month stay in Paris in 1801 have survived. They include Volta's own letters home, Brugnatelli's detailed diary, and scattered notes by scientists and amateurs like Marc Auguste Pictet, Count Rumford, Christoph Heinrich Pfaff, and Guillaume Robertson.[105] The reports allow us to trace two major threads of Volta's action in Paris. One was the quest for reward he had already begun before leaving Lombardy. The quest targeted Bonaparte in person qua reward dispenser, Sage—whose courses Volta had attended nineteen years before—, Berthollet, and Chaptal as major intermediaries in approaching the First Consul and securing his favors. The second goal Volta pursued had vaguer contours and required a variety of audiences and means to achieve it. It consisted in convincing the Parisian scientific public that Volta was right in the controversy with Galvani, and that his newly discovered apparatus was such a brilliant proof of ingenuity as to dispel the clouds that had overshadowed the reputation of its inventor in Paris since the times of the electrophorus and his first visit there. In pursuing both goals the support of Bonaparte proved very effective. The support of the scientific elite in control of the Institut National, on the other hand, was not a decisive factor.

Volta's impressive endeavors in Paris in 1801 are worth following in detail. They began on 26 September, when the *Journal de Paris* announced his arrival in the French capital, accompanied by his colleague and friend, the chemist Brugnatelli.[106] By 30 September Volta was a guest in Arcueil and had already met Claude Louis Berthollet, Antoine François de Fourcroy, Georges Cuvier, René-Just Haüy, Biot, and Robertson. On 2 October some of Volta's experiments were repeated at Fourcroy's after a lunch. The next day Volta and Brugnatelli were made members of the commission on galvanism previously appointed by the Institut. On the 4th and 6th he met and won over to his cause young Christoph Heinrich Pfaff, then in Paris, whom he soon sent to Van Marum in Haarlem with the purpose of having the performance of the big Teyler machine compared with the battery's in the decomposition of water. By 8 October, as already mentioned, in letters addressed to German journals Pfaff was advertising Volta's achievements in Paris to the German states.

On 6 October Volta had lunch at the Ministry of War and met there, among others, General Brune (who had been instrumental in obtaining the means that had brought Volta to Paris) as well as Monge and Berthollet. On the 7th he met Minister Chaptal, and was his guest for lunch on the 17th. Soirées were mostly spent attending Robertson's "Fantamagories": on 23 October Bonaparte's journal, the *Moniteur*, announced that Robertson had carried out public experiments confirming Volta's interpretation of the battery and disproving the galvanic interpretations previously endorsed by Robertson himself. By 3 November the list of Volta's new or renewed acquaintances in Paris included also Pierre-Simon de Laplace, Jean Nicolas Pierre Hachette, Charles-Bernard Désormes, Jean-Henri Hassenfratz, Louis Bernard

Guyton de Morveau, Louis Joseph Dumotiez, and Benjamin Thompson (Count Rumford).

On 6 November, the representative in Paris of the Cisalpine Republic (which Volta was now a citizen of), Ferdinando Marescalchi, introduced Volta to Bonaparte, with whom he would also have had lunch, had it not been for the delayed delivery of the ticket inviting him. The next day Volta, who had attended three previous sessions of the Institut's commission on galvanism, began reading a memoir in French that he had meanwhile polished up with the help of two other acquaintances of his in Paris: physicians Biron and Tourdes.

On that same day, 7 November 1801, Bonaparte was able to spend more than one hour at the Institut attending Volta's presentation. The length of the First Consul's stay was carefully noted down by another attendee, the methodical Rumford, whose report of the event conveyed a mixture of pride and jealousy at the attention Volta was being paid by Bonaparte.[107] The First Consul's interest in electricity, on the other hand, was genuine if generic. Like many amateurs in those days he regarded electricity and galvanism as among the major ingredients of a broad, vague, but compelling naturalistic worldview. Bonaparte once expressed this creed with the following words: "I believe that man is the product of the fluids of the atmosphere [including the electric], that the brain pumps these fluids and gives life, and that after death they return to the ether."[108]

At the end of the 7 November session at the Institut, the First Consul invited his colleagues to confer a gold medal on Volta. The next day Volta let Bonaparte know that he was ready to repeat his electrical experiments for him whenever he pleased.

The members of the Institut complied with Bonaparte's request to reward Volta with some hesitation. Many of them of course remembered the uneasy relationships Volta had had with Parisian natural philosophers since 1782, and they were aware of Volta's negative reactions to Coulomb's memoirs on electrostatics.[109] They realized, however, that in his latest initiative Bonaparte's amateur interests in science combined with his usual political goals. Those were indeed momentous days for France, and for Europe. Two days after Volta's first presentation at the Institut, Paris celebrated the (temporarily) restored peace, favorable to the French, with great fanfare. Volta, an Italian whose name was known by then to the cultured elites of most European capitals, who had served Austria for a quarter of a century, and had announced the battery to the British (alas!), but writing in French, was now conferring with his peers in Paris: Volta had all the symbolic characteristics needed to convey a "universal," "scientific" image of French ambitions in Europe.

Bonaparte also attended Volta's second presentation at the Institut, which took place the day after the big peace celebrations. Volta concluded his presentation on 13 November. The next day he was chosen, we know, to represent the Cisalpine Republic at the congress of the Italian regions under French influence that was to convene shortly in Lyon.[110]

Biot dated his final report to the Institut on Volta's apparatus 2 December. On the 4th Volta left Paris for Lyon; but Parisian newspapers continued to keep their readers informed about Volta's achievements. On 9 December Bonaparte's *Moniteur* reported that Van Marum and Pfaff in Holland had confirmed the common electrical (not galvanic) nature of the battery's effects. On the 20th, the journal once again published instructions on how to build a voltaic battery. In those same days the popular *Journal de Paris* hosted an exchange of letters on the battery, in which Robertson defended Volta's interpretation. As advertised every day in the same newspaper, by then Robertson had made of *physique expérimentale* a permanent attraction of his evening performance in Cour des Capucines. On Christmas Eve, 1801, the *Moniteur* published an abstract of Biot's report to the Institut (on which more below) on Volta's experiments.[111] Meanwhile, on 8 December, the Consuls of the French Republic had decreed a gratuity of 6,000 francs to be assigned to Volta.[112]

By the end of December 1801 Volta and Bonaparte would certainly have agreed that they had achieved their respective goals in dealing with the battery in front of the savants of the Institut, the Paris daily press, and, through them, Europe's cultivated elites.

## *Biot's Battery*

The chief scientific upshot of Volta's visit to Paris was the report that Biot wrote in the name of a committee appointed by the Institut and consisting of Laplace, Coulomb, Hallé, Monge, Fourcroy, Vaucquelin, Pelletan, Charles, Brisson, Sabatier, Guyton, and Biot himself.[113] The report was written after the committee, joined on the occasion by Volta and Brugnatelli, had met four times at Charles's, and concluded by endorsing Bonaparte's proposal that a gold medal be conferred on Volta.

As Geoffrey Sutton has shown, the main, difficult thrust of Biot's report was to assimilate Volta's views on metallic electricity with Coulomb's research tradition.[114] Haüy, in 1801 a leading personality within the Institut, had been endorsing that tradition—banned under the name of Aepinus—since 1787. As we know, the research program outlined by Coulomb and Haüy implied a substantial neglect of Volta's contributions and claims.[115] As a result of the 1801 report, the Paris philosophers who had rejected Volta's request for recognition for twenty years came under pressure (exerted on the one hand by the amateur Bonaparte, and on the other by the writings of young Biot) to grant the Pavian professor a place in the history of electricity. Volta was granted the place, however, at the cost of being assimilated into a tradition he could not easily identify with. Understandably, there is no proof that Volta especially liked Biot's report, or that he took pains to develop the elegant mathematical approach to the battery sketched out by Biot on the occasion.

Young Biot's first reactions to the voltaic battery had been in tune with the generic galvanic framework within which the battery was first reviewed in Paris.[116] The uncertain connections between the galvanic and the electric fluids, and the chemical effects of the battery, had claimed Biot's main attention in the earliest experiments he had carried out with the new apparatus.[117]

Inspired by Laplace, however, Biot soon introduced speculations on the speed of the electric fluid in metallic and wet conductors of different kinds, and in the summer of 1801 he sketched an interpretive model of the battery that was partly based on mechanical assumptions. He expressed them as follows:

1. The laws of motion of the galvanic fluid follow from the repulsive property of the molecules of which this same fluid is composed, and, under this regard, the laws are the same as those of electricity;

2. the diversity of galvanic phenomena observed in different apparatus is caused by the different proportions in which the quantity or the mass of the fluid is combined with its speed;

3. the galvanic fluid moves across the water with difficulty, and it glides over its surface speedily.[118]

As to the chemical effects of the battery, Biot wanted to keep them within the framework of the same electrical interpretation:

The chemical phenomena exhibited by galvanism cannot be regarded as distinguishing it from electricity, because the galvanic fluid is only displayed in our instruments with a great speed and a limited mass; while electricity, when we set it in motion by means of our batteries, is endowed with both a great mass and a great speed.[119]

Though many, important issues placed Volta and Biot worlds apart in scientific matters,[120] in the interpretation of the battery they shared two basic goals: to keep galvanism within the domain of electricity, and not to let the chemical phenomena of the battery undermine the efforts being made at a physical interpretation of electrical phenomena.

A hint in the direction of a mechanical and mathematical treatment of the battery had been offered to Biot by Laplace himself, who had observed that attractions and repulsions could be detected at the two ends of the instrument.[121] Further hints in that same direction derived from Biot's own speculations on the supposed speed the electric fluid acquired inside the battery, depending on the size and shape of the metallic discs being used. The measuring techniques necessary to substantiate this physical approach to the battery, however, simply did not exist. In the end, Biot's strategy when dealing with the battery in the report he finally submitted to the Institut was double, and somewhat inconclusive.[122] In the first part of the report he asserted the priority, in principle, of Coulomb's notions over the entire science of electricity, while in the second part he introduced a mathematical model of the battery that established no link with Coulomb's law.

Biot's mathematical model of the battery was based on simple algebra, as Volta himself had hinted at years before when illustrating his contact theory of electricity.[123] Biot adopted Volta's own contact explanation of the battery and focused on the notion of electric tension, regarded as proportional to the repulsive force of the molecules of the electric fluid, and to be measured with Coulomb's balance, not with Volta's electrometer.[124] He then assumed that the tension of each metallic disc inside the battery could be calculated depending on the number of metallic pairs making the battery; the battery being either isolated, connected to earth, or to a *condensatore*.

Algebra allowed Biot to calculate the (ideal[125]) distribution of tension inside the instrument. In Biot's treatment, $x$ being the tension generated in a disc of a single bimetallic pair of discs (zinc and copper) placed in contact with each other, the total tension of a battery of $n$ pairs is $2nx - n^2$. In an insulated battery, $2nx - n^2 = 0$, so that the tension of the disc at the top of the battery is $x = n/2$, and the tension of the bottom disc is $-n/2$. Discs placed at equal distances from the two ends of the battery are assumed to have the same degree of tension, one plus, the other minus.

Biot's quantitative approach to Volta's contact theory was much more sophisticated than Volta's. It can be regarded as an interesting example in the history of the mathematical treatment of electricity. In tune with his display of mathematical effectiveness and elegance, Biot—no doubt informed through Laplace and Coulomb of the story of Volta's earlier, unsuccessful claims for recognition pressed on the Parisian natural philosophers since the 1780s—took the opportunity offered by the report on the battery to reassert what was regarded in Paris as *the* winning strategy in the science of electricity: a strategy based on a mathematical approach that, Biot argued, first introduced by Dufay, had led to Coulomb's law via Franklin and Aepinus. Within Biot's quick historical survey produced for the occasion, Volta's merits were reduced to having "happily applied" the principles of Aepinus to the building of the electrophorus and the *condensatore*, and having shown, with the work on galvanism and the battery, that galvanism had to be kept firmly within the domain of electricity.[126]

In subsequent years Biot developed his quantitative approach to the battery further, and finally adopted Coulomb's balance when measuring electricity (as he had recommended but failed to do in 1801).[127] He also devoted considerable space to the mathematical treatment of the battery in his successful *Traité de physique experimentale et mathematique*.[128] However, Biot's mathematization of the battery did not exert an impact on his contemporaries comparable to the one exerted by Cavendish's memoirs on a previous generation of electricians, nor to that played by Poisson's treatment of electrostatics on scientists of the same generation as Biot.

Apparently, there was more to the battery than the mathematical physics of Biot could convey.

After 1801, at any event, the majority of the researchers interested in the battery focused their attention on its effects and how to improve them rather than on a mathematical treatment of what happened inside it. As to Volta, despite the considerable satisfaction he had got in Paris, and the ample role Biot had granted to his contact theory, he never mentioned Biot in his subsequent writings, either published or private.

### Volta's Battery, 3

Encouraged by the rewards obtained in Paris and the enthusiasm shown for his recent achievements by young physicists like Pfaff— who offered to expound Volta's views on electricity and his dismantling of galvanism in a purposely written treatise[129]—after 1801 Volta was led to new reflections on a possible explanation of what went on inside the battery. These further reflections remained in an imperfect form, and he never published them. They offer some interesting hints, however, on the kind of natural philosophy Volta continued to cultivate after the invention of the battery, while not committing himself to it publicly.

In these manuscript notes,[130] he recorded that he was no longer happy with the notion that a special, mysterious "impulse" somehow generated by the contact of two different conductors was responsible for the motion of the electric fluid inside the battery. He now preferred to develop an analogy between the bodies' capacity for heat and their capacity for the electric fluid. He had mentioned similar analogies between heat and electricity several times in his previous publications, and long before the invention of the battery. For example, he had sketched a similar analogy in connection with his experiments on the electricity observed in the phenomena of evaporation in 1782, and again when trying to refine his notion of electrical atmospheres in 1784. He had alluded to the analogy between electricity and heat—indeed, among all "radiant fluids"—in connection with his contact theory in a published, 1798 letter to Giovanni Aldini, where he however claimed to regard the principle on which the electromotive force acted as ultimately unexplained, or "arcane." Now (sometime after 1801) Volta developed the analogy between heat and electricity more in detail, in a piece written in the third person, as if intended for one of his occasional anonymous publications, aiming at a different explanation of the electromotive force. Volta developed the analogy as follows:

> Different bodies, depending on their attractions and affinities, have different capacities for electricity just like they have different capacities for caloric. So, precisely as, at the same temperature, different bodies retain larger or smaller quantities of caloric / caloric fluid—i.e., each one has its specific heat—so also with electricity at the same temperature, i.e., in a state of electrical equilibrium, each body retains a smaller or larger dose of the electric fluid depending on its specific capacity. If that is so, while two different metals, like for example copper and zinc, preserve the dose of the electric fluid pertaining to them with respect to the air surrounding them, or to all the conductors of the second class [that is, humid conductors] which they are in contact with, they will persist in a state of equilibrium, displaying no tension, nor any sign of electricity. But, notwithstanding the equilibrium in which these conductors of first [metallic]- and second [humid]-class will find themselves because the electric fluid is distributed among them in the right dose, that is, in proportion to their respective capacity, it will no longer be so between the metals when they are placed in contact. That is what Volta assumes to happen in any combination of two different metals, or first class conductors.

The analogy with specific heat allowed Volta to explain how a current developed inside the battery in the following terms:

> It can be assumed that copper and zinc, that were in a state of equilibrium with other, nonmetallic conductors, because they retained the electric fluid according to their respective capacity, when brought into contact will need quite different amounts of fluid for their respective capacities. So that, for example, copper will give 1 out of his 100 to zinc, the latter getting 101 and copper remaining with 99. They will thus find themselves satisfied,

and in equilibrium; but they will no longer be in equilibrium with the other, second-class conductors and with the environment, because one will lack, and the other will have as a surplus one-hundredth of the electric fluid. From that situation, tensions will arise, positive on one side, and negative on the other; tensions that will be eliminated by discharging the one in excess in the zinc, and by compensating copper with the one lacking. The equilibrium thus restored, . . . . charges and discharges will be renewed at every instant, and a continuous current will develop if communications will be continuous or, better, if the circle will be completed by means of a good conducting arc.[131]

As already noted, the interpretation we call "Volta's Battery, 3," remained a semiprivate speculation on the part of its author. As to the notion of a "capacity for electricity," specific to each substance and modeled on the notion of specific heat, it was not even mentioned in the most extensive, last discussion of the battery that Volta finally published (under the name of a pupil) in 1814. In this work ample room was given to a discussion of the capacity of the voltaic battery as compared to a battery of Leyden jars, but no mention was made of the notion of a specific capacity for electricity. However fond Volta might have been of a notion reminiscent of his juvenile speculations on microscopic attractions, he apparently did not deem it worth linking the destinies of the battery to that notion.

The latter circumstance, together with the fact that—as we by now know—Volta developed at least three different interpretations of his own invention, suggests that he was only moderately happy with his own interpretations of the battery. This, in turn, sheds some light on the complex relationships linking instruments, interpreters, and natural philosophies.

Together with the evidence mentioned when discussing Volta's path to the battery, the oscillations he showed in identifying the principles on which his battery worked suggest that Volta *knew*, somehow, that he had only a limited grasp of what was going on inside his instrument. Such an awareness was implicit in his first account of the battery to the Royal Society, all centered on the instrument itself, and skipping theory. The same awareness must have been reinforced by Nicholson's report of the chemical phenomena that accompanied the operations of the battery, which Volta had left unnoticed, and he felt it perhaps again amid the difficulties he met with his "fundamental experiment" in Paris in 1801.

Whatever Volta's own and the other interpreters' uncertainties, the most authoritative, public treatments of the battery were to remain Volta's own second interpretation ("Volta's Battery, 2") and Jean-Baptiste Biot's report to the Institut National.[132]

## A Name for All Purposes

Having described some of the attempts made to appropriate the battery by a number of interpreters in several countries, it may be instructive to trace the history of the names that were attached to the instrument being appropriated.

As it turns out, the names that circulated extensively were those that did *not* identify the instrument with a single interpretation, interpreter, or group. It was as if the experts agreed that it was reasonable to adopt a name for the instrument that allowed it to enjoy an amount of independence from the clash of the different interpretations. The most popular names adopted to designate the voltaic battery thus focused on its most obvious and least controversial features.

When first presenting the battery, to be sure, Volta had tried a different strategy, dictated by the need to be recognized as its inventor. He was well aware of the role that the names attached to new instruments played within expert and amateur communities. He had learned that from the time he had introduced his electrophorus, when the name imposed on the instrument had helped him to vindicate his merits. As he declared on presenting the battery to the Royal Society, "New instruments should be given new names, depending not only on their form, but also on their effects, or the priniciple on which they are based."[133] Accordingly, Volta had recommended two possible names for the battery. The first alluded to the organ of the torpedo fish, which had inspired him, and to Volta's own claims against galvanism. It was the name of "organe électrique artificiel," the organ of the fish being the "organe électrique naturel."[134] The second one alluded instead to the principle by which the battery worked according to its inventor: "appareil électro-moteur."[135] This was an elaboration of Volta's own notion (developed well before the invention of the battery) according to which different conductors, especially metals, when in contact had the power to set the electric fluid in motion.

Volta's own recommended names, however, were not adopted widely within expert and amateur circles: these needed a name that did *not* imply Volta's appropriation of the instrument.

When presenting the battery Volta had used a couple of other names in order to distinguish between the two forms of the instrument he described on the occasion. He used the phrase "appareil à colonne," or column apparatus, for the first battery he had built; and "appareil . . . à couronne de tasses," or crown of cups apparatus, for the other one, illustrated in his April 1st letter to the Royal Society. Early interpreters—who had built Volta's apparatus and replicated its basic ef-

fects, but were unwilling to commit themselves to Volta's interpretation of them—preferred similar names or phrases, describing the
instrument rather than its working principles, or its supposed role
within the controversy over galvanism.

Thus Nicholson, who in his July 1800 title cautiously mentioned "the
new *Electrical or Galvanic* apparatus of Sig. Alex. Volta," in that same
article called the instrument currently "the pile," or "the electric
pile."[136] Landriani, in a letter of August 1800 already mentioned, called
it the "apparato a colonna," or the column apparatus.[137] In those same
months the battery was called "l'appareil électrique ou galvanique" in
Paris and Geneva, while the Germans called it "Volta's Säule," or Volta's column. In the course of 1801, the French and English "pile," the
Italian "pila" (occasionally also "piliere"), and the German "*Säule*" prevailed, and Volta accepted it for convenience sake,[138] while continuing
to use his own "elettromotore" whenever possible.

Due to well-known interpretive uncertainties, the adjectives "galvanic" or "voltaic" (or "Volta's"), or both, continued to be attached to
the "pile" or "column" for several years, with Volta of course discretely
campaigning in favor of the latter. Meanwhile, another name emerged:
the battery.

Volta and all the early experimenters with the new apparatus noticed that the electricity produced by it resembled the electricity delivered by a battery of Leyden jars, that is, by a condenser made by connecting a number of Leyden jars together. The batteries of Leyden jars
were known to be able to retain and discharge large quantities of electricity at a weak charge; or, to use Volta's terminology, their electricity
displayed a high capacity and a low tension just like Volta's new apparatus. Since Franklin's studies in the 1740s, the batteries of Leyden jars
were known to any amateur and expert electrician, and they were described in all manuals and treatises devoted to the subject. In 1800
Volta had proudly reminded his readers of those earlier batteries in
connection with his battery, while emphasizing that, contrary to the
batteries of Leyden jars, his new device did not need to be charged
again and again with an external machine after discharge.[139]

Because of these earlier electrical batteries—and also because of the
common usage of the word battery, with its military origins, to denote
a set of similar pieces used for combined action[140]—experimenters
using Volta's device, made of many identical metallic pairs combined
together, began to call it "a battery." Humphry Davy used the expression "galvanic battery" in entries of his manuscript *Physical Journal*
from November 1800.[141] By 1803 "galvanic batteries," "Volta's batteries," and "voltaic batteries" obtained regular entries in British textbooks devoted to electricity.[142] Later names, like the English "trough,"[143]

the German *"Trog,"* and, later still, the English "cell,"[144] indicate that, for attaching a name to the apparatus, the form of the instrument continued to be regarded a safer ground than the controversial interpretations accompanying it.

The story of the names attached to the voltaic battery in different countries at the beginning of the nineteenth century points to the fact that, at a basic but important communication level, an instrumentalist attitude was shared by many among those interested in the new device. By providing some common, relatively unproblematic terms for describing the apparatus, the instrumentalist attitude made travel easier for the new device. Indirectly (but in an important respect) that same instrumentalist attitude favored the circulation of the conflicting interpretations of the new apparatus: some common, descriptive terms designating the basic features of the new apparatus were needed for the different appropriations of the battery to take place and their respective merits to be assessed across cultural frontiers.

## From Philosophic Instrument to Patented Device

By 1800, patent systems had been introduced in a number of countries.[145] Patents offered another, still infrequent, but relevant mediator between the inventor, a new instrument, and the people potentially interested in it.

There is no evidence, however, that Volta ever considered the idea of patenting his battery. The circumstance deserves attention because, although the step was rare among natural philosophers ca. 1800, and all the more so among university professors like Volta, the practice of protecting a new device with a patent was well known in the same circles Volta was at home in, and where his battery was first introduced and tried.

As we saw considering Volta's dealings with Bonaparte and the pro-French administrators of Lombardy, the battery *was* regarded as a potential, important source of financial reward for its inventor. The battery apparently enjoyed such a status because it was regarded—in expert and lay circles alike, both within and outside the particular reward system in which Volta acted in Lombardy—as proof of the inventor's ingenuity. On applying successfully to Bonaparte and the pro-French administrators of Lombardy for concrete benefits in the troubled situation of 1801, Volta was exploiting this implicit but very real set of values, agreed upon independently of any formal system of patents or other reward-conferring procedure intended specifically for natural philosophers or university professors like Volta.

As mentioned, on the other hand, within the mixed community of electricians to which Volta belonged recourse to a patent could occasionally occur. Some ascertained practical applications of the invention were needed for such a step to be taken, another requisite being of course that some profit could be expected from it. The well-known London instrument maker Edward Nairne, for example, in 1782 patented his improved, friction electrical machine pointing to its medical applications, and on the assumption that the practice of electrical therapy provided by then a market wide enough to make the patenting process worth undertaking.[146]

Utility, however, could be a flexible concept. In Paris, a year before the battery was announced and tested by Étienne Gaspard Robertson, Robertson himself had patented his *Fantascope*, the modified magic lantern that he used to produce apparitions during his ghosts show.[147]

Another electrician involved in the story of the battery, William Nicholson, was (among several other things) a London patent agent, he himself holding a number of patents.[148] Considering Nicholson's contribution to the discovery of the chemical effects of the battery, and his description of the battery as a powerful "machine," one may speculate that, had it not been for Volta and the traditional reward system within which the battery was first introduced, people like Nicholson would have patented the voltaic battery soon after its introduction. Volta, however, preferred other, for him more obvious, avenues to obtain recognition and reward. As a consequence—and because of the uncertainties over the useful applications to be expected from electrical apparatus circa 1800—several decades passed before batteries began to be patented in Europe and the United States.

Going through the patent records of Britain, France and the United States, one finds that (apart from Nairne's patent already mentioned) the earliest patents dealing with electrical apparatus were taken out in the mid-1830s, and still targeted medicine as their field of application. The early electrical patents taken after 1800 did not include improvements on the voltaic battery in their specifications. For several, combined reasons already hinted at the battery was regarded as falling outside the domain of the patent system. Only since about 1840, with the prospect of new, electromagnetic devices being introduced to generate motion, a rapid increase in the number of patents based on electricity developed, and soon afterwards patents focusing on improvements of the battery were introduced.

One such patent was the initiative of Frederick De Moleyns in England, in 1841. It is now known mainly for the system of electric, incandescent light described in the second part of the specifications, while a third part dealt with "the application of electricity to the obtainment of motion."[149] In his specification De Moleyns traced the history of the battery, since its first introduction by Volta, in a manner that offers hints on what must have been the received view of the matter among instrument makers and inventors. Until the 1830s, De Moleyns noted, many kinds of voltaic batteries were "in general use" in Great Britain. However, "the powers of all [these batteries] diminished greatly after a few minutes, *which limited their application to scientific research.*"[150] The situation changed with the improvements introduced by Antoine-César Becquerel, John Frederic Daniel, William Robert Grove, and De Moleyns himself. As a consequence of these improvements, the "electric effects [of batteries] are [now] so proportioned in regard to quantity and intensity that the electricity developed is peculiarly applicable to the production of *great electro-magnetic power.*"[151] Apparently, around 1840 the creation of new systems of motion based on electricity was the main goal perceived by people like De Moleyns as likely to push toward the improvement and patenting of new types of batteries. By the 1840s, in fact, in several countries patents had been taken out affecting all the fields in which the application of electricity was to develop during the nineteenth century: communication (telegraphy, above all), lighting, electrochemistry, and motion. This part of the story extends beyond the limits of the present study. Yet, it offers a glimpse on the unintended consequences deriving ultimately from Volta's own invention, and from the old reward system that procured him benefits and celebrity on the European scene.

### Conclusion

The early reception of the voltaic battery in several countries has been surveyed in the present chapter with the purpose of exploring how instruments, experts, and cultural contexts interacted in the assessment of a device that brought about a momentous breakthrough in the history of science and technology.

The events associated with the early reception of the battery discussed in this chapter seem to point to two distinct sets of circumstances. On the one hand, the rapid diffusion of the battery in expert and amateur circles, approaching the instrument from widely different backgrounds, indicate that the battery and its basic effects were comparatively easy to replicate and, in some obvious sense, the battery was regarded as being the same, basic instrument in London, Paris, Geneva, Vienna, Halle, Jena, Copenhagen, as well as in Como, where it had first been conceived and built. It is also clear however that—side by side with the rapid, relatively unproblematic diffusion of the instru-

ment—a wide range of different, sometime contradictory interpretations of the battery were put forward in various quarters, occasionally
by the inventor himself. As already pointed out, in the case of the battery easy replication did not carry uniformity of interpretation with it.
While every decently trained electrician in Europe could build a voltaic
battery and perform with it, there was little agreement on how to interpret it, apart from the vague perception that the instrument was new,
its effects sensational, and its consequences for contemporary science
important, if unpredictable. How are we to assess this dual set of
circumstances?

An easy way out of the difficulty, of course, would be to explain
away one of the two sets of evidence. We should refrain, however, from
underplaying either easy replication or disagreement over interpretation: both clearly played a part in the early history of the voltaic battery. We seem to need an interpretive framework compatible with both,
and the following, concluding remarks are aimed at sketching such an
interpretive framework.

To begin with, it is useful to recall how the battery was introduced
on to the scene in the first place. The path that led Volta to build the
battery has been reconstructed in detail in a previous chapter.[152] From
that reconstruction, the connection linking the instrument to the natural philosophy and interpretation cherished by its inventor has
emerged in all its complexity. As shown in that reconstruction, Volta
did not (in any reasonable sense of the word) deduce the battery from
his theory of contact electricity. The battery was, rather, the result of
an investigative process in which Volta's contact theory interacted with
a variety of other factors, including the intellectual and social pressures
deriving from the controversy over galvanism, the extraordinary manipulative ability Volta had acquired in over thirty-five years of dealing with weak electricity, and the hint coming from William Nicholson's intriguing model on how best to imitate the torpedo fish.

In that process, there was no single or privileged path leading Volta
from the contact theory (the core of his natural philosophy in the late
1790s) to the battery. Much less could the path be, as it were, deduced
or predicted. This is shown overwhelmingly by several circumstances:
first, the fact that it took more than four years for Volta to move from
a clear formulation of the contact theory to the battery; second, the fact
that he built the battery only after Nicholson's model of the torpedo
fish had intruded into his investigative path; and, third, the fact that
he did not bind the instrument entirely to the contact theory when he
first presented the battery to the public in the spring of 1800.

All this shows that the battery already enjoyed an amount of independence from the interpretations provided by natural philosophy in

the investigative experience and presentation strategies of its inventor. As pointed out above (under "Volta's Battery, 1"), the relative independence of his instruments from the interpretations provided by natural philosophy was something Volta was aware of, because he had experienced it when introducing his earlier machines, like the electrophorus. Volta also knew, on the other hand, that within the mixed community of expert and amateur electricians, natural philosophers and instrument makers he belonged to, and considering his position as a university professor, his claims as inventor rested mainly on his ability to provide evidence that *he possessed the theory explaining the instrument.* From his own experience with the electrophorus,[153] at any event, Volta knew that claims concerning the theory explaining an instrument were bound to be controversial, and open to troublesome, endless negotiations within that same community, while the instrument itself could enjoy untrammeled popularity in it.

All these circumstances may well explain why, when first presenting the battery, Volta emphasized the instrument and its effects rather than the theory he thought could explain it; and why he could adjust his presentation strategies to his peers' reactions and expectations, as he did shifting from *"Volta's Battery, 1"* to *"Volta's Battery, 2"*, in concomitance with the target of his quest for reward shifting from London to Paris.

Given the same circumstances, it comes as no surprise that Volta continued to modify his own interpretation of the battery after 1801, as described under "Volta's battery, 3," and that he decided eventually *not* to publish his latest speculations on the subject, based on the notion that each body must have a "specific capacity for electricity" just as it has a specific capacity for heat.

Other circumstances favored the relative independence of the battery from the author's interpretation of it, and, a fortiori, from the interpretations of others. Within the mixed community Volta belonged to, recognition and reward for the invention depended on well-administered publicity about the instrument, its construction, its effects, and its implications for natural philosophy. That publicity had to be addressed to a wide group of experts belonging to different cultural, social, and national contexts. That meant providing a description of the instrument appealing to many different audiences, rather than to a closely knit group of experts. That was what Volta pursued with his letters to the Royal Society, and with his propaganda trip to Paris and other European centers in the autumn and winter of 1801.

The decision of Joseph Banks in London to pass the first of Volta's letters on to Carlisle, a physician, who in turn passed it to Nicholson, a chemist, electrician, natural philosopher, and instrument maker (who did not belong to the Royal Society), confirms that from the earliest steps of its public career the battery was not perceived as belonging to a closely knit group of experts. The subsequent news leaks about the device and its performance that took place in London—from the circles inside and around the Royal Society to those of the Royal Institution, and hence to the London and Paris daily press—further encouraged the public career of the battery independently of the interpretations recommended by its author, or by any self-selected group of experts. As a result of those several news leaks—and because of other, more cogent circumstances—for many months (the time needed for the slow *Philosophical Transactions* and the *Philosophical Magazine* to publish Volta's spring 1800 paper) the battery was often built, and its functioning and effects assessed around Europe, independently of the presentation with which the author had thought fit to accompany it. Under the circumstances, that entirely different interpretations were soon attached to the battery seems an obvious consequence.

The degree of independent life the voltaic battery enjoyed during its earliest steps on the public scene was apparently augmented by its simplicity and by the easy availability of the ingredients needed to build it. In subsequent years, no doubt, with the introduction of different formats, greater sizes, and chemical improvements, the battery became a highly sophisticated device, and its use was a complex enterprise on the part of accomplished natural philosophers like Davy and Becquerel, and technologists like Cruickshank, Daniel, and Grove, supported by institutions that, in several countries, invested considerable amounts of money in building big and improved batteries. The fact, however, that only in 1841 De Moleyns, in England, thought it fit to patent improvements in battery construction and regarded all earlier batteries as "voltaic batteries" that he considered to be "in general use," points to the circumstance that the basic identity of the instrument was, until then, agreed upon despite the considerable technical developments that had taken place in the four decades following its first introduction.

The agreed-upon identity of the voltaic battery, as pointed out, went hand in hand with an impressive array of diverse *appropriations* of the instrument from different quarters. These responded to different needs, depending on the intellectual and social contexts of the persons or groups interested in the new instrument.

For Volta himself, because no controversy developed over his claim to be regarded as the inventor, appropriating the battery was above all a matter of intellectual pleasure, as well as part of the social strategy aimed at obtaining reward for his new achievement. Those who read Volta's second presentation of the battery, or saw him performing with the battery during his European trip of 1801, could nurture no doubts about his main intellectual ambition in connection with the instrument. Volta regarded the battery as a decisive blow in the battle to show that there was no special animal or galvanic electricity, but only the same plain, physical electricity everywhere in nature. This broad, natural philosophical, physicalist tenet was, according to Volta, what conferred general meaning to the battery and gave an appropriate measure of his achievement in building it.

The many who disagreed with Volta over his interpretation of the instrument agreed in any case on the broad intellectual implications that the battery was perceived, however confusedly, to entail. All the early appropriations of the battery discussed in this chapter seem to have worked perfectly well—that is, according to the intellectual and social needs of each different circle of interested people—side by side with a recognition of the basic identity and "philosophical" worth of the instrument itself.

Nicholson, for example, when he emphasized the chemical phenomena accompanying the battery that had gone unnoticed by Volta, was at the same time appropriating part of the battery (its chemical effects), claiming originality and hence recognition over it, and (implicitly) vindicating the role his model of the torpedo fish had had in leading Volta to the battery. Nicholson at any event implicitly agreed that what made the battery special (and his own contribution to it worth) were the broad implications the instrument was supposed to have for the natural philosophies involved.

Because there was a general, if vague, understanding of the unexplored but momentous philosophical implications of the battery, widespread disagreement about which particular philosophy explained it better did not detract from what was universally a generous appreciation of Volta's achievement in inventing it, and of the merits of Nicholson and Carlisle in discovering its chemical effects.

In the present chapter, the diversity of the interpretations that accompanied early experiments with the battery has been exemplified through the works of Ritter, Ørsted, and Robertson (before his conversion to Volta's explanation). Especially in the case of the galvanic and/or chemical interpretations suggested by Ritter and Ørsted, it is hard to imagine explanations more at odds with the philosophy pursued by the inventor of the battery throughout his entire career. Yet, the mechanisms of appropriation used by Ritter and Ørsted seem to have worked along lines similar to those allowing people like Nicholson and Biot, much closer to Volta in their interpretations, to pursue their own appropriations of the new instrument.

Considering the remarkable degree of popularity the battery enjoyed beyond expert circles, some form of appropriation of the battery appealed to a wide variety of nonexpert cultivators of the sciences as well. If appropriating the battery within expert circles meant to contribute original explanations and/or new evidence and technical improvements, among lay audiences it was linked to other, arguably not-too-dissimilar intellectual and social pleasures. Judging from the early press reports describing the battery and from the fascination journalists and lay audiences felt for the water, electricity and galvanism business in London, Paris, Geneva, or Vienna, what made the battery popular was not so much the instrument *per se*, but its vague, and yet intriguing, connections with broad speculations concerning the nature of life and matter. Such speculations apparently appealed to many cultivated people scattered in Europe circa 1800. To them, the easy replication and the popular effects of the battery added wonder to an already sensational subject.

Judging again from the early reception of the battery among lay audiences in several countries, an understanding of the broad implications the instrument had for natural philosophy and an adhesion to the cultural and social rituals involved in its recognition were widespread commodities in Europe around 1800. It was the significant amount of shared, if uncertain, understanding of the instrument and its implications, and the similar social rituals associated with the assessment of knowledge production, that allowed Volta to search for and obtain recognition as inventor of the battery outside his country, allowing him to spend in Lombardy the rewards he obtained in London, Paris, and Vienna.

That same amount of shared understanding, and the not too dissimilar rituals associated with the reward of intellectual achievement across Europe's troubled borders, allowed Bonaparte to use the prestige of Volta and his battery to please the cultured elites of several European countries at a time, leaving a lasting impression of the intriguing connections between science, power, and public culture on the imagination of subsequent generations of scientists and lay audiences alike.

For several reasons mentioned in the present chapter, the early history of the voltaic battery does not fit easily either into a traditional, realist interpretation of scientific instruments, or into a constructivist view of science conceived as the product of strictly local cultures. A combination of both seems to be needed if we want to understand the geography of the new continent opened up by the invention of the battery.

To understand this geography, we should grant the following premises, and no more than them. First, that circa 1800 circles of expert and lay people scattered in several European countries subscribed to cultures that shared bits and pieces of Enlightenment notions associating achievement with natural philosophy, "useful knowledge," instruments, and machines. Second, that Volta himself subscribed to such notions, in the particular combination that valued natural philosophy and instruments most, while also valuing their possible useful applications to the professions and industry. Third, that through the specialist and daily press, as well as through correspondence and travel, members of those circles constituted loose but effective networks of people, often crossing Europe's national borders, interested in assessing innovation in natural philosophy and new instruments like the battery. Fourth, and last, that within such networks instruments like the battery were subjected to repeated and varied assessments, in the course of which frequent permutations of the mentioned Enlightenment notions and their ranking took place, leading to widely different interpretations of what was nonetheless regarded (and named) as the same basic instrument.

This, and no more than this is needed to explain why in little more than one year after spring 1800 the same simple instrument, invented by Volta as a device imitating the electric fish and showing the principle of contact electricity at work, was interpreted instead by Nicholson as a chemical machine prone to industrial applications, by Ritter and Ørsted as an icon of the Romantic philosophy of nature, by Robertson as yet another attraction provided by galvanism to his ghost show, by Biot as an example of Laplacian mathematical physics, and by General Bonaparte as a symbol of achievement whose celebration could convey a reassuring image of French political ambitions over Europe.

Indeed, in order to follow the early steps of the voltaic battery on the European scene one must admit an amount of diversity, shared (if vague) notions, and unintended consequences that neither the realists nor the social constructivists have been willing so far to accommodate in their views of the history of science and technology.

# The Scientist as Hero

## VOLTA AND THE USES OF
## PAST SCIENCE IN THE INDUSTRIAL ERA

The example set by Bonaparte's celebration of Volta before Europe's cultured elites had an impressive follow-up. The present chapter surveys a few representations produced in the nineteenth and early twentieth centuries to convey a notion of Volta's achievement to wider audiences. From the battery's earliest appearance on the public scene in 1800 until the huge celebrations to commemorate the centennial of Volta's death, held in 1927 in Como, the battery, together with its inventor, were used in ways that offer interesting clues to how science and technology were being granted a prominent place in public culture. The social uses of past science discussed in the following pages, however, also reveal the difficulties that the public, and scientists themselves, found in developing a specific set of values against which to assess scientific and technological achievement. Far from hindering the celebrations, the lack of a specific set of criteria for assessing achievements like the battery probably contributed to their very popularity. Yet, it did not help to dispel the several ambiguities that affected the images of science and technology conveyed to the public during the industrial era.

Ambiguities were apparent in the contrast between the universal virtues attributed to the hero being celebrated and the overwhelmingly local interests of the people supporting the celebrations; between the disinterestedness granted scientists and technologists and the usefulness exalted in the products of their work; between the public merit of their achievements and the individual character of their genius; between the moral worth ascribed to scientific and technological endeavor and the secular nature recognized in that same endeavor.

These conflicting themes contributed to the (temporary) magic of the rituals performed to celebrate the scientist as hero in the industrial era, but they also caused the rapid disappearance of the magic once the celebrations were over.

## Admitted to "Galileo's Tribune"

Soon after 1815, when the Congress of Vienna had restored the dozen "legitimate" (and mostly foreign) sovereigns that Napoleonic wars had temporarily swept from the Italian peninsula, the times were ripe for reflection on what was increasingly being construed as the "national" Italian tradition in the sciences. In those days Italian achievements in the sciences (since the time of Galileo, and before) were exalted the more willingly as they contrasted with the frustrations Italians had suffered in other, especially political arenas. What was often referred to as "the Italians' primacy in the sciences" thus acquired the role of a precious substitute for lack of primacy in other fields. Scientific excellence, on the other hand, appeared or at any rate was presented as universal enough to guarantee that the vindication of the Italians' merits in the sciences did not carry strictly political implications with it, thus posing no threat to the recently restored foreign powers. Given the circumstances, the universal, "civic" virtues attributed to science by the Enlightenment tradition made science a more appealing substitute for the missing political achievements than the artistic and literary merits Italians were usually credited with.

Among the peninsula's cultured elites, the theme of an Italian tradition in the sciences appealed especially to those, like the Tuscans, who felt closer to the sources of that tradition. They now appeared willing to include leading figures from other Italian regions in it. This was the program pursued by Vincenzio Antinori, a Tuscan nobleman with some scientific training, who in Florence in 1816 edited a collection of Volta's scientific papers in two volumes.[1] The dedication of the volumes—containing the publications of a scientist who had served Austria for most of his life—to Ferdinand III, Grand Duke of Tuscany and Imperial Prince of Austria, reassured everyone that the addition of a Lombard (Volta) to the Parnassus of Italian (Tuscan) science did not depart from the track of dynastic legitimacy recently enforced in the peninsula by the great European powers.

The line pursued by Antinori and his patrons, aimed at reinvigorating the notion of an Italian tradition in the sciences, reached a climax in 1841 when, in view of a meeting of scientists from several Italian states due to convene in Florence, it was decided to erect what was called "Galileo's Tribune."[2] The place was conceived as a chapel, but devoted to a lay cult: the cult of Galileo, who was supposed to remind the visitors, from his "tribune," of the great past, and hopefully also the persistent vitality of Italian science. The paintings and the statues enriching the place were in tune with the message. The inclusion in the "Tribune" of a fresco depicting Volta in the act of presenting the battery to Bonaparte was conceived (like the Florentine edition of Volta's works already mentioned) to show that Italian glories in physics were not only the old ones, that they could occasionally come from outside Tuscany, and that they were worth the attention of public authorities, represented by Bonaparte. Sovereigns and public authorities indeed played a prominent (according to many, overwhelming) role in the annual, itinerant meetings of Italian scientists that had began two years before in Pisa, also on the initiative of a grand duke of Tuscany, Leopold II.[3]

Thanks to Antinori, who cultivated electricity and had a passion for the history of science, the otherwise conventional fresco by Niccola Cianfanelli[4] dedicated to Volta adorning Galileo's Tribune (fig. 8.1, p. 260) contained a quite accurate description of the apparatus (the electroscope-*condensatore*, the pile, and the crown of cups) that Volta had presented in Paris in 1801. The scientists assembled were also diligently represented, with the Italian(Piedmontese)-born Lagrange standing out among them, as required by the main message Italians expected from Galileo's Tribune.

For the rest, the painting shows that scientific instruments (that, as we know, during Volta's lifetime lay and expert audiences had had some difficulties in accommodating within their notions of scientific endeavor) were now—in the 1840s—regarded as the chief symbol able to convey to wide audiences the notion of achievement in the sciences. As shown by another visual representation of Volta's scientific achievements discussed below, the shift in the public attitude toward scientific instruments during the central decades of the nineteenth century was linked to the parallel diffusion of other impressive machines, like the steam engine, emblematic of the technological feats of the industrial era.

## Secular Saint in the Positivist Calendar

During the 1830s and 1840s, celebrating Volta and the battery appealed to many, outside Italy as well. The most exceptional such celebration was perhaps the one conceived by the founding father of positivist philosophy, Auguste Comte. Although Comte regarded electricity as the least developed among the branches of physics and inclined toward a chemical interpretation of the battery, in his major philosophical work he decried Volta, as inventor of the battery, "immortal."[5] Consistently, when he later embarked on one of the most bizarre undertakings of his career, the *Positivist Calendar*, he reserved a prestigious place for Volta in it.

8.1. Niccola Cianfanelli, Volta presenting the battery to Bonaparte and the Institut, Florence, "Galileo's Tribune," 1840s. The fresco was commissioned by Leopold II, Grand Duke of Tuscany (*Museo della Specola, Florence*).

The *Positivist Calendar*, conceived after the revolutionary ferments of 1848, was meant by Comte to encourage a "respectful appreciation of the diverse services of all our predecessors," and scientists and inventors had a prominent place in it, as required by the principles of positivist philosophy.[6] The day assigned to Volta in the calendar revealed a high regard for the Lombard: he was to be celebrated on the sixth day (Saturday) of the month devoted to "Modern science," the other weekdays being reserved for Copernicus, Kepler, Huygens, Jacques Bernoulli, James Bradley, and the Sunday to Galileo. Volta thus enjoyed posthumously in Comte's calendar the company of the natural philosophers that he had aspired (with mixed success) to join in public recognition during his lifetime. It may also be noted that, for reasons perhaps connected with criteria of distributive justice and other such technicalities, in Comte's calendar Coulomb shared a somewhat less prestigious company than Volta, his celebration recurring in the month devoted to "Modern industry."[7]

## "The Triumph of Science"

A high regard for science and technology spread impressively from the 1840s to the 1870s in Italy (since 1861, an independent country) if we judge from another painting celebrating Volta, conceived in 1876, and compare it with the one in Galileo's Tribune. While that earlier celebration of Volta's achievements was the combined initiative of a man who had himself some competence as a scientist, and of a sovereign, and was destined to a public space devoted to science, this later one (fig. 8.2, p. 262) was commissioned by a wealthy lawyer, having no scientific credentials, and was intended for his private palazzo. Volta (standing, leaning toward the table, and touching the top of his column battery) is central in this representation of scientific achievement, although he has to share the stage with a high number of colleagues (thirty-eight), belonging to different ages and assembled as if in a timeless international congress. The original is a fresco, painted in a room of Palazzo Orsini, in Genoa, and it was commissioned by Tito Orsini, an expert in commercial and maritime law, and a member of the Italian Senate.[8] The painter, Nicolò Barabino,[9] having spent part of his life in Florence was well aware of the earlier, Florentine fresco mentioned above.

The title of Barabino's painting is *The Triumph of Science*. The overall conception is extremely ambitious, the architecture of the scene evoking no less a model than Raphael's *School of Athens* in the Vatican Palaces. The genre is academic and symbolic. Science is represented as a

8.2. Nicolò Barabino, *Il Trionfo della Scienza*, fresco in the Palazzo Orsini, Genoa, 1876–80s. The painting was privately commissioned by Tito Orsini, a wealthy lawyer (*Private collection, Genoa*; photo: Ottavio Tomasini)

woman, a geometry book in her hand, surrounded by light: the light of an enlightened age. At her feet you see a figure laying: it is obscurantism defeated. All around science is a crowd of natural philosophers, including Galileo and Newton, but also Columbus, Gutenberg, Stephenson, Watt, and Montgolfier.

When this painting was made the age of electricity applied to industry was clearly in sight. Accordingly, the impact of the battery is conveyed more clearly than in the former fresco: the battery is shown on a table packed with other, obvious symbols of nineteenth-century technological achievements, such as the steam locomotive. Bonaparte is still there (although in the back stage, center-left, with the typical hat) to mediate between the state and the achievements of the natural philosophers and technologists. Power and its rhetoric are also there: they spread, as it were, from the high vaults of the building down to the attire of the "philosophers."

Were we to submit this painting to the analyses of iconology, it would turn out that there was continuity in the representation of philosophical and scientific achievement; a continuity that went on from the Renaissance, through the Baroque and the Enlightenment, to well into the nineteenth century. The reasons for this continuity are not difficult to grasp. Well into the industrial age people like this painter and his patron lacked a specific set of criteria with which to assess and eventually celebrate achievements like the battery and the science of electricity. Being in doubt about the distinctive agendas and merits of scientists and inventors, to celebrate them artists and patrons had recourse to the symbols used to celebrate achievement, and power, in general. Scientific instruments, on the other hand, were being granted an increasingly prominent place (side by side with the classical "geometry book" placed in the hands of the woman representing science) among the icons conveying the goals pursued by the scientists; but instruments like the battery had to share the stage with the machines produced by the industrial era.

A lack of specific criteria with which to assess, and eventually celebrate scientific and technological achievement in front of the general public is transparent also in the last episode to be discussed in this chapter: the big international congress set up in Como on the first centennial of Volta's death.

## In the Nobel Laureates' Era

From 11 September 1927, a group of sixty-one physicists from fourteen countries met for seven days in Como, to celebrate the first centenary celebrations of Volta's death.[10] The group included twelve Nobel Prize winners: Niels Bohr, James Franck, Max von Laue, Max Planck, Francis W. Aston, William L. Bragg, Ernest Rutherford, Guglielmo Marconi, Hendrik A. Lorentz, Pieter Zeeman, Arthur H. Compton, and Robert A. Millikan. The group also included three men who were to join the Nobel laureates later: Max Born, Werner Heisenberg, and Enrico Fermi.

For the Volta celebrations, a huge villa overlooking the lake of Como was bought and kept open from May to September. A special journal, *Voltiana*, was published in thirty-three issues over ten months. Exhibits of the electric, telephone, and silk industries (the strong engine of Como's economy) were organized and put on display. To lure people from all over Italy, special train fares were arranged and a postal stamp series launched. The iconography adopted throughout was in keeping with the themes and rhetoric of the fascist regime that had come to power in Italy five years earlier. Alessandro Volta was portrayed in the pose of an ancient roman. Volta's electric battery was made to look like the bundle of elm branches containing an axe that was the symbol of the fascist regime. The fascist bundle figured large on the special search-light built on a mountain near Como to symbolize the "light of science" spreading out from Volta's birthplace. Mussolini himself, together with Guglielmo Marconi, was chosen as honorary president of the committee in charge of the celebrations. The journal of the celebrations liked to portray Mussolini in his military dress, the supposedly menacing air slightly softened by the "flou" technique cherished by photographers in those days. The involvement of Mussolini should have included his reading a special message to the nation on Volta's anniversary, exalting Italy's scientific achievements. The text of the message survives, though it was never delivered, for reasons we shall see below.

As the "manifesto" of the celebrations announced, side by side with the congresses and exhibits a wide range of leisure and sporting activities took place. These included art, flower, horticultural, hunting and fishing exhibits, dog shows, and international competitions for mandolin players, musical bands, and choirs. A piano piece by Giacomo Puccini, which had been commissioned by the association of Italian telegraph operators on the anniversary of the electric battery in 1899, was reprinted on the occasion. Puccini's piece in honor of Volta was appropriately entitled "The Electric Shock," and it was written in the tempo of a "little triumphal march."

The budget for these massive celebrations reached the then huge sum of 7,291,568 lire.[11] The sum corresponded to 13 percent of the total annual expenditure of the Italian state for universities, or the budget of the Fermi group in Rome for twenty years.[12] On the revenue side—

if one excludes the effects of the celebrations on the Como tourist industry—the only lasting, scientific product of the massive financial outlay was the two-volume proceedings of the physicists' congress, with their impressive list of Nobel Prize winners as contributors.[13] Indirect benefits included a much longer list of historical books and articles elicited by the celebrations, most of which, however, were of an ephemeral nature and value.

The sheer scope and size of the 1927 Volta celebrations invite questions about the uses of past science that were current among specialist and lay audiences in the 1920s, questions such as the following: First, exactly what role did past science play in the celebration machine and in the rhetoric being used throughout? Second, what "virtues" were ascribed to the "hero" being celebrated? Were these virtues peculiar to Volta—to the hero-scientist—or could they have been shared by other herolike figures, such as a philosopher or an artist? A third and more concrete question: exactly what kind of benefits did the celebrating people—lay and professional—expect and draw from the celebrations?

The aims that moved the promoters of the celebrations, as we shall soon see, were overwhelmingly national, and often strictly local. Yet the prominent part played by foreign scientists in the celebrations invites additional questions relating to the crossnational uses of science and its past. Let us start with an outline of the national scene.

Italian science between the two world wars has been the subject of considerable historical work.[14] To outline the context in which the Volta celebrations of 1927 took place, a few general comments will suffice. During the 1920s and 1930s, Italy slowly passed from a mainly agricultural country to a predominantly industrial one. Science based industries, like the electrical and chemical industries, played a key role in this long overdue change, but they had to rely mainly on the know-how of foreign firms. The science policy pursued by the fascist regime after 1922 seems indeed to have put great stock in applied research as a consequence of increasingly nationalistic and military concerns. At an institutional level, important initiatives were taken, such as the establishment of a national research council in 1927. However, the amount of attention, institutional care, and financial support given science by private individuals and the state fell well short of reaching the critical mass necessary for securing the badly needed takeoff of Italian science between the wars.

Against this gloomy background, the mammoth 1927 celebrations of Volta appear astonishing, not to mention paradoxical. How did the lavish celebrations square with the neglect of research facilities throughout the peninsula that any serious observer of the day complained of? When I first posed the question I thought that the answer should be sought in Rome: I assumed that the huge "propaganda" venture centered around Volta must have been just one of the many ventures Mussolini and his entourage resorted to to stir up national pride and secure foreign attention. Research into the archives proved me wrong. I found that the promoters and unrelenting supporters of the initiative were in fact strictly local figures, representing local interests. These figures were close to being the least conspicuous personalities of all the people involved in the celebrations. Yet, it was they who engineered the concerted effort needed to call national and international attention, *via* Volta, to the town of Como.

The chief character in the story was a now forgotten man, Carlo Baragiola. At the time, Baragiola was the leading fascist authority in Como and a member of the senate in Rome, as his father had been. He liked to call himself "an industrialist." In fact, his conspicuous family riches came from land and from the transport business in the Como region. The high marks of his career were two wounds suffered and two decorations won during the First World War, and a third wound suffered during the skirmishes of the civil war that accompanied the fascist seizure of power. While all this qualified him as a war and a fascist hero, it did not guarantee him fame beyond the regional boundaries.[15] As to his possible connections with Volta, I was unable to find in his biography any link whatsoever tying him to science, except the fact that his father had invested part of the family capital in the electric industry.

The second figure behind the celebrations was another forgotten name, Enrico Musa. A former commercial traveler from southern Italy, Musa had settled in Como and accumulated considerable wealth as a silk industrialist there. He had also held office in the local administration of Como on several occasions and had distinguished himself for his philanthropic initiatives.[16] The special tie linking Musa to Volta was an amateur interest in the eighteenth-century scientist, an interest that had led Musa to edit the ephemeral journal published during the Volta celebrations of 1899, without adding anything lasting to the literature.

The two major figures behind the celebrations thus represented the local fascist elite and the silk industry of Como. However tenuous their ties with science in general and with Volta in particular, they were extremely ambitious and determined to make the Volta business a national and international concern. With this goal in mind, the two men set up an impressive array of honorary and executive committees for the celebrations, including people ranking from the king of Italy to the local journalist. The full list of committee members comprised no less than 370 people, whose duties ranged from that of just being on the list (the members of the honorary committee) to that of helping organize the mandolin competition. The promoters were very successful in obtaining the adhesion and support of authorities, both political and economic, at a local level. They were much less successful with national authorities and firms.

The campaign for national support began with a meeting with Mussolini himself.[17] Once some rather vague but encouraging statements from Mussolini had been obtained, the Como group used them for lobbying the ministries in Rome and the leading electric and telephone firms. The immediate results were promises of financial support, which the Como group interpreted as encouragement for their lavish projects. The funds they eventually raised from national firms, however, amounted to very little compared to their expectations, the result being that the Como group came very close to bankruptcy just a few days before the opening of the big physicists' congress, which was supposed to be the grand finale of the celebrations. None of the Nobel Prize winners who gathered in Como on the evening of September 9th ever suspected that the fees they had been promised were being paid by a loan a local bank had granted at the very last minute against a guarantee signed by Como's mayor.[18]

The expectations of the organizers were that, attracted by the name of Volta, big electrical companies would flock in, bringing with them generous support. The expectations had been nurtured by top representatives in the Italian government. They watched with a mixture of admiration and jealousy the success companies such as Edison were enjoying at that time in Italy. As a top government official remarked during a meeting with Como organizers, companies such as Edison of Italy were prospering thanks to the monopolistic regime they operated in with the government's consent. The line of argument was that foreign investors benefiting from the situation had almost a duty to reciprocate the favor by contributing to the celebrations of a national hero such as Volta.[19] There were precedents supporting the argument: the Edison company had contributed money for the publication of the seven-volume national edition of Volta's works, for which government support had long been overdue. The same company had given the money needed for the building of a special room at the Lombardy Institute of the Sciences, where Volta's manuscripts were, and still are, kept.[20] Such celebrations of national glory were, apparently, the main cultural dividend Italian politicians could come up with as an argument for their contributing to the success of international firms such as Edison in Italy.

The chief Italian executive of Edison had a different view of how best to celebrate the past glories of Italian science. He stubbornly resisted the pressures on him to get involved in the Como celebrations.

He suggested instead the creation of a large national electrical engineering institute, to be dedicated to Volta. Of the people mentioned so far, the Edison executive was the only one having a genuine perspective on Italy's recent technological achievements and failures. His name was Giacinto Motta and he had an impressive record as a businessman, a technician, and a teacher. While proud of the rapid expansion of the electrical industry in Italy—which he himself had helped launch—he was well aware that this success was almost entirely based on imported technology. He was also convinced that, to ensure long-lasting success, a broad-based system of technological education was required in Italy.[21]

The failure to raise funds from companies like Edison can be put down to the ephemeral nature imposed on the celebrations by its Como promoters. Their political leanings, which placed them at the far right of the fascist movement, was probably another reason. Political reasons must have caused rivalry with people such as Motta, who was a well-known liberal, and who for a while advocated Mussolini's removal from power.[22] The political leanings of the Como promoters must have caused embarrassment for Mussolini himself, who at the time was moving toward the center of the political spectrum to consolidate his power after the early, "revolutionary" phase of his regime. What is certain is that the Como promoters experienced growing resistance in Rome. Mussolini decided in the end not to deliver to the nation the message on Volta solicited by Como, and he wrote a big "NIENTE" (nothing) in blue pencil on the text prepared for broadcasting. He wrote an equally big "NO" in blue pencil on the last-minute, pressing request from Como for money to reimburse the physicists due to convene in a few days time.[23]

Let us turn to the big event itself, the physicists' congress, its wide international participation, and the results achieved during its six-day working session. These results are of course a major part of the "uses" of past science at work during the 1927 celebrations.

The program of the congress had been drawn up by Quirino Majorana, at the time President of the Italian Physicists Association and head of the Physics Department at the University of Bologna.[24] Majorana had since prewar times been personally acquainted with a number of leading foreign physicists, a fact that enabled him to convene a truly international gathering. In this, Majorana was decidedly successful: the Como congress was one of the first international conferences after the First World War to which scientists from Germany and the former Austrian empire were admitted again. He was less successful in drawing up the program of the conference.

Keeping in mind the celebrative purposes of the conference, Majorana chose to devote four full days of the meeting, from Monday to Thursday, to themes linked either with the heritage of Volta's work or with classical physics in general. Only on Friday were scientists at long last allowed to discuss hot current issues, the theme being "Theories of Matter and Radiations." The high point of this session, and of the entire conference, was Niels Bohr's paper on "The Quantum Postulate and the Recent Development of Atomic Theory."[25] Most likely in view of the fifth Solvay conference due to meet the following month, Bohr had decided to open in Como what he called "a general discussion" on the state of quantum theory. The time and occasion were well chosen. In the few years and months preceding the Como conference a number of new, outstanding contributions to quantum theory had been made. As to the occasion, in Como Bohr was able to address directly three generations of physicists who had been responsible for the impressive developments in the field: the older generation of scientists like Planck and Lorentz; the middle generation including Rutherford, Max Born, and Niels Bohr himself; and the youngest generation including Heisenberg and Fermi. Lorentz found in Bohr's paper "la clarté et la simplicité merveilleuses qui lui sont tout à fait particulières" (the marvelous clarity and simplicity absolutely characteristic of him). It raised a lively discussion, the only one recorded in the proceedings. Born, Kramers, Heisenberg, Fermi, and again Heisenberg in answer to Fermi, joined in the discussion. As Lorentz noted, everybody felt that too little time had been left for discussion, though to be sure discussion went on informally, as Lorentz hinted in his final report on the conference.[26]

Taking Lorentz's hint seriously, I have studied a plan, which has survived in Como, showing how scientists and dignitaries—118 in all—were seated around the huge table arranged for the dinners during the conference.[27] Incidentally, the table was shaped like the letter M in honor of Mussolini. The conclusion I have drawn from the seating arrangements is that fruitful discussions could hardly have taken place during such dinners. Bohr's seat was near Cavalier Musa, one of the two promoters of the celebrations, who knew absolutely nothing of contemporary physics. Planck was sitting near Como's town mayor, whose only scientific merit was that he had saved the conference from bankruptcy. Sitting at the top of the table, the old generation of physicists could not mix with the middle and younger generations, who were seated along the sides of the M-shaped table. Only Eddington and Born, who had the good fortune to be seated opposite each other, are likely to have had any kind of fruitful exchange during those dinners.

Bohr's paper and a few other highlights notwithstanding, the scientific value of the Como conference consisted mainly in being a kind of dress rehearsal for the fifth Solvay conference due to meet the following month. The absence in Como of scientists like Einstein (who had been invited but did not come), Schrödinger, and Dirac, all of whom were to take part in the Solvay meeting,[28] detracted from the value of the Italian conference. Even Bohr's paper seems to have owed its impact on the scientific community more to its publication in *Nature*,[29] than to the presentation in Como or to its later, belated publication in the proceedings of the conference.[30] That said, the fact remains that the assembly of eminent physicists meeting in Como in the name of Volta was memorable in the annals of science, and the conference was by far the most serious event in the mammoth celebrations. However, some of the questions casting doubts on the uses of science past seem to apply to the physicists' conference too. That is why the following, concluding remarks will address these issues with reference to the entire 1927 undertaking, and will point out a few analogies with the cases discussed in the previous paragraphs.

## Conclusion

At the origin of the Como's initiative, as we have seen, there was a high degree of local pride, local interest, and local ambition involved. However, judging from the promoters in Como, and leaving aside their obvious career, financial, and political interests, there was also a major cultural motive at work behind their initiative. The motive was the attempt made by these common men to share some of the characteristics attributed to the great man, the "hero" being celebrated. The hero was presented as a symbol of genius, fame, success, and disinterestedness. Such virtues were not peculiar to Volta or indeed to any individual scientist as such. However, they were the kind of virtues that rub off on anyone who joins in celebrating them. Such a psychological mechanism apparently played a major role in stimulating the promoters and attracting the lay masses to Como to celebrate Volta.

Virtues of the kind mentioned cannot be self-ascribed without losing their credibility. Universality of assent and judgment are required. That is one reason why the Como promoters earnestly looked for national and international approval. They felt it their duty to involve in the celebrations the experts of the field their hero belonged to, that is, the physicists. It should be noted, however, that the money spent on the physicists' congress amounted to barely 7 percent of the total budget of the celebrations.[31] The renowned physicists invited to join in the feasts and to confirm universality of assent were expected to give their blessing to the entire operation, but they were not its central focus.

Among the experts themselves, of course, the uses of past science followed different, more sophisticated routes, and yet the physicists gathered in Como who actually commented on Volta came out with remarks that had much in common with the themes of the lay audiences. Achievement, dedication, disinterestedness were for the physicists, too, the virtues of the hero being celebrated. The most frank among the remarks on Volta made by the physicists were those in which they revealed their genuine amazement at the enormous changes that physics had undergone since Volta's days. Put in this perspective, celebrating past science meant for the physicists celebrating scientific progress and *current* achievements. For the rest, the historical contributions of the physicists did not go beyond adding minor details to well-known aspects of Volta's work or ascribing back to Volta the credit for this or that discovery.

Common only to the experts, to be sure, was the goal of bestowing, with the assent of foreign experts, part of the prestige from past Italian science on its contemporary representatives. This naturally was one goal pursued by the Italian scientists who organized the international congress, together with the other goal of showing to the learned world that Italy was eager to conquer again a decent position in the international scientific community. If one takes the comments made by foreign scientists recorded in the proceedings at face value, these easy goals were attained. The achievement, however, was the result of hospitality mechanisms, rather than of the appeal to past scientific glories.

More concrete benefits for Italian science stemming from the conference are hard to detect. Majorana certainly reinforced his personal position among Italian physicists. He was later to be awarded the then prestigious Mussolini prize. Scientists like Fermi, however, hardly needed the Volta celebrations to further their work and career opportunities. Even the vague international spirit activated by the conference was to be soon curtailed by the autarchy policy of the fascist regime, which affected scientific circles as well.

To conclude, can we say that past science played a distinct, special role in the celebrations of scientific achievement discussed in this chapter? The answer is negative. Past science—if by that we mean the record and meaning of past intellectual and technical breakthroughs—seems to have played no major role. It was used as a tool by which the (often vague) cultural identity of the people involved in the celebrations was affirmed. Even if we grant (as I am personally inclined to) that the enterprise we call science and technology has peculiar features that place it somewhat apart from other cultural systems of interest to

the historian and the anthropologist, its distinctive features seem to
reduce or dissolve when science and its past are used in the way they
were in the cases under discussion. It would make little sense for the
historian to ignore that in such contexts science and scientists are mere
symbols of general values that pertain to culture at large. In dealing
with such values, and the uses they are put to, the tools of the general
historian seem to be more appropriate to the task than the specific tools
of the historian of science and technology.

# Conclusion: Science, Technology, and Contingency

## ENLIGHTENMENT LEGACIES

*Electricity reveals new worlds. They are pigmy worlds, you might think;*
*but who knows that one day giants too may amuse themselves with them.*
—*G. Beccaria to R. Boscovich,*
*22 April 1768*[1]

## Invention

The voltaic battery has been compared to such diverse inventions as the steam engine, the telescope, and the microscope. While making these two last comparisons in the mid-twentieth century, George Sarton noticed that the "creativity" of the voltaic battery was, however, unsurpassed by the telescope and the microscope.[2] Sarton's remark was reminiscent of Michael Faraday's mid-nineteenth-century statement, according to which the battery was, above all, a "magnificent instrument of philosophic research."[3] Faraday's and Sarton's comments were similar in turn to those often made by Volta's contemporaries, to the effect that the battery opened the way to "an ample field of experimental speculation."[4] Sarton, however, also mentioned that the battery "opened to man a new and incomparable source of energy," alluding to the big impact that the steady current first provided by the voltaic battery had on chemistry and electromagnetism, and through them on subsequent industrial civilizations. The question that remained unanswered by those commentators, and has been there throughout much of this book, is: Where did those novel, "philosophic," and at the same time very useful things associated with the battery come from?

In the preceding chapters the study of Volta and the battery has been conducted on the assumption that, to understand scientific work and its productions, we must place them in proper historical context. The halo of myth that tends to be associated with the notions of "invention," "discovery," and "creativity" when they are referred to the individual scientist alone has been reduced as a consequence.[5] Those too often vague notions have been substituted by handier historical objects: Volta's intellectual and social background in Austrian Lombardy; his training as a natural philosopher and civil servant, and the connected ambitions; his day-to-day activities as an investigator in his chosen field of expertise; his interaction with the community of eighteenth-century electricians; and his dealings with patrons who shared his Enlightenment ideals and engaged in the imitation-competition game that several governments played in their well-advertised support of fashionable natural philosophy.

Volta's investigations leading to his inventions have thus been found to conform to a pattern that can be described as competitive imitation. When inventing the electrophorus Volta imitated and improved on Aepinus's sulphur and cup apparatus, urged on by the controversy with which he was involved with Beccaria, and seconded by his own ideas about electrical atmospheres, which had helped him to define his place within the expert community in the early 1770s. In the case of the battery, he imitated and improved on Nicholson's model for the torpedo fish, this time urged on by the controversy over galvanism, and seconded by his contact theory of electricity, which had identified his stance within the controversy throughout the 1790s. In both cases the products of Volta's pursuits—his inventions—embodied something of the models being imitated, and something else deriving from the ongoing competition with other expert electricians. The inventions thus contained much that had been known for some time, and something else that was unquestionably new and the result of Volta's own endeavor. The latter, however, was itself the outcome of intense interaction with other experts, and thus part of a pursuit that Volta's competitors had themselves contributed to shape.

The novelty that contemporaries perceived, and we can still see in Volta's instruments, was the result of this process of competitive imitation and invention, in which the goals of the inventor interacted with those pursued by the other actors involved in the game: novelty was as much the product of this interaction (unpredictable in its detailed outcome) within the expert community, as of Volta's own goals, ideas, and creativity, and their interaction with the material world.

Important though the contextualization of invention is, we should not lose sight of some other broader, if ill-defined, questions that seem to have puzzled Volta, his contemporaries, and historians of science and technology when assessing the merit and status of instruments like the battery. The uncertainties apparent in the assessments quoted above should themselves be the object of historical reflection. Uncertainties affecting contemporary and later assessment of what the battery was about can in fact be regarded as pointing to some long-term developments of the enterprise we now call science and technology; long-term developments not clearly perceived by Volta's contemporaries, or later commentators.

In these concluding pages I will be developing a few reflections on invention and the legacies of the Enlightenment from such a long-term perspective, focusing on the fortunes and assessment of the battery. The aim is to answer broad questions like: How did the old concerns of natural philosophers combine with the new opportunities that Enlightenment society provided for experts specializing in fields like electricity? What exactly were the links between the emphasis on "utility," proper to Enlightenment culture, and the introduction of instruments like the battery, which had only limited useful applications, and yet paved the way for technological developments that would modify an entire civilization? How did it happen that a culture that liked to portray itself as "enlightened," but had developed no electrical industry, ended up with an invention that would make of electric lighting an everyday marvel?

My survey of these questions falls under six headings: useful knowledge, the quantifying spirit, investment, value assessments, contingency, and Enlightenment legacies. The goal is to review the main notions that have emerged from the present study of the economy of invention of late eighteenth-century science, sketching in some final considerations on the interrelations of science, technology, and cultures in the early age of electricity.[6]

## "Useful Knowledge" and Unintended Consequences

In the 1760s, when young Volta began cultivating electricity and mature Beccaria expressed the view that one day electricity would be turned into something deserving the attention of giants, only scant evidence supported the expectation. It rested mainly on Franklin's lightning rod, and on the controversial claims of medical electricity. The then recent fuss about Symmer's electrified socks, on the other hand, could well have confirmed the view of many lay observers that electricians were wasting their time dealing with pigmy worlds and irrelevant problems. All the more so because in the meantime other branches of physics were busy in ostensibly cyclopean ventures, like measuring the meridian and assessing the shape of the earth, to the satisfaction of those enlightened rulers and learned elites who delighted in the philosophers' well-publicized achievements.

Those expectations electricity aroused in Beccaria in the 1760s, and that sustained Volta throughout his career, still rested on similarly uncertain terrain around 1800, when the battery was introduced. Yet, judging from the reactions to Volta's new instrument by technologists like William Nicholson and generals like Bonaparte, the battery *was* perceived as capable of somehow fulfilling those long-cherished expectations, which finally turned into reality in the middle decades of the nineteenth century. The history of the multifaceted connections linking Enlightenment culture, instruments like the battery, and what we now call science and technology, is in an important respect the history of the unpredicted avenues through which the expectations nurtured by eighteenth-century natural philosophers like Franklin, Beccaria, and Volta came to be fulfilled. The story told in the present book shows, in fact, that those avenues were interspersed with elements of both deliberate intention and chance: intention took the form of the powerful, guiding forces provided by the Enlightenment notions of "useful knowledge" and "the quantifying spirit"; chance took the form of unintended consequences.

Let me take "useful knowledge" first. Enlightened natural philosophers like Volta were imbued with notions implying that knowledge, especially natural philosophy, must be useful.[7] Happily, they had an idea of usefulness broad enough to accommodate the fact that, for a long while, the science of electricity gave little sign of having any major useful application. It was above all the curiosity of natural philosophers, sustained by the powerful Enlightenment expectation that knowledge *must* be useful, that nurtured the growth of studies in electricity in the period spanning from Franklin to Volta, and beyond. As for the battery itself, we know that when it was first presented to the Royal Society in London and the Institut National in Paris, its claims to utility were on the whole modest. In 1800, of course, not one of the thousand devices now needing an electric battery had yet been introduced: by our own, early-twenty-first-century standards the voltaic battery was almost a *useless* machine.

The present study of Enlightenment and the battery also shows that the useful applications of the battery developed in fields that scientists themselves had *not* anticipated. This leads to my second point: chance, in the form of unintended consequences. It has been seen throughout this book that Volta expected his battery to be applied above all to medicine; though other contemporaries, living in what was quickly becoming the age of steam, expected electricity and electromagnetism to be applied to motion, mechanics, and engines of various kinds. Contrary to their expectations the early successful applications of the vol-

*[margin annotation:]* battery "useless" at first

taic current developed instead in fields like chemistry and communications, with the telegraph.

How are we to account for the natural philosophers' inability to predict *how* and *where* their "useful knowledge" was going to become really useful? How do unintended consequences arise in a field like science and technology that we have been taught to regard as a well-disciplined field of human endeavor? And note that even today, in what we regard as a mature technological age, scientists and technologists do not perform much better than their late-eighteenth-century predecessors in predicting the useful applications of scientific and technological research.

I think the answer to these questions is quite simple, but it requires some significant changes in our current view of the enterprise we call science and technology. We must accept the notion that diversity (as well as discipline) acts powerfully within science and technology just as in other fields of comparatively free human endeavor: we must recognize that in science and technology too diversity is an endless source of unintended consequences. The example of Volta and the battery is revealing.

The inventor of the battery, we know, was from Lombardy, then an Italian province of the Austrian Empire, and he had his deep personal and cultural roots there. Through frequent travel and correspondence, however, he had spent thirty-five years prior to his major invention in close touch with a wide and varied network of natural philosophers, instrument makers, amateurs, and patrons from at least seven different European countries. Despite being Italian and a subject of the Austrian Empire for fifty-five years, in 1800 he decided to announce his major invention first to the British, the people he liked most as a natural philosopher, a citizen of the Republic of Letters, and a traveler. He addressed the Royal Society of London, however, using the French language, since that was then the "lingua franca" of the natural philosophers, as well as of the aristocracy and the merchant classes throughout Europe. At the same time, he made personally every possible effort to win followers and secure rewards for his invention in at least five different countries. The battery was thus exposed to a multiplicity of different cultural traditions, milieus, and interests, just as its inventor had been. No surprise, then, that the public career of the battery followed paths that the inventor himself was unable to anticipate and control.

The unintended consequences observed in the early history of the voltaic battery confirm that diversity—side by side with the discipline nurtured by the Enlightenment notions of "useful knowledge" and "the quantifying spirit" (see below)—plays as crucial a role in scientific culture. Emphasis on diversity and on unintended consequences, may seem surprising to those (especially in the natural sciences and technology) who are supposed to be taming diversity as part of their job. Yet, the story of the battery overwhelmingly shows that discipline *and* diversity are major assets also in that part of culture we call science and technology.

## "The Quantifying Spirit"

A major characteristic of the science of electricity in the second half of the eighteenth century was the effort made to quantify electrical phenomena. The works of Aepinus, Cavendish, and Coulomb, as well as Volta, stand out as emblematic in this respect. They belong to a pervasive, increasingly successful tradition that historians have singled out and explained as the product of "the quantifying spirit" of the eighteenth century. Some by now classical historical studies of the science of electricity during the Enlightenment have accordingly emphasized that tradition as a major trend.[8] These studies have often implied that, by endorsing and advancing "the quantifying spirit," the best electricians of the late Enlightenment were joining in a common tradition and were paving the way for the sort of standard mathematical physics that became customary in the nineteenth century and beyond. Caution was occasionally expressed in those historical studies to the effect that, in reality, the quantifying spirit of the eighteenth century did "include the passion to order and systematize as well as to measure and calculate," and that, above all, " 'mathematical' or 'geometrical' did not mean the same thing to all parties."[9] These important qualifications notwithstanding, those historical studies could encourage the view that a unilinear march linked the works of Aepinus to those of Cavendish, Volta, and Coulomb, and that the works of Biot and Poisson stemmed from these, to the direct accretion of the later formulations of a fully mathematized science of electricity.

The case of Volta and his contributions to the science of electricity discussed in this book invites caution and additional qualifications to the picture offered by the classical accounts of the quantifying spirit in the eighteenth century. Caution seems mandatory in the light of the diverse cognitive strategies and the conflicts that we have observed at work especially in the interaction between Volta and Parisian natural philosophers in 1782, Volta and Coulomb later in the 1780s, and Volta and Biot later still. According to the reconstruction developed in this book (chapters 4 and 7, above), the diversity of the goals pursued and the means adopted by people like Volta, Coulomb, and Biot were just as important as the similarities in the language, and in the experimental and mathematical tools that allowed them to interact, albeit with the difficulties that were manifest to all the parties involved.

Volta's momentous contributions to the science of electricity in the 1780s and 1790s, despite his ignorance of the calculus and opposition to Coulomb, and his invention of the battery are sufficient proofs that the quantifying spirit nurtured by electricians could assume several distinct forms, and that the quantifiers themselves pursued quite different agendas. The decisive role that sophisticated measuring techniques, developed in the hunt for weak electricity stimulated by galvanism, played in the invention of the battery, and the negligible role mathematization had in that same invention and in its early impact, seem to indicate that around 1800 (and beyond) the quantifying spirit harbored more diversity in it than we may be prepared to concede. Certainly, if we judge from the voltaic battery's spurious origins in the controversy over galvanism and the "imitate the electric fish" program, the quantifying spirit paid more rewarding dividends when, as in Volta, it managed to accommodate diversity of interests and widespread curiosity, extending into topics regarded by others as being on the verge of scientific unorthodoxy.

Judging from the story of Volta and the battery, there was more diversity at work in the science of electricity around 1800, and in the quantifying spirit itself, than the mathematical physics of Coulomb and Biot conveyed.

## Investment

As we know, when Volta built his first batteries in the fall of 1799 the University of Pavia was closed. After having taught there for twenty-one years, Volta had been suspended in the spring, like many other professors. The suspensions had been decided by the Austrian government on regaining control over Lombardy after a three-year French occupation. As usual in those years, the situation was soon to be reversed. Bonaparte regained control over Lombardy less than one year after the announcement of the battery, and he soon accorded Volta some very special rewards, for reasons that included strictly political motivations and a genuine (if amateur) interest in scientific and technological developments. Having employed Volta as a professor for over twenty years, however, the Austrian government of Lombardy could rightly claim to have been a long-term and munificent investor in Volta's abilities. The questions are: In what exactly had the Austrians been investing over the years, and what role did Volta's ability as an inventor of new, intriguing machines play in the motives behind the investment?

At turning points in his career Volta had recourse to ingenious instruments that he had invented or improved. He had first caught his patrons' attention with a curious, if not very effective, electrostatic machine made entirely of wood. He then came up with the electrophorus, an electric pistol, a eudiometer, an "electric lamp" (i.e., a gas lamp lit by a spark), etc. If we draw up a balance sheet of the transactions between Volta and the Austrian government, however, we find that the instruments he invented played quite a marginal role. It seems clear that the government was investing in Volta, first and foremost, as a teacher, then as a zealous public servant, and only thirdly as an inventor of new machines. Going through the records, one has the impression that the government, while generally nurturing expectations about utility, simply did not know what sort of useful applications to expect from the kind of physics Volta was practicing. The educational system was, admittedly, interested in utility. A well-established rhetorical tradition celebrated utility and useful instruments. On these grounds, fields traditionally given credit for their usefulness, like medicine and natural history, were granted higher status than physics. The physics professor at Pavia, despite Volta's prestige, effort, and achievements, earned about 20 percent less than the anatomy and natural history professors, and 25 percent less than the clinical professor.[10] Furthermore, while the physicians were being urged by administrators to improve their scientific collections, in the case of physicists they themselves had to press for more funds and facilities in a field that was displaying unprecedented, if little understood, vitality.

One could say that, in the case of Volta, the Austrian government of Lombardy made no specific investment in invention; not at least in invention as distinct from the pursuit of knowledge and education. Which leads to another point: value assessments.

## Value Assessments

In Volta's own writings, one finds mention of batteries varying in value from 36 lire to 2,000 lire.[11] Thirty-six lire was the sum needed to buy in Milan twelve popular books. Two thousand lire corresponded to about one-third of Professor Volta's yearly salary. The variation in price depended largely on the actual size of the battery. In fact, one has the impression that the number of discs and the quality of the metals employed were the sole criteria for setting its price. In a sense, for Volta and his contemporaries the value of the battery was the value of the metals used to build it. As we know from chapter 7, above, Volta was not interested in patenting the battery and, as his letters to the Royal Society made plain, easy replication was one of the chief goals he was pursuing on first describing the new instrument. Replication by others was indeed a requisite for the kind of recognition as a natural philosopher that Volta expected from his colleagues. Moreover, the battery did not lend itself to the sort of embellishment other scientific instruments traditionally enjoyed (though the great elector of Saxony was presented with a battery made of silver and gold).[12]

The estimated cost of the battery, however, was only a small part of the value attributed to it. Value assessment, in a broader sense, should include the rewards the battery earned its inventor. Volta's own earliest documented request for a reward mentioned three points. First, he had discovered "a new principle," implying the demise of the opposite principle of animal electricity. Second, the apparatus he had invented could be "of interest" to physicians and chemists alike. Third, the invention had won fame and honor for its author and the institution with which he was associated.[13] The battery itself and its potentially useful applications did not figure very highly in Volta's own perception of his achievement.

What about other people's value perceptions? For an answer to this question we may turn to Volta's most famous and generous patron after 1800: Bonaparte. The first of the rewards he accorded to Volta, it will be recalled, was a gold medal of the Institut. The statement accompanying this medal was blatantly political, rather than scientific. It proclaimed that Volta, who had personally presented the battery in Paris in 1801, had been the first foreign scientist to address the Institut after the signing of the Lunéville peace treaty with Austria. The reward was intended to show that the Institut was eager to welcome contributions from foreign scientists in times of restored peace, and to reassure European elites about the nature of Bonaparte's ambitions. Scientific issues figured more prominently in the reasons behind the second of the rewards accorded to Volta in Paris. Conferring on Volta a 6,000-franc gratuity, his colleagues at the Institut mentioned, first, Volta's long-lasting devotion to what they called "the useful truths."[14] They then pointed to the fact that he had chosen the Institut as the appropriate place for "revealing the secrets of galvanism": again, no specific mention was made of the apparatus that we—from our own present-day, overly technological perspective—regard as Volta's main achievement.

The uncertainties manifest in both Volta's and his contemporaries' appraisals of achievements like the battery confirm that the community of electricians around 1800 was made up of a variety of actors, endorsing a wide range of criteria for the assessment of scientific merit and value. Volta, we know, was part of a mixed community in which the traditional figure of the natural philosopher, to which he as a university professor aspired, sat side by side with independent lecturers, inventors, instrument makers, physicians, and amateurs of many kinds. Individual sectors of this variegated community presumed they knew how to rank the members and the value of their contributions, and they thus contributed to shaping the ambitions and the investigative styles of the electricians seeking recognition, including Volta.

As shown in chapters 1, 4, and 5, above, several distinctive features of Volta's work were produced by his adjusting his ambition to be regarded as a genuine natural philosopher to the more modest role of inventor of intriguing machines that some of his colleagues assigned him. Similar adjustments, on the other hand, could affect the perceived theoretical dignity (and "value") of the notions Volta was introducing in the science of electricity. The dubious status that Volta's contemporaries ascribed to the concepts stemming from Volta's laboratory practice and machines, can explain Volta's stunted attempts at imposing his own conceptual system on the science of electricity, even after the big impact of the battery.

Around 1800, apparently, the values and concerns of traditional natural philosophy still held sway in the appraisal of theoretical worth and expert status within the community of electricians. Those traditional values coexisted at variance with the new emphasis on utility, ingenuity, efficacy, and machines promoted by Enlightenment literature, and pursued within the reformed educational and technical institutions created by governments and patrons inspired by that literature. The uncertain status enjoyed by instruments like the battery, and by Volta's own midrange notions of tension, capacity, "actuation," and "contact" electricity, was a product of this mixed system of values.

In the case of Volta, the mixed system of values just described was the product of geography, as well as intellectual and social contexts. Living in a provincial setting where no expert could seriously compete with him in his major field of expertise, Volta felt part of a cosmopolitan network in which different ranks of competence and different research traditions could merge more pervasively than in any single, well-established, and controlled research center. In the case of Volta, the variety of intellectual stimuli deriving from this combination of provincialism and cosmopolitanism was added to the stimuli proper to the mixed system of values governing the community of late-eighteenth-century electricians. Navigation through this complex reality was admittedly difficult, as can be seen by the occasional frustrations from which Volta suffered. Yet, clever and determined investigators like him could find in this navigation the right combination of stimuli, freedom, and constraints that led to achievements like the battery.

The economy of invention described so far was not particularly concerned with scientific instruments *per se*. In the system of directives and rewards that the Austrians, the French, and later the Italians themselves pursued in Lombardy, the inventions of physicists like Volta were assessed against traditional values such as the pursuit of knowledge and education. As shown in chapters 7 and 8, above, instruments like the battery were rewarded and celebrated, first and foremost, as generic symbols of achievement. As such they could be attributed functions, according to need, which went beyond knowledge and education, and could easily serve diplomacy and politics. Within a not strictly utilitarian system of symbols and rewards, on the other hand, recompense could be no less concrete than in a system dominated by utilitarian concerns. Experts like Volta, at any event, shared with their patrons and audiences this traditional system of appreciation of their work.

## Contingency

From what has been suggested under the rubrics "investment" and "value assessments," it might be assumed that, while the Austrian government of Lombardy and Bonaparte were not sure where enterprising physicists like Volta were heading, the physicists themselves knew. This was only partly the case, however. Several levels of contingency have emerged from the present study concerning Volta's science of electricity, the origins of the battery, and the latter's career on the public scene.[15] The fact that the contingency elements detected throughout were intertwined with events displaying long-term trends within the science of electricity and/or Volta's own personal endeavors should not induce to lose sight of them. The battery was indeed invented by a natural philosopher who, around 1800, was among the top handful of competent electricians in Europe, and who had then been practicing the science of electricity for some thirty-six years. These circumstances, however, should not conceal from view the important signs of unpredictability that marked Volta's own path to the battery and its early public career.

As discussed in detail in chapter 6, there is no evidence that Volta was pursuing the construction of an instrument like the battery, showing his long-cherished theory of contact electricity at work, until a few months before his building the instrument. Despite what was by then Volta's excellent and long-standing record as an electrician, the battery

was the outcome of a comparatively quick and unexpected turn of
events. This is shown by the important, precipitating effect played by
Volta's reading of Nicholson's suggestions (unrelated to the theory of
contact electricity) on how best to imitate the electric fish by means of
an electrical and mechanical device. The role of Nicholson in Volta's
path to the battery is a sign of contingency not to be forgotten, as it
reminds us that Volta's purpose while building the battery—however
strange it may sound to physicists and technologists today—was to
create an apparatus that could imitate the electric fish. In Volta's view,
such an apparatus would deal a mortal blow to animal electricity and
galvanism. As emphasized in chapters 6 and 7, above, there was no
single or privileged path leading Volta from his contact theory of electricity—the core of his natural philosophy in the 1790s—to the battery.
Much less could his path to the battery be, as it were, deduced from or
predicted by theory alone as some historians have tended to assume.

The battery was *not* inscribed in any predictable chain of events dictated by the science of electricity around 1800. The contingencies nurtured by Volta's own widespread curiosity, and by his frequent dealings with the mixed population of expert and amateur electricians
from several countries he was in touch with, both played a role in the
first introduction of the new machine. Once built, moreover, the impact
and significance of the instrument went immediately beyond the controversy over galvanism, and far beyond Volta's own program and intentions in building it.

In fact, observing Volta's demeanor in expert and amateur circles of
electricians in the twenty-five years prior to the invention of the battery, one is impressed by the diversity of the strategies adopted in the
pursuit of knowledge of electrical phenomena within those circles.
Throughout Volta's own career, an ability to come to terms with this
diversity and exploit it for his own benefit was a major achievement
on his part. By dealing with British, French, German, Austrian, Dutch,
Belgian, as well as Italian colleagues at a time, Volta knew that the philosophers' republic was far from constituting a single community. It
was more like an informal federation of different constituencies,
molded according to different hierarchies of recognized expertise, as
well as by national and social differences. As shown especially in chapters 3 and 4, above, the diversity of this mixed community was just as
important in shaping Volta's science of electricity, as were the proclaimed universality of the scientific enterprise and the common goals
pursued by expert electricians. It was precisely the diverse, competitive, occasionally conflicting interests of the different constituencies of
the republic of natural philosophers that offered an outsider like Volta
his best opportunities to innovate and to elicit reward.

When compared with the strong research programs pursued within a few, closely knit groups of experts, like those Volta met with in Paris in 1782, and again in 1801, Volta's program for the science of electricity may indeed appear weak, loosely knit, and eclectic. Yet, it had its peculiar strengths. One such strength, as already emphasized, was curiosity: a curiosity unrestricted by the usually severe rules of exclusion enforced by strong research programs and expert groups. Volta was usually able and willing to compare his notions and results with others, whether they were the independent lecturers and instrument makers he liked to mix with in London, or well-established Parisian authorities like Lavoisier or Laplace, or physicians suspected of being on the verge of scientific unorthodoxy like Galvani, or mathematical physicists à la Coulomb and Biot, or even amateurs and quacks, like Robertson.

According to the present study, the intertwined issues of curiosity, diversity, and contingency are basic but important ones. They contain a lesson for historians. The lesson is that perhaps the single major strength of the science of electricity as practiced around 1800 was its diversity: a diversity nurtured by different cultural traditions merging somehow, in the work of people like Volta, under the Enlightenment banners of "useful knowledge" and the "quantifying spirit." Historians of science should recognize rather than downplay that diversity. If we want to recapture the wealth of interests, goals, and actors that inhabited the republic of natural philosophers around 1800, we must resist the homogenizing efforts imposed by several generations of physics textbooks, by normative philosophies of science, and by unilinear histories of science. Diversity and contingency were just as important as the Enlightenment ideal of "useful knowledge" and the "quantifying spirit" in bringing about the battery.

Diversity and contingency are even more manifest in the ways the battery was appropriated soon after its introduction. These are well illustrated by the rapid transformations the battery underwent from a strictly physical instrument, as it was conceived by its inventor in the fall of 1799 in Lombardy, into a mainly chemical apparatus, as in the hands of Nicholson the following spring in London, and into a mathematised physical instrument, as it was for Biot (inspired by Laplace) toward the end of 1801 in Paris.

These rapid changes, unforeseen by the inventor, can be explained of course in terms of the different communities of experts, driven by different disciplinary cultures. They were no doubt a consequence of the "contingency of community" affecting the life of human artifacts, often remarked upon by social constructivist interpreters of science and technology.[16] Within the perspective adopted in the present book, however, that combines elements of realist and constructivist approaches, the rapid changes in the battery's significance in 1800–1801 offer evidence also of another level of contingency, acting at the intersection between instruments and natural phenomena.

Within the realist and constructivist perspective adopted here, it goes without saying that Volta, who in the battle against galvanism regarded chemical explanations as unacceptable and pursued strictly physical interpretations of electrical phenomena, "could not even see," and at any rate downplayed the chemistry of the battery. It also goes without saying that Nicholson, on the other hand, had several good reasons for seeing the chemical phenomena of the battery that he quickly saw. Within the perspective adopted here, however, the chemical developments unforeseen by Volta and soon perceived by Nicholson reveal a level of contingency that goes beyond the experts' community: it is the level of contingency that pertains to the interaction—itself potentially rich with unexpected developments—between the experts and a material world that, with instruments like the battery, they try to intercept and bend to their needs.

The several degrees of contingency thus detected in the story of the battery and its early public career may seem surprising, especially comparing the accounts usually offered by historians of science and technology. This leads to our final point: Enlightenment legacies.

## Enlightenment Legacies

As we have seen throughout this book, around 1800 physicists like Volta, who had grown up within the values of classical natural philosophy and the Enlightenment, could not fully grasp what was involved in products like the battery, created by the pragmatic ethos that they had themselves contributed to shape. In Volta's writings and career one finds only hints at the ties that (in the eyes of later generations) were to connect his work and instruments to the enterprise we call science and technology. With hindsight, one such hint is revealing, however. On the very same day when Volta ordered via his bookseller in Leipzig a subscription to Nicholson's journal, which put him on the track to the battery, he also ordered a new edition of the first book ever published to contain the word *technology* in its title. It was Johann Beckmann's *Anleitung zur Technologie* ("Instructions on Technology").[17] Another similar episode is that, when at the University of Pavia they decided (some twenty years after the introduction of the battery) to add a course of "Tecnologia" to the syllabus—to be combined, as it seemed appropriate, with the chair of natural history—they had Volta sitting in the search committee, and they appointed a pupil of his for the job.

Better than Volta's own figure and career, however, it was the steady current provided by the battery, with its applications to chemistry and electromagnetism, that was to give to later generations a measure of what traditional natural philosophy, pushed by the utilitarian emphasis and the quantifying spirit of the Enlightenment, and further combined with the commitments of natural philosophers turned professors, inventors, zealous civil servants, entrepreneurs, and scientists, could achieve. That we live in societies still molded, via instruments like the battery, by the kind of mixed scientific and technological endeavors that pushed people like Volta and his fellow late-eighteenth-century natural philosophers is something often granted, though—we have contended throughout this book—rarely understood in its proper terms.

Despite the difficulties involved in the assessment of the work of people like Volta, or of instruments like the battery, the major interpreters of the Enlightenment have been well aware of the continuing influence linking the legacies of the Enlightenment discussed in this book with later industrial societies.[18] The interpreters of the Enlightenment, however, have systematically removed from view the amount of diversity and unintended consequences that have acted (and apparently continue to act) in the supposedly well-disciplined world of science and technology typical of our industrial societies. The removal of diversity and contingency has had momentous consequences. It has led both the supporters and the critics of the Enlightenment to perpetuate two contrasting, but symmetrical misrepresentations of its heritage. The supporters of the Enlightenment have tended to impose on their predecessors the notion (which not all of them shared[19]) that the enterprise we call science and technology is a strictly rational, ordered, and purposeful enterprise: a model to be imitated in all other human endeavors if we want to control disorder, enforce efficiency, and favor "progress." The critics of the Enlightenment, on the other hand, have exposed that same view of science and technology as a diversity-free and contingency-free enterprise, as if it confirmed the vicious ambition of the founding fathers and followers of the Enlightenment to repress diversity and control society in the name of elites deceitfully presenting themselves as pursuing rational policies modeled on the natural sciences.

Within the perspective adopted throughout this book, recognition of the many levels of contingency that affect the enterprise we call science and technology is a necessary emendation of the conception of science and technology handed on to us by the Enlightenment tradition.

Recognition of the role played by diversity and contingency in the history of science and technology should make us appreciate another, connected circumstance. Contrary to the messages conveyed by several forms of scientism, normative philosophies of science, and unilinear histories of science, since the early times of the age of electricity the main instrument we have had to encourage science and technology is to let a plurality of experts, actors, and differently interested people interact freely both among themselves and with natural phenomena. As often remarked throughout this book, science and technology is a highly imitative and competitive game, which involves a constant effort at appropriating and reinventing somebody else's results and achievements: this is what Volta did with Franklin, Nollet, Beccaria, Galvani, and Nicholson; and this is what Nicholson, Davy, Ørsted, and Faraday in turn did with Volta. Breakthroughs like the battery occasionally take place in the middle of this imitative and competitive game, providing novelty and unintended consequences to further fuel the exchange.

It is within this imitative, competitive, entangled game, shaped by diverse interests and unintended consequences, that the Enlightenment mentality emphasizing useful knowledge and the quantifying spirit managed to harvest the huge dividends symbolized by the battery and its aftermath in industrial societies. In ensuring these fruits, Enlightenment notions and directives have acted as a kind of sieve: an informal system for sorting out whatever was potentially useful amid the new knowledge acquired. The sieve, however, was half-blind: none of Volta's contemporaries apparently could (just as nobody can now) see with certainty which among the new knowledge acquisitions and the often useless machines would turn out to be useful, nor where, nor how. Through such a process—in which the intended pursuit of useful knowledge and the quantifying spirit have been merging constantly with the unintended consequences of diversity and invention—some of the useless machines produced by people like Volta did become useful, mostly in fields unforeseen by the inventor.

If there is a moral to be drawn from the story told in this book, then it is that we should adopt a view of science and technology able to accommodate both the intended pursuit of useful knowledge and the quantifying spirit, and the unintended consequences of invention. Whether we like it or not, the success of the enterprise we call science and technology seems to depend no less on the diversity of the interests and goals of the people involved, and on the contingency attached to diversity, than on the discipline adopted in its pursuit. Whatever "progress" we are willing to recognize in the products brought about by the legacies of the Enlightenment, we must recognize that those same legacies could not (and cannot) enforce any strictly oriented direction in the history of science and technology, whatever some followers of the Enlightenment may wish, and its detractors complain.[20]

If so, the main strategy available to foster science and technology—and the most valuable among the legacies of the Enlightenment for industrial societies—consists in encouraging the free circulation of people, ideas, instruments, and practices in a wide variety of different cultural contexts. Because, as that public officer assured Volta as he was on the point of leaving for a trip abroad to circulate the battery, scientific pursuits benefit immensely from the "fast communication of enlightenment."[21]

# N O T E S

## Introduction

1. With qualifications. For a survey of Enlightenment historiography, from a perspective akin to the one adopted in the present book, see Roy Porter, *Enlightenment: Britain and the Creation of the Modern World* (London: Allen Lane, The Penguin Press, 2000), especially the Introduction and chapter 1. See also Dorinda Outram, *The Enlightenment* (Cambridge: Cambridge University Press, 1995); G. Ricuperati (ed.), *La Reinvenzione dei Lumi* (Florence: Olschki, 2000); and the literature discussed in chapter 9, below.

2. John L. Heilbron, "Experimental Natural Philosophy," in G. S. Rousseau and Roy Porter (eds.), *The Ferment of Knowledge: Studies in the Historiography of Eighteenth-Century Science* (Cambridge: Cambridge University Press, 1980), pp. 356–87, 368. See also John L. Heilbron, *Electricity in the Seventeenth and Eighteenth Centuries: A Study in Early Modern Physics* (Mineola, New York: Dover Publications, 1999), pp. 134, 500; idem, "The Measure of Enlightenment," in Tore Frängsmyr, John L. Heilbron, and Robin E. Rider (eds.), *The Quantifying Spirit in the Eighteenth Century* (University of California Press, 1990), pp. 207–42.

3. A famous motto circulated by the French economist and administrator Turgot asserted that Franklin had "snatched the lightning from the sky and the scepter from the tyrants" ("Eripuit coelo fulmen sceptrumque tyrannis"), thus combining an appreciation of Franklin's science of electricity with his political endeavors. See I. Bernard Cohen, *Science and the Founding Fathers: Science in the Political Thought of Jefferson, Franklin, Adams, and Madison* (W. W. Norton & Company, 1995), pp. 184–86. See also Jessica Riskin, "Poor Richard's Leyden Jar: Electricity and Economy in Franklinist France," *Histor. Stud. Phys. Biol. Sci.*, 28 (1998): 301–36.

4. For a discussion of the conflicting interpretations of the Enlightenment, and substantial new contributions towards an understanding of the sciences in enlightened Europe, see William Clark, Jan Golinski, and Simon Schaffer (eds.), *The Sciences in Enlightened Europe* (Chicago: The University of Chicago Press, 1999), esp. the Introduction, and Dorinda Outram, "The Enlightenment our Contemporary" (ibidem, pp. 32–40).

5. A survey of the controversies in Ian Hacking, *The Social Construction of What?* (Cambridge, Mass.: Harvard University Press, 1999). Important new contributions, often blurring the realism-constructivism opposition, in Lorraine Daston (ed.), *Biographies of Scientific Objects* (Chicago: The University of Chicago Press, 2000). Hints at new equilibria between constructivist and realist perspectives also in: Mario Biagioli, "From Relativism to Contingentism," in Peter Galison and David J. Stump (eds.), *The Disunity of Science: Boundaries, Contexts, and Power* (Stanford: Stanford University Press, 1996), pp. 190–206; Bruno Latour, *Pandora's Hope: Essays on the Reality of Science Studies* (Cambridge, Mass.: Harvard University Press, 1999); Mary P. Winsor, "The

Practitioner of Science. Everyone Her Own Historian," *J. Hist. Biol.*, 34 (2001): 229–45.

6. As (loosely) defined by Jed Z. Buchwald and Sylvan S. Schweber, "Conclusion," in Jed Z. Buchwald (ed.), *Scientific Practice: Theories and Stories of Doing Physics* (Chicago: The University of Chicago Press, 1995), pp. 345–51. A form of pragmatic realism is advocated also by Andrew Pickering, *The Mangle of Practice: Time, Agency, and Science* (Chicago: The University of Chicago Press, 1995), pp. 180–85.

7. An excellent, sympathetic, but critical survey of constructivism is Jan Golinski, *Making Natural Knowledge: Constructivism and the History of Science* (Cambridge: Cambridge University Press, 1998).

## Chapter 1
## The Making of a Natural Philosopher

1. For an assessment of scientific biographies as a genre: R. Yeo and M. Shortland (eds.), *Telling Lives: Studies of Scientific Biographies* (Cambridge: Cambridge University Press, 1995). A survey of some major contributions to the genre, and some penetrating remarks, in T. Söderqvist, "The Architecture of a Biographical Pathway," *Hist. Stud. Phys. Biol. Sci.*, 25, part 1 (1994): 165–75.

2. F. Venturi, *Settecento riformatore*, V, *L'Italia dei lumi (1764–1790)*, I (Torino: Einaudi, 1987), pp. 425–834.

3. The best biography is still C. Volpati, *Alessandro Volta nella gloria e nell'intimità* (Milano: Treves, 1927). See also C. Volpati, *Scritti voltiani*, ed. V. Lucati (Como: Comune di Como, 1974); J. L. Heilbron, "Volta," *DSB*, 14, 69–82; F. Venturi, *Settecento riformatore*, V: 725–41.

4. See Joseph II's report in F. Valsecchi, *L'Assolutismo illuminato in Austria e in Lombardia* (Bologna: Zanichelli, 1931–34), 2:323–24. For a survey of the history of Lombardy in the eighteenth century, see D. Sella and C. Capra, *Il Ducato di Milano dal 1533 al 1796* (Torino: UTET, 1984).

5. C. Cantù, *Storia della città e della diocesi di Como* (Florence: Le Monnier, 1856), 2:218ff. For a penetrating description of Como in the eighteenth century: F. Venturi, *Settecento riformatore*, 5:699–747. For an eighteenth-century survey of key figures in Como's life: G. B. Giovio, *Gli uomini della comasca diocesi* (Modena, 1784). See also B. Scolari, "Como nell'età di Volta," *Periodico della Società Storica Comense*, n.s. 7 (1951): 71–94; S. della Torre, "Lo sviluppo di Como tra 700 e 800," in *Omaggio a Volta nel centocinquantenario della morte* (Como: Nani, 1978), 71–84.

6. S. Bettinelli, *Opere*, 8 vols. (Venice, 1780), 2:249.

7. B. Caizzi, *Industria, commercio e banca in Lombardia nel XVIII secolo* (Milan: Banca Commerciale Italiana, 1968), pp. 116–19, 239, 248. On the technology involved in the silk industry, see C. Poni, "All'origine del sistema di fabbrica: tecnologia e organizzazione produttiva dei mulini da seta nell'Italia settentrionale (Sec. XVII–XVIII)," *Rivista storica italiana*, 88 (1976): 444–96.

8. See M. Romani, "La ripresa dell'agricoltura nel fervore delle riforme (1760–1786)," in *Storia di Milano*, 16 vols. and Index (Rome: Enciclopedia Italiana, 1953–66), 12, 525–43.

9. On Lombard nobility in the eighteenth century, see the fundamental essay by J. M. Roberts, "Lombardy," in A. Goodwin (ed.), *The European Nobility in the Eighteenth Century* 2nd ed. (London: Black, 1967), pp. 60–82. See also: U. Petronio, *Il Senato di Milano: Istituzioni giuridiche ed esercizio del potere nel Ducato di Milano da Carlo V a Giuseppe II* (Rome: Giuffrè, 1972); F. Pino, "Patriziato e decurionato a Milano nel secolo XVIII," *Società e storia*, 5 (1979): 339–78; E. Rotelli, "Gli ordinamenti locali della Lombardia preunitaria 1755–1859," *Archivio storico lombardo*, 100 (1974): 171–234.

10. "Catalogo de' regolari di Città, & Diocesi," Archivio Storico della Diocesi di Como, MS dated November 1767.

11. Gattoni in *VE*, 1:1, and Z. Volta, *Alessandro Volta* (Milano: Civelli, 1875), p. 36.

12. Z. Volta, *Alessandro Volta*, p. 37; C. Volpati, *Alessandro Volta*, p. 202.

13. G. G. Brenna, "Pranzi di carnevale in casa del conte Filippo Maria Volta," *Omaggio a Volta nel centocinquantenario della morte* (Como: Nani, 1978), pp. 85–91.

14. Filippo Volta was no longer alive on 28 June 1756, when Nicolò Stampa dictated his will mentioned below (ASC, Notarile, Gio. Batt. Stampa, Gravedona).

15. *VE*, 1:27–28.

16. Antonio Maria Odescalco, "Diverse notizie spettanti al Rv.mo Capitolo della Cattedrale di Como," BCC, MS 2.3.19, pp. 483, 493, and *VE*, 1:68.

17. See *VE*, 3:88.

18. See *VE*, 3:229, 233.

19. *VE*, 2:412.

20. Maddalena Inzaghi to Carlo Manzi, 27 July 1757, 25 Aug. (?) 1758, 2 July 1760, Dec. (?) 1764, Volta Family Papers in VMS.

21. *VE*, 2:144.

22. ASC, Notarile, Gio. Batt. Stampa, Gravedona, 28 June 1756.

23. ASC, Notarile, Gio. Batt. Stampa, Gravedona, 13 July 1756.

24. ASC, Notarile, Giacomo Filippo Clerici, 12 August 1794 (no. 349, encl.).

25. A list in *VE, Indici*, 2:792.

26. *VE*, 2:61. See also *VE*, 2:141.

27. See the letters exchanged between Volta and his Jesuit professor of philosophy, partly published in *VE*, 1:17–33. The originals are kept in ASM, PADD, 9. On Lombard educational institutions in the eighteenth century, see A. Annoni, "Cultura e scuola nell'epoca del Volta," in *Conferenze Voltiane* (Como: Società Storica Comense, 1978), pp. 69–124.

28. See Gattoni in *VE*, 1:3, and Z. Volta, *Alessandro Volta*, p. 69, who disagree, however, on the name of the seminary to which Volta was sent on leaving the Jesuit College.

29. E. Chinea, *L'istruzione pubblica e privata nello Stato di Milano dal Concilio tridentino alla riforma teresiana (1563–1773)* (Florence: La Nuova Italia, 1953).

30. *Ibidem*, 12. The plan is discussed in detail in A. Annoni, "Cultura e scuola nell'epoca del Volta."

31. Z. Volta, *Alessandro Volta*, p. 48.

32. See the printed "Catalogus personarum, et officiorum provinciae medio-lanensis Societatis Jesu, Anno 1761, . . . ," kept in ASSI, and the MS signed "Med. 69, Mediol. Cat. 1761–1764," also in ASSI. On the Jesuits in Como: G. Rocchi, *Profilo storico della chiesa di S. Ananzio e del Gesù in Como* (Como: Noseda, 1968). See also: *I gesuiti a Milano* (Milano: San Fedele, 1984); S. Di Bella, *Chiesa e società civile nel Settecento italiano* (Milano: Giuffré, 1982).

33. For surveys of the educational system of the Jesuits, see A. P. Farrell, *The Jesuit Code of Liberal Education* (Milwaukee: Bruce, 1938), and especially F. de Dainville, *L'éducation des Jésuites (XVIe–XVIIIe siècles)* (Paris: Les Editions de Minuit, 1978), where pp. 311–423 are devoted to science teaching. On physics and the Jesuits, see J. L. Heilbron, *Electricity in the Seventeenth and Eighteenth Centuries* (Mineola, N.Y.: Dover Publications, 1999), pp. 180–92. Valuable recent assessments include Mordechai Feingold, ed., *Jesuit Science and the Republic of Letters.* Cambridge, Mass.: MIT Press, 2002.

34. The printed catalogue mentioned in note 32, above, says that Girolamo Bonesi was born on 14 November 1701 and that he had entered the Society of Jesus on 13 October 1742. According to the same source, from 1763 he taught at the Jesuit College in Cremona. If I am right in decoding the system adopted to keep the evaluation of the professors covered, Bonesi's superiors in the College of Como judged that "judicium" and "prudentia" were more prominent among Bonesi's qualities than "ingenium," though they credit him "talenta" (ASSI, MS Med. 69, Mediol. Cat. 1761–1764).

35. *VE*, 1:19; Gattoni in *VE*, 1:3. On animal souls, see K. Thomas, *Man and the Natural World* (London: Allen Lane, Penguin Books, 1983), 137–42.

36. Gattoni, loc. cit.

37. Farrell, *The Jesuit Code of Liberal Education*, p. 343.

38. "Indice della Libreria nel Coll.o degl'Exgesuiti di Como . . ." (1774), ASM, MS Studi, Parte antica, folder 230. On the circulation of foreign physics (and especially Newtonian) textbooks in eighteenth-century Italy, see P. Casini, *Newton e la coscienza europea* (Bologna: il Mulino, 1983), pp. 173–227.

39. Bonesi to Volta, 1 August 1761, and Bonesi to Volta, 28 September 1761, ASM, PADD 9.

40. Bonesi to Volta, 20 August 1761, ASM, PADD 9.

41. On Como's seminaries, see R. della Bella, "Il seminario della diocesi di Como e la sua biblioteca" (Università Cattolica di Milano, "Laurea" thesis 1969–1970). See also P. Gini, "L'età posttridentina, secc. XVII–XVIII," in *Storia religiosa della Lombardia, Diocesi di Como*, ed. A. Caprioli, A. Rimoldi, L. Vaccaro (Brescia: La Scuola, n.d.), pp. 101–14. On the teaching imparted in the seminary of Santa Caterina, see the Archivio del Seminario di Como, Seminario Santa Caterina, MS folder 2, E, 10.

42. A list of the books kept in the library of the seminary of Santa Caterina is found in "Inventario di crediti, debiti e mobili," MS dated 27 July 1760, in Archivio del Seminario di Como, Seminario Santa Caterina, folder 2, B, 2.

43. A list (drawn in 1774) in ASM, Studi, Parte Antica, folder 230. Volta's interest in the library is mentioned in *VE*, 1:436.

44. On Gattoni see Volpati, *Alessandro Volta*, pp. 5–15.

45. MS published in *VE*, *Aggiunte*, pp. 119–35. See also: *Il poemetto didascalico latino di A. Volta con versione italiana di Zanino Volta* (Pavia: Fusi, 1899).

46. *VE, Aggiunte*, p. 120.

47. F. Algarotti, *Dialoghi sopra l'ottica neutoniana* (1752), ed. E. Bonora (Turin: Einaudi, 1977), p. 146.

48. *VE, Aggiunte*, pp. 123, 132, 135.

49. Ibid., p. 123.

50. Ibid., p. 124.

51. Ibid., pp. 126–28.

52. Ibid., p. 122.

53. Saepe etiam nulla foris accedente favilla
    Materie in pingui per sese accenditur ignis,
    Lucentesque globi flammarum sponte cientur:
    Haud aliter gigni veteres, multique recentum
    Sulphureo ex habitu meteora ignita docentes.
    Tali crediderant conflatum ab origine fulmen,
    Sed falso: nam postquam illuxit clarior aetas
    Monstravitque novas geniis nostratibus artes,
    Heu quantum Sophiae vultus mutatum ab illo est!

Ibid., pp. 132–33. Cf. Lucretius on lightening, *De rerum natura*, book 6, verses 160–422.

54. *VE*, 1:44. A survey of Volta's poetical compositions in A. Longatti, "Il Volta poeta," *Omaggio a Volta nel centocinquantenario della morte*, pp. 47–59.

55. Information drawn from Volta's MS (not autograph), "Idee sulla maniera d'insegnare per le classi inferiori e superiori delle regie scuole . . ." (1775), published in *VE*, 1:440–69.

56. Published in *VE, Aggiunte*, pp. 136–45.

57. *VE, Aggiunte*, pp. 146–52.

58. Giovio in *VE*, 2:263.

59. Bonesi to Volta, 23 July 1761 and 24 July 1761, ASM, PADD 9, and *VE*, 1:18.

60. *VE*, 1:7–8.

61. *VE*, 1:475–93.

62. *VE*, 1:480–84.

63. The MS is published in *VE*, 1:431–34.

64. *VE*, 1:433.

65. *VE*, 1:433–34.

66. *VE*, 5:504–8.

67. Volta to Giacomo Ciceri, 6 January 1815, *VE*, 5:289–92.

68. P. Vaccari, *Storia della Università di Pavia* (Pavia: Il Portale, 1948), pp. 112–14. See also: E. Rota, *Il giansenismo in Lombardia* (Pavia: Fusi, 1907); A. C. Jemolo, *Il giansenismo in Italia prima della rivoluzione* (Bari: Laterza, 1948).

69. *VE*, 3:131–35.

70. Volpati, *Llessandro Volta*, p. 193ff.

71. Ibid., 185ff.

72. *VE*, 5:290–91.

73. *VE*, 4:7.

74. Examples in *VE*, 1:488, and II: 129–30.

75. *VE*, 3:102.

76. Details on the letters in *VE*, 1:33–34; an excerpt from the letter to Nollet in *VO*, 3, 38 n.

77. *VE*, 1:35.

78. *VE*, 1:35–36.

79. *VE*, 1:40.

80. *VO*, 3:19–20.

81. *VE*, 1:42.

82. (Como, 1769), now in *VO*, 3:21–52. See chapter 3, p. 156ff.

83. A. Volta, *Novus ac simplicissimus electricorum tentaminum apparatus* . . . (Como, 1771), now in *VO*, 3:53–76.

84. *VE*, 1:51–52. On Frisi see ch. 2, p. 81ff.

85. *VE*, 1:53–54, 56, 66.

86. Paolo Frisi, "Piano del regolamento di studj dell'Università di Pavia e delle Scuole Palatine di Milano," Biblioteca del Politecnico di Milano, MS vol. 34, no. 10.

87. Ibid.

88. C. Barletti, *Nuove esperienze elettriche, secondo la teoria del Sig. Franklin e le produzioni del P. Beccaria* (Milan, 1771).

89. V. Cappelletti, "Barletti," *DBI*, 6:401–405, 40.

90. *VE*, 1:59–60.

91. *VE*, 1:62, and *VO*, 6:1–4.

92. According to Volta's own interpretation and phrase, in *VE*, 1:64.

93. *VE*, 1:66.

94. A. Volta, "Idea di uno stabilimento di scuole pubbliche per la città di Como," MS published in *VE*, 1:435–39.

95. *VE*, 1:68.

96. *VE*, 1:130–35.

97. *VE*, 1:67.

98. *VE*, 1:72.

99. *VE*, 1:73; the MS is published in *VE*, 1:440–69. A very good discussion of Volta's plan in Ada Annoni, "Cultura e scuola nell'epoca del Volta," pp. 103–16.

100. *VE*, 1:77.

101. A. Volta, "Articolo di una lettera del Signor Don Alessandro Volta al Signor Dottore Giuseppe Priestley, Como, 10 giugno 1775," *Scelta di opuscoli interessanti* (September and October 1775), now in *VO*, 3:93–108.

102. *VE*, 1:96.

103. *VE*, 1:100–101.

104. Bonesi to Volta, 23 July 1761, ASM, PADD 9, no. 1.

105. Bonesi to Volta, 7 August 1761, ASM, PADD 9, no. 11.

106. *VE*, *Aggiunte*, p. 158.

107. G. Parini, *Il giorno*, ed. D. Isella (Milan and Naples: Ricciardi, 1969). On the economic, social, and cultural functions of villas in the Enlightenment, see Dianne Suzette Harris, "Lombardia illuminata": The Formation of an Enlightenment Landscape in Eighteenth-Century Lombardy (Ph.D. dissertation, University of California at Berkeley, 1996), especially chapters 3–5.

108. *VE*, 1:47–48.

109. *VE*, 1:58.

110. Volta to Bertola, 24 July 1788, Uppsala University Library, Waller Collection.

111. *VE*, 1:467–68.

112. *VE, Aggiunte*, pp. 91–92.

113. *VE*, 2:182, 185, 308.

114. *VE, Indici*, 2:888–89.

115. *VE*, 3:87, 89.

116. Professor Luigi Valentino Brugnatelli, a chemist and a close friend of Volta, was reported (by an anonymous witness, and after Volta's death) to have said that Volta was the kind of person who attended masses in the mornings and brothels at night. See the MS comments on page 48 of the copy of *Lettere inedite di Alessandro Volta* (Pesaro, 1834) in BCC, classmark 146-4-286.

117. Brugnatelli's report in *VE*, 4:529; see also, ibid., 486, 488.

118. *Lichtenbergs Briefe* (Leipzig: Weicher, 1901–4), 2:153–54.

119. *VE*, 3:14.

120. *Storia di Milano*, 12:311.

121. *VE*, 3:40, 44–46.

122. *VE*, 3:123.

123. *VE*, 3:41, 124.

124. VE, 3:44, 109, 159–61, 167–68.

125. *VE*, 3:51–52.

126. *VE*, 3:57.

127. *VE*, 3:79.

128. *VE*, 3:109–110.

129. *VE*, 3:125–26.

130. *VE*, 3:111.

131. *VE*, 3:101, 123, 158, 169, 211.

132. *VE*, 3:164–65, 167. ASC, Notarile, Giacomo Filippo Clerici, 12 August 1794.

133. *VE*, 3:232.

134. *VE*, 3:191.

135. *VE*, 4:84–86.

136. *VE*, 4:168–69; *VE*, 5:3, 21, 42, 43.

137. *Lichtenbergs Briefe*, 2:150.

138. Penetrating insights into the world and sociology of musicians in the age of Volta in N. Elias, *Mozart: Zur Soziologie eines Genies* (Frankfurt: Suhrkamp, 1991).

139. *VO*, 3:6. Similar skills were employed in the assessment of the thickness and quality of silk threads in the silk industries of Como: Carlo Poni, "Standard, fiducia e conversazione civile: misurare lo spessore e la qualità del filo di seta," *Quaderni storici*, 32 (1997): 717–34.

140. *VO*, 1:154.

141. *VO*, 5:69. See also *VE, Indici*, 659–60. For reflections on similar topics see S. Schaffer, "Self Evidence," *Critical Inquiry*, 18 (1992): 327–62. See also S. Schaffer, "The Consuming Flame: Electrical Showmen and Tory Mystics in the World of Goods," in J. Brewer and R. Porter (eds.), *Consumption and the*

*World of Goods* (London: Routledge 1993), pp. 488–526. Additional reflections in Jan Golinski, *Making Natural Knowledge: Constructivism and the History of Science* (Cambridge: Cambridge University Press, 1998), pp. 182–85.

## Chapter 2
## Enlightenment Science South of the Alps

1. See: F. Venturi, *Settecento riformatore*, 5 vols. (Turin: Einaudi, 1968–1990); G. Ricuperati, *L'organizzazione della cultura nell'Italia del '700: Istruzione e accademie* (Turin: Tirrenia, 1976); G. Ricuperati, "Giornali e società nell'Italia dell'Ancien Régime (1668–1789)," in *La stampa italiana dal '500 all"800* (Bari: Laterza, 1976). See also D. Carpanetto and G. Ricuperati, *Italy in the Age of Reason, 1685–1789* (London: Longman, 1987).

2. Major examples in: P. Casini, "Boscovich," *DBI*, 13:221–30; R. Pasta, *Scienza politica e rivoluzione: L'opera di Giovanni Fabbroni (1752–1822) intellettuale e funzionario al servizio dei Lorena* (Florence: Olschki, 1989); V. Ferrone, *La Nuova Atlantide e i lumi: Scienza e politica nel Piemonte di Vittorio Amedeo III* (Turin: 1988); M. Cavazza, *Settecento inquieto: Alle origini dell'Istituto delle Scienze di Bologna* (Bologna: il Mulino, 1990); Massimo Mazzotti, "Maria Gaetana Agnesi: Mathematics and the Making of the Catholic Enlightenment," *Isis*, 92 (2001): 657–83; and, on an earlier period, Giuseppe Olmi, *L'inventario del mondo: catalogazione della natura e luoghi del sapere nella prima età moderna* (Bologna: il Mulino, 1992), and Paula Findlen, *Possessing Nature: Museums, Collecting, and Scientific Culture in Early Modern Italy* (Berkeley: University of California Press, 1994).

3. Major examples in: Renato G. Mazzolini, *The Iris in Eighteenth-Century Physiology* (Bern: H. Huber, 1980); P. Casini, *Newton e la coscienza europea* (Bologna: Il Mulino, 1983); W. Bernardi, *Le metafisiche dell'embrione: Scienze della vita e filosofia da Malpighi a Spallanzani (1672–1793)* (Florence: Olschki, 1986); M. Pera, *La rana ambigua* (Torino: Einaudi, 1986; Amer. trans., Princeton: Princeton University Press, 1992); W. Bernardi, *I fluidi della vita: Alle origini della controversia sull'elettricità animale* (Florence: Olschki, 1992); Luca Ciancio, *Autopsie della terra: illuminismo e geologia in Alberto Fortis (1741–1803)* (Florence: Olschki, 1995); Mario Morselli, *Amedeo Avogadro: A Scientific Biography* (Dordrecht: Kluwer, 1984); Marco Ciardi, *L'atomo fantasma: Genesi storica dell'ipotesi di Avogadro* (Florence: Olschki, 1995).

4. See however: U. Baldini, "L'attività scientifica nel primo Settecento," in *Storia d'Italia, Annali 3, Scienza e tecnica nella cultura e nella società dal Rinascimento a oggi* (Turin: Einaudi, 1980), pp. 465–545; and P. Redondi, "Cultura e scienza dall'illuminismo al positivismo," ibid., pp. 677–811.

5. *VE*, vols. 1, 2, 3.

6. D. Carpanetto and G. Ricuperati, *Italy in the Age of Reason, 1685–1789*, p. 152.

7. P. Frisi, "Stato odierno della letteratura," Biblioteca del Politecnico di Milano, Coll. Frisi, MS vol. 35, no. 7.

8. F. Griselini, *Dizionario delle arti e de' mestieri*, 18 vols. (Venice, 1768).

9. R. L. Kagan, "Universities in Italy, 1500–1700," in D. Julia, J. Revel, R. Chartier (eds.) *Histoire sociale des populations étudiantes*, Tome 1 de la série "Les universitées européennes du XVIe au XVIIIe siècle" (Paris: EHESS, 1986), pp. 152–86. An excellent, general survey of the teaching of natural philosophy and related topics in European universities during the eighteenth century is Laurence Brockliss, "Curricula," in *A History of the University in Europe*, general editor Walter Rüegg, vol. 2, *Universities in Early Modern Europe (1500–1800)*, editor Hilde de Ridder-Symoens (Cambridge: Cambridge University Press, 1996), pp. 563–620.

10. Baldini, "L'attività scientifica nel primo Settecento," pp. 526–29.

11. M. Roggero, "Professori e studenti nelle università tra crisi e riforme," *Storia d'Italia. Annali 4, Intellettuali e potere*, ed. C. Vivanti (Torino: Einaudi, 1981), pp. 1039–81.

12. A. Annoni, "Cultura e scuola nell'epoca del Volta," in *Conferenze Voltiane* (Como: Società Storica Comense, 1978), pp. 69–124, on pp. 80, 89.

13. Roggero, "Professori e studenti nelle università tra crisi e riforme," p. 1079.

14. R. Rettaroli and F. Tassinari, "Studenti e docenti dell'Ateneo tra VIII e IX Centenario," in *Lo studio e la città. Bologna, 1888–1988*, ed. W. Tega (Bologna: Nuova Alfa, 1987), pp. 284–88, on p. 284.

15. L. Simeoni, *Storia dell'Università di Bologna: L'età moderna* (1940; reprint, Bologna: Forni, n.d.), p. 101.

16. G. C. Mor, *Storia della Università di Modena* (Florence: Olschki, 1975); A. Visconti, *La storia dell'Università di Ferrara* (Bologna: Zanichelli, 1950), L. Pepe, "Scienziati e stabilimenti scientifici a Ferrara," *Museologia scientifica*, 3 (1986): 113–19; A. Fiocca and L. Pepe, "L'Università e le scuole per gli Ingegneri a Ferrara," *Annuario dell'Università di Ferrara, Sc. Mat.*, 32 (1986): 125–66.

17. V. Ferrone, "Riflessioni sulla cultura illuministica napoletana e l'eredità di Galilei," in *Galileo e Napoli*, ed. F. Lomonaco, and M. Torrini (Naples: Guida, 1987), pp. 429–48.

18. For the amount and variety of consulting in which academicians and professors were involved, see Ferrone, *La Nuova Atlantide e i lumi*, and *Tra scienza e società. 200 anni di storia dell'Accademia delle Scienze di Torino* (Turin: Allemandi, 1988).

19. *Memorie di matematica e fisica della Società Italiana* (Verona, the 7 vols. 1782–1794); *Miscellanea Philosophico-mathematica*, then *Mélanges de philosophie et de mathématique*, then *Mémoires de l'Académie Royale des Sciences* (Turin, the volumes pertaining to the years 1770–1791; *Scelta di Opuscoli interessanti* (*Opuscoli scelti* from 1778) (Milan, 1775–1794); *Biblioteca fisica d'Europa* (Milan, 1788–1791); *Giornale fisico-medico* (Milan, 1792–1795); *Annali di chimica e storia naturale* (Pavia, 1790–1795); *Saggi scientifici e letterarj dell'Accademia di Padova* (the 4 vols. published in 1786, 1789, and 1794, parts 1 and 2); *Memorie della Reale Accademia di Scienze belle lettere ed arti in Mantova* (1 vol., 1795); *De Bononiensi scientiarum et artium Instituto atque Academia commentarii* (the 2 vols. published in 1783 and 1792); *Atti dell'Accademia delle Scienze di Siena detta de' Fisiocritici* (the 4 vols., 1771–1794); *Atti della Reale accademia delle scienze e belle-lettere di Napoli* (the vol. published in 1788, covering the period since the foundation).

20. This is especially the case of the *Giornale de' letterati* (Pisa), the *Antologia romana* (Rome), the *Giornale enciclopedico* (Venice), the *Nuova raccolta di opuscoli scientifici e filologici* (Venice). See B. Dooley, *Science, Politics, and Society in 18th-Century Italy. The* Giornale de' letterati d'Italia *and its World* (New York: Garland, 1991).

21. Giovanni Aldini (1762–1834), Carlo Allioni (1728–1804), Carlo Amoretti (1741–1816), Pietro Arduino (1728–1805), Prospero Balbo (1762–1837), Carlo Barletti (1735–1800), Biagio Bartalini (1746–1822), Giambattista Beccaria (1716–1781), Teodoro Massimo Bonati (1724–1820), Camillo Bonioli (1729–1791), Luigi Bossi (1758–1835), Luigi Valentino Brugnatelli (1761–1818), Giovanni Brugnone (1741–1818), Gabriello Brunelli (1728–1797), Michele Francesco Buniva (1761–1834), Francesco Buzzi (1751–1805), Antonio Cagnoli (1743–1816), Leopoldo Marcantonio Caldani (1725–1813), Tommaso Caluso di Valperga (1737–1815), Stanislao Canovai (1740–1812), Sebastiano Canterzani (1734–1819), Marco Carburi (1731–1808), Giovacchino Carradori (1758–1818), Gregorio Casali (1721–1802), Filippo Cavolini (1756–1810), Angelo Giovanni Cesaris (1749–1832), Vincenzo Chiminello (1741–1815), Giovanni Francesco Cigna (1734–1790), Domenico Cotugno (1736–1822), Giovanni Dana (1736–1801), Giovanni Fabbroni (1752–1822), Nicola Fergola (1753–1824), Felice Fontana (1730–1805), Gregorio Fontana (1735–1803), Alberto Fortis (1741–1803), Vittorio Fossombroni (1754–1844), Francesco Franceschinis (1756–1840), Paolo Frisi (1728–1784), Luigi Galvani (1737–1798), Giovanni Antonio Giobert (1761–1834), Giuseppe Maria Giovene (1753–1837), Michele Girardi (1731–1797), Giuseppe Luigi Lagrange (1736–1813), Antonio Maria Lorgna (1735–1796), Michele Malacarne (1744–1816), Gian Francesco Malfatti (1731–1807), Petronio Matteucci (1740–1800), Carlo Mondini (1729–1803), Carlo Luigi Morozzo (1744–1804), Pietro Moscati (1739–1824), Andrea Mozzoni (1754–1842), Gian Francesco Napione (1748–1830), Barnaba Oriani (1752–1832), Ermenegildo Pini (1739–1825), Saverio Poli (1746–1825), Lorentino Presciani (1721–1799), Tarsizio Riviera (1759–1801), Esprit-Benoit Nicolis de Robilant (1724–1801), Michele Rosa (1731–?), Pietro Rossi (?–1804), Pietro Rubini (1760–1819), Girolamo Saladini (?–1813), Giuseppe Angelo Saluzzo (1734–1810), Giambattista San Martino (1739–1800), Antonio Scarpa (1747–1832), Giovanni Antonio Scopoli (1723–1788), Paolo Spadoni (1764–1826), Lazzaro Spallanzani (1729–1799), Simone Stratico (1733–1824), Giuseppe Toaldo (1719–1797), Giambattista Vasco (1733–1796), Giambattista Venturi (1746–1822), Alessandro Volta (1745–1827), Leonardo Ximenes (1716–1786).

22. *DBI.*

23. For the period discussed here, the repertoire used most frequently is E. De Tipaldo, *Biografia degli Italiani illustri*, 10 vols. (Venice, 1834–1845). It must be supplemented, however, by several other regional and local repertoires found among the 5,000 titles listed in C. Manzoni, *Biografia Italica* (Osnabrück: Biblio Verlag, 1981).

24. Kagan, "Universities in Italy 1500–1700," loc. cit.

25. See ch. 5.

26. Research under way on eighteenth-century Italian women scientists by Marta Cavazza, Paula Findlen, and Beate Cheranski is going to change this side of our picture soon.

27. R. Darnton, *The Business of the Enlightenment* (Cambridge, Mass.: Belknap, 1979), p. 315.

28. Frisi, "Stato odierno della letteratura."

29. Darnton, *The Business of the Enlightenment*, p. 317.

30. Ibid., p. 314.

31. A. T. Villa, *Caroli Comitis Firmiani Vita* (Milan, 1783), p. 41n.

32. *Bibliotheca Firmiana*, 6 vols. (Milan, 1783).

33. D. Barsanti and L. Rombai, *Leonardo Ximenes* (Florence: Osservatorio Ximeniano, 1987), 201–25.

34. M. Piacenza, "Note biografiche e bibliografiche e nuovi documenti su G. B. Beccaria," *Bollettino storico bibliografico subalpino*, 9 (1904): 209–354, 347–54.

35. Casini, *Newton e la coscienza europea*, p. 178.

36. T. L. Hankins, "Lalande," *DSB*, 7:579–82; D. M. Simpkins, "J. E. Smith," *DSB*, 12:471–2; "Walker, Adam," *DNB*, 20:499; "Young, Arthur," *DNB*, 21:1272–78.

37. J.-J. Le Français de Lalande, *Voyage en Italie* (Yverdon, 1787), 1:xj.

38. Ibid., 2:64–66.

39. A. Walker, *Ideas Suggested on the Spot in a Late Excursion through Flanders, Germany, France, and Italy* (London, 1790), p. 192.

40. Loc. cit.

41. Ibid., p. 410.

42. A. Young, *Travels during the Years 1787, 1788, and 1789*, 2nd ed. (London 1794), 1:256.

43. Ibid., 1:261–62.

44. Ibid., 1:213.

45. J. E. Smith, *A Sketch of a Tour on the Continent in the Years 1786 and 1787* (London, 1793), 3:65.

46. Ibid., 1:265.

47. Ibid., 3:114.

48. Ibid., 1:316.

49. Contrast this situation with the situation in Britain, that won the admiration of Italian travelers like Volta (chapter 5, below): R. Porter, *English Society in the Eighteenth Century*, rev. ed. (London: 1990), esp. ch. 8, esp. p. 314. See also Larry Stewart's penetrating *The Rise of Public Science: Rhetoric, Technology, and Natural Philosophy in Newtonian Britain, 1660–1750* (Cambridge: Cambridge University Press, 1992).

## Chapter 3
## The Electrophorus

1. *VO*, 3:99; *Encyclopedia Britannica*, 3d ed. (1797), 6:424. The author of the article was probably James Tytler (see ibid., 1:xv).

2. John L. Heilbron, "Volta," *DSB*, 13:69–82.

3. F. K. Achard, "Expériences sur l'electrophore avec un théorie de cet instrument," in *Nouveaux mémoires de l'Académie Royale des Sciences et Belles-lettres* (Berlin, 1776), pp. 122–34.

4. J. L. Heilbron, *Electricity in the Seventeenth and Eighteenth Centuries* (Mineola, New York: Dover Publications, 1999), pp. 421–26, 462. On Volta and Coulomb, chapter 4, below.

5. R. W. Home, Essay Review of Heilbron, *Electricity*, in *Ann. Sci.*, 38 (1981): 477–82, on p. 479.

6. *VO*, 6:4.

7. For a criticism of the notion of "discovery," with specific reference to the history of electricity, see Michael Ben-Chaim, "Social Mobility and Scientific Change: Stephen Gray's Contribution to Electrical Research," *Brit. J. Hist. Sci.*, 22 (1990): 3–24, esp. 21. For some perceptive remarks (and bibliography) on the notions of "discovery" and "invention" in the historiography of science, see Lorraine Daston, "Introduction," in Lorraine Daston (ed.), *Biographies of Scientific Objects* (Chicago: University of Chicago Press, 2000), pp. 1–11.

8. Nollet's comment was quoted by Volta in 1769, *VO*, 3:23n. For Volta's later descriptions of his own letters of 1763, see Volta to Beccaria, *VE*, 1:35–36, and *VO*, 3:38n.

9. R. Symmer, "New Experiments and Observations Concerning Electricity," *PT*, 51 (1759): 340–89. On Symmer: J. L. Heilbron, "Robert Symmer and the Two Electricities," *Isis*, 67 (1976): 7–20, and J. L. Heilbron, "Symmer," *DSB*, 13, 224–25. On Franklin's theory and its circulation: B. Franklin, *Experiments*, ed. I. B. Cohen (Cambridge, Mass.: Harvard University Press, 1941); I. B. Cohen, *Franklin and Newton* (Philadelphia: American Philosophical Society, 1956); A. Pace, *Benjamin Franklin and Italy* (Philadelphia: American Philosophical Society, 1958); Heilbron, *Electricity*, pp. 324–72; *Aepinus's Essay on the Theory of Electricity and Magnetism*, ed. R. W. Home (Princeton: Princeton University Press, 1979), pp. 77–87.

10. The story is told in J. A. Nollet, *Lettres sur l'électricité* (Troisième Partie, Paris, 1767), pp. iii–iv, xii. See also G. F. Cigna, "De novis quibusdam experimentis electricis," *Miscellanea Taurinensia* (1763–65), 3:31–72. On Cigna: U. Baldini, "Cigna," *DBI*, 25:479–82.

11. *VE*, 1:42.

12. "Je verrai avec bien du plaisir votre nouveau système sur les causes de l'électricité, quand vous le ferez paroitre: je serai surpris, si vous tirez de l'attraction Newtonienne des explications physiques des phénomènes de ce genre; il me semble, qu'en laissant subsister les loix, qu'on attribue à cette espèce de vertu, il est bien difficile de rendre raison des principaux faits: personne jusqu'à présent n'a osé l'entreprendre; il sera glorieux pour vous de l'avoir fait avec succès" (*VO*, 3:23n).

13. *VO*, 3:37–38 and 38n.

14. On the Franklin square's relation to the Leyden jar, see Heilbron, *Electricity*, p. 317.

15. I. Newton, *Philosophiae Naturalis Principia Mathematica* 3d ed. (1726, with variant readings), ed. Alexandre Koyré and I. B. Cohen (Cambridge: Cambridge University Press, 1972), 431. On "Newtonianism" and magnetism see

R. W. Home, " 'Newtonianism' and the Theory of the Magnet," *Hist. Sci.*, 15 (1977), 252–66.

16. *VO*, 3:38n.

17. *VE, Aggiunte*, pp. 119–35.

18. On this tradition see R. Fox, *The Caloric Theory of Gases from Lavoisier to Regnault* (Oxford: Clarendon, 1971), p. 13.

19. *VE, Aggiunte*, p. 122.

20. Ibid.

21. In 1764 Volta knew the French translation edited by Thomas François Dalibard: B. Franklin, *Experiences et oservations sur l'électricité*, 2d edition (Paris, 1756); see *VE, Aggiunte*, p. 133. In the French edition just quoted, the passages on electric and common fire were in vol. 2, p. 88.

22. Fox, *Caloric Theory*, p. 17. See also W. M. Sudduth, "Eighteenth-Century Identifications of Electricity with Phlogiston," *Ambix*, 25 (1978): 131–47.

23. *VO*, 3:3–16.

24. This was done by presenting threads, suspended from a conductor repeatedly put into contact with the electrified body, subsequently to a rubbed glass tube and to a rubbed sulphur tube: if threads were repelled by the glass tube (minus) and attracted by the sulphur tube (plus), the body's electricity was minus, and vice-versa *(VO*, 3:3–4).

25. *VO*, 3:6–8.

26. Johan Carl Wilcke had described his research on the subject in 1757. See Heilbron, *Electricity*, p. 387.

27. *VO*, 3:12.

28. *VO*, 3:9. On the electrostatic machine: W. D. Hackmann, *Electricity from Glass: The History of the Frictional Electrical Machine, 1600–1850* (Alphen aan den Rijin: Sijthoff and Noordhoff, 1978).

29. *VO*, 3:9–10.

30. *VE, Aggiunte*, pp. 19–20.

31. *VO*, 3:19.

32. *VE*, 1:40–41.

33. On the Peking experiment, Aepinus and Beccaria: Heilbron, *Electricity*, pp.405–6, 409–10, and Home in *Aepinus's Essay*, pp. 130–31.

34. As reported in G. B. Beccaria, *Experimenta, atque observationes quibus electricitas vindex late constituitur, atque explicatur* (Turin, 1769), p. 45.

35. Beccaria's text, with English translation, is reprinted in B. Franklin, *Papers*, ed. L. W. Labaree et al. (New Haven: Yale University Press, 1959–), 14:41–57.

36. See *VO*, 3:24 and *VE*, 1:43.

37. Beccaria's formulation of the theory proceeded by stages, and so must have proceeded Volta's understanding of it. See note 72, below.

38. For convenience sake, we refer here to the fullest (and amended) account of the theory of "vindicating electricity" expounded by Beccaria in *Elettricismo artificiale* (Turin, 1772), pp. 400ff. An English translation of this book, at the initiative of Franklin (see A. Pace, "Beccaria," *DBI*, 7:469–71), was published in London in 1776. A detailed discussion of Beccaria's theory in Heilbron, *Electricity*, pp. 408–12.

39. Beccaria, *Elettricismo artificiale*, pp. 400–1, 409.

40. Ibid., pp. 421–22.

41. *VO*, 3:24.

42. (Como, 1769), reprinted in *VO*, 3:21–52.

43. See for example Franklin, *Experiences* (French translation), 1:5 and 209–11; 2:10–11.

44. R. W. Home, "Franklin's Electrical Atmospheres," *Brit. J. Hist. Sci.*, 6 (1972): 131–51; Heilbron, *Electricity*, pp. 334–37.

45. Cohen, *Franklin and Newton*, pp. 468–71; Home, *Aepinus's Essay*, pp. 84, 98–99.

46. Home, *Aepinus's Essay*, p. 86 f.

47. G. B. Beccaria, *Dell'elettricismo artificiale e naturale* (Turin, 1753), pp. 101–10.

48. See chapter 2, above, pp. 107ff.

49. *VO*, 3:24–25.

50. See P. Casini, *Newton e la coscienza europea* (Bologna: il Mulino, 1983), pp. 207–24.

51. *VO*, 3:25.

52. Volta (*VO*, 3:25) quoted Musschenbroek's *Essai de physique* and *Nouveau Cours de Chymie*.

53. Home, "Newtonianism," p. 263; Heilbron, *Electricity*, pp. 1–2.

54. An example of this eclectic tradition is the manual by Aimé-Henri Paulian, *Dictionnaire de physique* (Avignon, 1761), that circulated in Jesuit colleges and elsewhere in the 1760s.

55. R. J. Boscovich, *Philosophiae naturalis theoria redacta ad unicam legem virium in natura existentium* [1758] (2nd ed., Venice, 1763). References below refer to the second edition, as reprinted, with English translation, by J. M. Child (ed.) (Chicago and London: Open Court, 1922). On Boscovich see P. Casini, "Boscovich," *DBI*, 13:221–30. On Boscovich's impact on Priestley and on other chemists: A. Thackray, " 'Matter in a Nutshell': Newton's *Opticks* and Eighteenth-Century Chemistry," *Ambix*, 15 (1968): 29–53, 49–51.

56. Casini, "Boscovich," p. 228.

57. J. M. Child in Boscovich, *Philosophiae naturalis*, pp. xii–xiii.

58. Casini, "Boscovich," p. 229.

59. *VO*, 3:31.

60. Boscovich, *Philosophiae naturalis*, pp. 360–63.

61. Ibid., pp. 362–63.

62. Loc. cit.

63. Loc. cit.; cf. *VO*, 3:29.

64. *VO*, 3:28.

65. *VO*, 3:31.

66. *VO*, 3:30.

67. *VO*, 3:32–34.

68. *VO*, 3:35.

69. *VO*, 3:40.

70. *VO*, 3:39–40.

71. *VO*, 3:40–41.

72. *VO*, 3:23, 52. It is worth reminding that, at the time of *De vi attractiva ignis electrici*, Volta knew Beccaria's essay of 1767 and the essay "De atmosphera electrica," *PT*, 60 (1770): 277–301, not the 1769 *Experimenta*. See A. Volta, *Opere scelte*, ed. M. Gliozzi (Torino: UTET, 1967), pp. 50n and 83n.

73. Heilbron, "Robert Symmer and the Two Electricities," p. 15.

74. Disagreement with Beccaria is very cautiously expressed in 1769 (*VO*, 3:45–47); much more emphatically so after the discovery of the electrophorus (*VO*, 3:138–40n).

75. According to Beccaria's comment in a letter to Volta, reported by Volta in Volta to Landriani, 3 June 1775, *VO*, 3:85. Beccaria alluded to his *Elettricismo artificiale e naturale*.

76. Beccaria, *Elettricismo artificiale*, pp. 407–9. Beccaria quoted also Cigna in his support, ibidem, p. 409. Volta's reactions in *VO*, 3:139n ff.

77. *VE*, 1:64.

78. Ibid.

79. *VO*, 3:85.

80. *VE*, 1:52.

81. Ibid.

82. *VE*, *Aggiunte*, p. 121.

83. A. Volta, *Novus ac simplicissimus electricorum tentaminum apparatus: seu de corporibus eteroelectricis quae fiunt idioelectrica. Experimenta, atque observationes* (Como, 1771), in *VO*, 3:53–76. On the role of instruments in Volta's work: Bellodi, Giuliano, Brenni, Paolo, "The 'Arms of the Physicist': Volta and Scientific Instruments," in Fabrio Bevilacqua and Lucio Fregonese (eds.), *Nuova Voltiana: Studies on Volta and His Times*, vol. 3 (2001), pp. 1–40. On the social uses of electrical instruments in the age of Volta: Lissa Roberts, "Science Becomes Electric. Dutch Interaction with the Electrical Machine during the Eighteenth Century," *Isis*, 90 (1999): 680–714.

84. *VO*, 3:58.

85. *VO*, 3:69.

86. *VO*, 3:71.

87. *VO*, 3:55–56, 59–60.

88. *VO*, 3:58.

89. *VO*, 3:56n.

90. *VO*, 3:55.

91. *VE*, 1:54, 56.

92. *VE*, 1:52.

93. *VE*, 1:118.

94. On the Brera schools see Frisi's report, "Stato della letteratura," discussed in ch. 2, above, pp. 81ff.

95. *VE*, 1:118; *VO*, 3:91; *VE*, 1:79. The translation of Franklin's *Experiments* (an abridged version from the French edition) bore the title *Scelta di Lettere e di Opuscoli del Signor B. Franklin* (Milan, 1774).

96. *VO*, 3:159.

97. *VE*, 1:277–78.

98. *VE*, 1:410, 472.

99. *VE*, 1:61–62.

100. *Scelta di opuscoli interessanti sulle scienze e sulle arti.*

101. R. De Felice, "Amoretti," *DBI*, 3:9–10.

102. Quoted in *VE*, *Indici*, p. 667.

103. On Barletti see: V. Cappelletti, "Barletti," *DBI*, 6:401–5.

104. *VE*, 1:66–74.

105. Volta to Firmian, 23 March 1775, *VE*, 1:75, and the accompanying project report. See ch. 1, above, p. 32.

106. *VE*, 1:100–101.

107. A. Volta, "Articoli relativi ai bisogni delle scuole di Como" (1776), *VE*, 1:471–73. See also *VE*, 1:84–86.

108. *VE*, 1:119–20.

109. Ibid.

110. *VE*, 1:472.

111. *VO*, 3:77–80. See W. Ammersin, *Brevis relatio de electricitate propria lignorum* (Lucerne, 1755), discussed also in Nollet, *Lettres sur l'électricité*, vol. 2 (Paris, 1760).

112. *VE*, 1:59.

113. "L'idée de votre machine, faite de carton, me frappa. C'est pourquoi je la fis construire et fus bien surpris d'en voir les effets, quoiqu'ils soient beaucoup au dessous de ceux de nos globes de verre. Neanmoins je peux bien supposer, que si on la construisse en un meillieur maniere, sa force seroit plus grande; sur tout si on introduiroit plusieurs plaques de carton, ou de bois (qu'il seroit assez aisé de faire) quand un frottoir suffiroit pour deux plaques" (*VE*, 1:59–60).

114. Volta to Priestley, May 1772, as reported by Volta in May 1776, *VO*, 3:140–41n.

115. *VO*, 3:139–40.

116. *VO*, 3:141.

117. *VO*, 3:141n.

118. *VE*, 1:61–62,

119. This is the judgement expressed also by Mario Gliozzi, "Consonanze e dissonanze tra l'elettrostatica di Cavendish e quella di Volta," *Physis*, 11 (1969): 231–48, on p. 233. Beccaria's book was *Elettricismo artificiale*. The other was Carlo Barletti, *Physica specimina* (Milan, 1772).

120. Joseph Priestley, *The History and Present State of Electricity* (London, 1767), pp. 232–33. Frisi very likely lent the French translation (Paris, 1771) of Priestley's book to Volta on 9 July 1771 (*VE*, 1:52). Many circumstantial references to Priestley's book on electricity, and to the French translation, are found in Volta's correspondence from 1771 onward. See: *VO*, 3:77; *VE*, 1:59; *VE*, 1:59–60; *VO*, 3:95; *VE*, 1:116; *VO*, 6:155.

It is very unlikely that Volta did not come across Priestley's description of Wilcke's and Aepinus's experiments, which were reported in vol. 1, 423–24, of the French translation, in a section devoted to the "two electricities," one of Volta's own main concerns in those years. Later, Volta would claim that he did not know Wilcke's and Aepinus's experiments when working on the electrophorus. However, he put the claim, literally, *in brackets*, and we interpret it as meaning that he did not know Wilcke's and Aepinus's works firsthand (*VO*, 3:138).

121. For Priestley, see note 120, above. For Barletti: C. Barletti, *Physica specimina*, p. 12n. Barletti described Aepinus's sulphur and cup experiment in detail, as reported in F.U.T. Aepinus, *De similitudine vis electricae atque magneticae* (St. Petersburg, 1758).

122. Priestley, *History*, p. 233.

123. Ibid. Priestley took the experiment from F.U.T. Aepinus, *Tentamen theoriae electricitatis et magnetismi* (St. Petersburg, 1759), p. 66.

124. Cf. Home, *Aepinus's Essay*, p. 130n, and Heilbron, *Electricity*, p. 416.

125. See William Henley, "Experiments and Observations on a New Apparatus, Called, A Machine for Exhibiting Perpetual Electricity," *PT*, 66 (1776): 513–22, on p. 513.

126. *VO*, 3:123; and F. K. Achard, "Expériences sur l'électrophore avec un théorie de cet instrument," *Nouveaux mémoires de l'Académie Royale des Sciences et Belles-lettres* (Berlin, 1776), pp. 122–34, on p. 131.

127. Cigna, "De novis quibusdam experimentis," p. 50; Priestley, *History*, p. 290.

128. Priestley, *History*, p. 290.

129. *VO*, 3:93–108.

130. *VO*, 3:102.

131. *VO*, 3:140n.

132. *VO*, 3:97, 171.

133. *VO*, 3:127; *VE*, 1:119; *VO*, 3:144.

134. *VO*, 3:102.

135. *VO*, 3:113.

136. *VO*, 3:108; Joseph Klinkosch to Count Kinsky, 15 January 1776, in VMS, F10; M. Landriani in *Scelta di Opuscoli* (1776), no. 19, pp. 73–86; E. Darwin, "Commonplace Book," MS in Darwin House, Downe, Kent; W. Nicholson, "Observations on the Electrophore," *Journal of Natural Philosophy, Chemistry, and the Arts*, 1 (1797): 355–59.

137. *VO*, 3:108.

138. Priestley, *History*, p. 422.

139. *VO*, 3:121.

140. *VO*, 3:99. Volta knew how effective the "imposing name" had been in securing the success of the machine (*VO*, 3:137).

141. *Observations sur la physique*, 7 (1776): 442.

142. *VO*, 3:108.

143. A. Volta, *Osservazioni sulla capacità dei conduttori elettrici* (1778), *VO*, 3:199–229.

144. G. Polvani, *Alessandro Volta* (Pisa: Domus Galilaeana, 1942), pp. 89–110; Heilbron, *Electricity*, pp. 415–19; Home, *Aepinus's Essay*, p. 130n.

145. Barletti, *Physica specimina*, pp. 12n and 82, quoted Aepinus's *De similitudine vis electricae atque magneticae*, and *Tentamen* (the latter perhaps known to him through Priestley's *History*: see *VE*, 1:121). He later gave some extracts from Aepinus to Volta, *VE*, 1:120–21. Fromond lent Aepinus's book on the

tourmaline [*Recueil de differents mémoires sur la tourmaline* (St. Petersburg, 1762)] to Volta in 1775 (*VE* 1:86–87).

146. See M. Piacenza, "Note biografiche e bibliografiche e nuovi documenti su G. B. Beccaria," *Bollettino Storico-Bibliografico Subalpino*, 9 (1904): 340–54.

147. *Scelta di Opuscoli*, 2 (1775): 117–18.

148. *VO*, 3:89–91.

149. *VE*, 1:79; *VO*, 3:112.

150. *VO*, 3:156–57n; *VE*, 1:89–92; *VE*, 1:106–107; *VE*, 1:147.

151. *VO*, 3:106n.

152. *VO*, 3:112n.

153. *VO*, 3:176n.

154. *VO*, 3:118.

155. *VE*, 1:103.

156. *Scelta di opuscoli interessanti*, 1 (1775): no. 9, 91–107; no. 10, 87–113 (*VO*, 3:95–108).

157. *VE*, 1:100; *VE*, 1:98; *VE*, 1:68.

158. *VE*, 1:102; *VO*, 3:145n; and VMS, F 10.

159. J. Ingenhousz, "Electrical Experiments, to Explain How Far the Phenomena of the Electrophorus May Be Accounted for by Dr. Franklin's Theory of Positive and Negative Electricity," *PT*, 68 (1778): 1027–48, esp. 1031.

160. Ibidem, p. 1029; *VE*, 1:119.

161. Henley, "Experiments and Observations."

162. Ibidem, p. 513; *VE*, 1:160.

163. Priestley to Volta, 25 April 1776, *VE*, 1:123, and VMS, G1. That there was no earlier answer is confirmed by Volta to Landriani, 27 January 1776, *VO*, 3:159.

164. Magellan to Volta, after 25 April 1776, *VE*, 1:124 (and *VE*, *Aggiunte*, pp. 78–79).

165. *VO*, 3:113.

166. "Sur l'électrophore perpétuel de M. Volta," *Observations sur la physique*, 7 (1776): 501–508.

167. *Observations sur la physique*, 7 (1776): 442.

168. Achard, "Expériences sur l'electrophore," p. 134.

169. *VO*, 3:146.

170. *VO*, 3:138–139.

## Chapter 4
## Volta's Science of Electricity

1. Priestley (and Magellan) sent several of Priestley's offprints and vol. 2 of his *Experiments and Observations on Different Kinds of Air* to Volta in the spring of 1776. See *VE*, 1:122–23, 124 and *VE*, *Aggiunte*, pp. 78–79. See also Home, Roderick W., "Volta's English Connections," in Fabio Bevilacqua and Lucio Fregonese (eds.), *Nuova Voltiana: Studies on Volta and His Times*, vol. 1 (2000), pp. 115–32.

2. See Volta's works published in volumes 6 and 7 of *VO*. On Volta's chemistry see the papers by Ferdinando Abbri, Bernadette Bensaude-Vincent, Raf-

faella Seligardi, Marco Beretta and Frederic L. Holmes in Fabio Bevilacqua and Lucio Fregonese (eds.), *Nuova Voltiana: Studies on Volta and his Times*, vol. 2, (2000).

3. See J. L. Heilbron, *Electricity in the Seventeenth and Eighteenth Centuries* (Mineola, New York: Dover Publications, 1999), p. 419n.

4. See R. J. Haüy, *Exposition raisonnée de la théorie de l'électricité et du magnétisme d'après les principes de M. Aepinus* (Paris, 1787), p. 54: "Nous devons observer ici, que c'est M. Aepinus qui, le premier, a employé un appareil construit sur le même plan que l'Electrophore.... On voit, par cet exposé, combien il restoit peu à faire, pour arriver de cet appareil à l'électrophore." See also J. B. Pujoulx, *Paris à la fin du 18e siècle* (Paris, 1801), p. 374: "Il y a environ vingt ans qu'un savant italien vint en France publier, comme le fruit de ses observations, plusieurs expériences très-importantes, qui complétaient en quelque sorte la théorie de l'électricité. Il fut accueilli par les hommes les plus instruits; et voilà que quelques années plus tard on découvre des Mémoires lus à l'Académie de Pétersbourg en 1759, et imprimés dans cette ville en 1760, dans lesquels on retrouve toutes les observations et les expériences du physicien italien. Si les savans eux-mêmes y sont pris, comment se fier aux réputations!"

5. De Luc and an unnamed electrician in Paris, as reported by De Luc to Volta; both in *VE*, 2:163.

6. *VO*, 3:266–67.

7. *VO*, 3:315. On Bellisomi's collection of instruments, see Lalande, *Voyage en Italie* 2nd ed. (Paris, 1786) 1:331.

8. Volta to De Luc, 30 March 1784, Uppsala University Library, MS in the Waller Collection: "Le mémoire [of 1778] devoit être incessamment suivi d'un autre que j'annoncois sur la capacité des conducteurs combinés, ou conjugués, comme j'aime à les appeller; mais les expériences m'ayant conduit au condensateur, dont je m'occupai longtemps, ce second memoire ne fut fini qu'en 1780, et rouloit presque entièrement sur ce meme [*sic*] condensateur. . . ." Of this letter, only the minute of a part—dealing with chemistry—was known and published in *VE*, 2:196–200, under the date 20 March 1784. I found the original, including eight dense pages on electricity and containing crucial information on Volta's work, in the Waller Collection, to which I had been directed by Marco Beretta.

9. *VO*, 3:159.

10. *VO*, 3:172 and 178.

11. See note 3, above.

12. *VO*, 6:261.

13. See chapter 1, above.

14. *VE*, 1:222.

15. *VO*, 6:224.

16. *VO*, 6:261. Not known in the form sent to Senebier, the article can probably be identified with the copy preserved in VMS, I, 1.

17. *VO*, 6:261.

18. *VE*, 1:275.

19. Ibidem.

20. *VO*, 3:229.

21. *VE*, 1:236. Volta's memoir "Osservazioni sulla capacità de' conduttori elettrici ... al Signor De Saussure," originally published in *Opuscoli*, is now in *VO*, 3:199–229.

22. *VO*, 3:209. On Volta's concepts: Jürgen Teichmann, "Volta and the Quantitative Conceptualisation of Electricity: From Electrical Capacity to the Preconception of Ohm's Law," in Fabio Bevilacqua and Lucio Fregonese (eds.), *Nuova Voltiana: Studies on Volta and His Times*, vol. 3 (2001), pp. 53–80.

23. *VO*, 3:201–3.

24. See ch. 3, above.

25. *VO*, 3:205.

26. *VO*, 3:203, 217. In what follows I abstain from comparing Volta's notion of tension with the notions of electromotive force and electric potential developed by physicists during the central decades of the nineteenth century, when the "volt" was gradually adopted as a practical unit of potential difference. Comparisons of that sort became routine after the volt was included in the system of electric units adopted at the International Congress of Electricians in Paris in 1881. For some brilliant exercises generated by the (partial) identification of Volta's tension with electric potential see Heilbron, *Electricity*, pp. 449–57. On the introduction of the notion of potential: Elizabeth Garber, "Siméon-Denis Poisson: Mathematics versus Physics in Early Nineteenth Century France," in *Beyond History of Science: Essays in Honor of Robert E. Schofield*, ed. Elizabeth Garber (Bethlehem: Lehigh University Press, 1990), pp. 156–76. On the scientific and commercial context that led to the adoption of electric units in late Victorian Britain, see Simon Schaffer's excellent "Accurate Measurement Is an English Science," in M. Norton Wise (ed.), *The Values of Precision* (Princeton: Princeton University Press, 1995), pp. 135–72. See also Bruce J. Hunt, "The Ohm Is Where the Art Is: British Telegraph Engineers and the Development of Electrical Standards," *Osiris*, 9 (1993): 48–63. On the Paris Congress of electricians: Paul Tunbridge, *Lord Kelvin: His Influence on Electrical Measurements and Units* (London: Peter Peregrinus, IEE, 1992), pp. 34–40.

27. *VO*, 3:202.

28. *VO*, 3:215–16.

29. *VO*, 3:206, 209.

30. See especially, VMS, L, 3. On the "sphere of activity" see VMS, I, 5, and VMS, I, 9.

31. *VO*, 3:206.

32. *VO*, 3:209.

33. *VO*, 3:206.

34. *VO*, 3:213.

35. Ibidem.

36. See ch. 3, above, p. 147ff.

37. *VE*, 1:306–8.

38. G. B. Beccaria, *Elettricismo artificiale* (Turin, 1772), p. 174ff.

39. Ibid., p. 173.

40. Ibid., pp. 176–80.

41. Ibid., p. 190.

42. H. Cavendish, "An Attempt to Explain Some of the Principal Phenomena of Electricity by Means of an Elastic Fluid," *PT*, 61 (1771): 584–677; H. Cavendish, "Some Attempts to Imitate the Effects of the *Torpedo* by Electricity," *PT*, 66 (1776): 196–225.

43. *VO*, 3:210n.

44. On the complex reactions towards Aepinus's adoption of the notion of action at a distance see R. W. Home in *Aepinus's Essay on the Theory of Electricity and Magnetism*, ed. R. W. Home (Princeton: Princeton University Press, 1979), pp. 195–97.

45. *VO*, 3:233–34. Cf. *Aepinus's Essay*, pp. 285–86.

46. M. Gliozzi, "Consonanze e dissonanze tra l'elettrostatica di Cavendish e quella di Volta," *Physis*, 11 (1969): 231–48, and J. L. Heilbron, *Electricity*, pp. 454n, 478.

47. *VO*, 1:11.

48. See Gliozzi, "Consonanze," p. 234. By 1775 Volta read English: *VO*, 3:84.

49. Heilbron, *Electricity*, p. 484.

50. *VE*, 1:312.

51. See ch. 5, below.

52. *VE*, 2:49.

53. *VE*, 2:4ff.

54. *VE*, 2:58.

55. The MS is preserved in RSA, "Letters and Papers," VIII, no. 242.

56. See RSA, "Journal Book" (Copy), vol. 30.

57. *VO*, 3:306.

58. *VE*, 2:471.

59. *VO*, 3:312–77.

60. RSA, "Journal Book," 30:509–13.

61. See the letters by Cavallo preserved in BL, MSs Additions 22,897/898.

62. RSA, "Journal Book," 30:518.

63. See Heilbron, *Electricity*, pp. 487–88.

64. *PT*, 72 (1782): 237–80.

65. *PT*, 72 (1782), Appendix, xxix. On the role of Lavoisier in these experiments see Marco Beretta, "From Nollet to Volta: Lavoisier and Electricity," *Rev. Hist. Sci.*, 54 (2001): 29–52.

66. See chap. 5, below.

67. *VE*, 2:163. On De Luc see R. P. Beckinsale, "Deluc," *DSB*, 4:27–29.

68. ". . . je trouvai encore, que, Mr De La Place excepté, et un peu Mr Lavoisier, tout le reste ne vous avoit point senti, et que celui qui se pique le plus d'être Electricien, et auquel pour ensi dire on se rapporte sur ce qui concerne l'Electricité, rioit de mon Enthousiasme, en levoit presque les epaules.

"Pour le coup je fus poussé au bout. Plus je sentois, moins j'exprimais; rhuminant seulement aux moiens de frapper quelque grand coupe sur ces têtes dures, et annonçant que ce ne seroit plus dans un simple Memoire au *Journal de Physique* que j'attaquerois les Préjugés.

"En effet, de retour à Londres je me mis à considérer mes Expériences, faites sur vos traces, avec les yeux tels que je les devois supposer (d'après ce que j'observois) à ceux qu'il importoit le plus de convencre; et je compris qu'ils ne seroient point frappés, qu'ils resteroient froids. Il falloit découvrir une Cause générale, fondamentale, au travers de bien des causes étrangères, qui la modifient; et quoique je ne perdisse jamais le fil de cette cause dans toutes ces modifications, quoique des Esprits comme celui de Mr De La Place l'y eussent surement retracée, je compris que je ne devois pas les supposer communs ces Esprits, puisque vous même vous aviez si peu excité l'attention en opéran sous bien des yeux" (*VE*, 2:164).

69. Cf. *VO*, 4:66.

70. R. J. Haüy, *Exposition raisonnée de la théorie de l'électricité et du magnétisme d'après les principes de M. Aepinus*, cit.

71. "Il est bien à remarquer qu'aucun Physicien François n'ait rien connu jusqu'à present aux athmospheres électriques. Voila pourquoi (Mr. de la Place excepté, cette excellente tête, et Mr Lavoisier auxquels j'ai plus particulierement expliqué mes idées), on ne m'a pas senti à Paris, malgré les expériences capitales que je montroit avec mes disques" (Volta to De Luc, 30 March 1784, Uppsala University Library, MS in the Waller Collection). On the only person Volta mentions approvingly among his acquaintances in Paris in 1782—the artisan named Billaum or Billaux (*VO*, 3:298, 375)—see M. Daumas, *Scientific Instruments of the Seventeenth and Eighteenth Centuries and Their Makers* (Engl. trans., London: Portman, 1972), p. 290.

72. *PT*, 72 (1782), Appendix, xxxi.

73. On the scientific circles to which Volta was admitted in London in 1782, see ch. 5, below. See also Richard John Sorrenson, *Scientific Instrument Makers at the Royal Society of London, 1720–1780* (Ph.D. dissertation, Princeton University, 1993), and David Philip Miller, "The Usefulness of Natural Philosophy: The Royal Society and the Culture of Practical Utility in the Later Eighteenth Century," *Brit. J. Hist. Sci.*, 32 (1999): 185–201. On Cavallo, Paola Bertucci, *Sparks of Life: Medical Electricity and Natural Philosophy in England, c. 1746–1792* (D. Phil. dissertation, University of Oxford, 2001), chapter 5. For Adam Walker's acquaintance with Volta, see A. Walker, *Ideas Suggested on the Spot in a Late Excursion through Flanders, Germany, France, and Italy* (London, 1790), p. 384.

74. *PT*, 72 (1782), Appendix, xxxi.

75. Volta to De Luc, 30 March 1784, Uppsala University Library, MS in the Waller Collection: ". . . d'ailleurs je suis sûr qu'en formant [with my scattered writings] un corps de doctrine comme je l'ai dans la tête, vous les présenterez au public beaucoup mieux que je ne sçauroie [*sic*] faire moi même, étant pourvu d'une tête beaucoup plus systematique que la mienne, et ayant le don d'écrire que je n'ai pas."

76. A copy of Volta's rare *De vi attractiva ignis electrici*, published in 1769, accompanied Volta's letter to De Luc of 30 March 1784, as mentioned in the letter itself.

77. "Mais personne que je sache n'a tenté de soumettre à l'action des athmospheres électriques un si grand nombre de phénomènes, comme j'ai fait. Voila peut-être la seule chose qui m'est due. En désduisant du même principe la théorie de la charge et de la decharge de Leyde, de l'Electrophore, du Condensateur, des pointes, la théorie électrique en general, qui étoit en butte à de fortes objections de la part de ceux qui ne voyaient pas l'ensemble, qui parois-

sait au moins fort imparfaite, devient on ne peut plus lumineuse et satisfaisante; et je ne doute point qu'elle ne resorte victorieuse et triomphante maniée par vous, Monsieur." (Volta to De Luc, 30 Mar. 1784, Uppsala University Library).

78. ". . . vous avez du trouver, *non par hazard, mais à priori* et l'Electrophore et le Condensateur" (*VE*, 2:207, emphasis added).

79. Volta to De Luc, 30 Mar. 1784, Uppsala University Library.

80. "Dans toutes vos spéculations sur l'objet qui nous occupe [electrical movements], vous liez deux choses, que je sépare au contraire toujours; savoir la *cause* des *Mouvemens électriques*, et l'*état* des Corps même qui se *meuvent*. C'est de ce dernier objet seulement que j'avois l'honneur de vous parler dans ma dernière lettre, et je vais maintenant vous expliquer pourquoi je pense qu'il faut toujours le traiter à part.

"Je remarque d'abord que le *Phénomène* à examiner n'est au fond que les *mouvemens* des *Corps libres*, résultant d'une Electrisation quelconque. Ces *Corps* paroissent *s'attirer* ou se *repousser*, ils *s'approchent* en un mot, ou *s'écartent*, suivant leur différens *états électriques*; et la question de *théorie* à cet égard consiste uniquement à déterminer, quels sont les *états* de ces Corps, quand il se meuvent d'une ou d'autre manière. En un mot il s'agit de déterminer les *Lois* des *attractions* ou *répulsions* dans les Corps *que nous voyons se mouvoir*.

"Sans doute qu'ensuite on peut, et l'on doit même s'occuper, de la *Cause* de ces *Mouvemens*: mais c'est là une Question *Systématique*, toute différente de la première, et à la quelle la détermination préalable de celle-ci est absolument indispensable" (*VE*, 2:204). Despite the methodological awareness displayed in this letter, De Luc himself was often tempted by similar speculations; see J. A. De Luc, *Idées sur la météorologie* (London, 1786), 2:249ff.

81. *VE*, 2:206, 208.

82. *VO*, 5:51.

83. "Il est aisé de concevoir que l'action de l'électricité [$Q$], est en raison composée de son intensité [tension = $T$]; et de la capacité [capacity = $C$] des conducteurs. Ainsi un petit conducteur fortement électrisé donnerà une étincelle à une grande distance, mais elle ne produirà pas un grand effet. Et au contraire un conducteur trés grand foiblement électrisé, donnerà une étincelle à un moindre distance qui produira un grand effet. Il faut donc bien distinguer, l'*intensité* de l'électricité d'avec sa quantité. La quantité est la somme de toutes les électricités contenues dans tous les points de la surface du corps électrisé. L'*intensité* est la force que chacun de ces points exerce, pour se defair [*sic*] de l'électricité dont il jouit, et se retablir en equilibre" (VMS, O, 23). These lecture notes were later developed and absorbed in Volta's then unpublished "Lezioni compendiose di elettricità," now in *VO*, 4:389–457. On Volta's law see Heilbron, *Electricity*, pp. 454–55.

84. "L'électricité d'un corps se fait sentir à tous les corps environans, bien au delà de la distance à la quelle la transfusion du fluide électrique peut se faire, témoin les mouvemens d'attraction et de répulsion qui ont lieu à quelques pieds de distance, tandis que l'aigrette ou l'étincelle électrique ne peut se porter qu'à quelques pouces. Les corps donc qui sont presentés au corps élec-

trisé hors la portée de l'étincelle ne laisse pas de se ressentir de l'état d'électri-cité de celui ci sans pourtant lui en prendre" (VMS, O 23, 5).

85. T. Frängsmyr, J. L. Heilbron, R. E. Rider, eds., *The Quantitying Spirit in the Eighteenth Century* (Berkeley, Los Angeles, London: University of California Press, 1990), esp. pp. 8–9. See also T. S. Kuhn, "The Function of Measurement in Modern Physical Science," *Isis*, 52 (1961): 161–93, now in T. H. Kuhn, *The Essential Tension* (Chicago: University of Chicago Press, 1977), pp. 178–224.

86. *PT*, 66 (1776): 206.

87. Ibidem, 209.

88. VMS, L, 3, published in *VO*, 3:165–67.

89. VMS, I, 9; VMS, I, 23; VMS, I, 28; VMS, I, 29. See *VO*, 3:245–58.

90. VMS, I, 13.

91. See F. Massardi, "Concordanze di risultati e formule emergenti da ma-noscritti inediti del Volta con quelli ricavati dalla fisico-matematica nella riso-luzione del problema generale dell'elettrostatica," *Rendiconti dell'Istituto Lom-bardo*, 2nd series, 56 (1923): 293–308; and Heilbron, *Electricity*, p. 456.

92. VMS, I, 9.

93. For Aepinus see Home in *Aepinus's Essay*, p. 115. As far as Volta is con-cerned, see his laboratory notes and drafts of papers kept in VMS, I, 1–55, amounting to a total of about 400 pages.

94. "Extrait d'un mémoire lu à l'Académie des Sciences, par M. Coulomb . . .," *Observations sur la physique* (1785), 116–17; see *VO*, 5:78n. The memoirs of Coulomb were published in Académie des Sciences, *Mémoires de mathématique et de physique* (1785–1788). On Coulomb and electricity see C. S. Gillmor, *Cou-lomb and the Evolution of Physics and Engineering in Eighteenth-Century France* (Princeton: Princeton University Press, 1971), pp. 175–221; Heilbron, *Electricity*, pp. 468–73; Christine Blondel and Matthias Dörries (eds.), *Restaging Coulomb: Usages, controverses et réplications autour de la balance de torsion* (Florence: Olschki, 1994). On Volta and Coulomb: Lucio Fregonese, "Two Different Scien-tific Programmes: Volta's Electrology and Coulomb's Electrostatics," in Chris-tine Blondel and Matthias Dörries (eds.), *Restaging Coulomb*, pp. 85–98.

95. Heilbron, *Electricity*, pp. 473–77; Peter Heering, "The Replication of the Torsion Balance Experiment: The Inverse Square Law and Its Refutation by Early Nineteenth-Century German Physicists," in Christine Blondel and Mat-thias Dörries (eds.), *Restaging Coulomb*, 47–66.

96. On Coulomb's instrumentalism, Heilbron, *Electricity*, pp. 474–75.

97. Ibidem, 475.

98. *VO*, 5:78–79.

99. *VO*, 5:55–56.

100. *VO*, 5:79–80.

101. *VO*, 5:77–81. See also VMS, I, 7 (published in *VO*, 5:94–105), VMS, I, 8, and VMS, I, 15.

102. Volta's main publication in this period were the nine *Lettere sulla mete-orologia elettrica*, published from 1788–1790. The first seven letters were also published in German translation from 1793–1799. See *VO*, 5:29–307.

103. *VO*, 5:35–36.

104. *VO*, 5:37–38.

105. *VO*, 4:74, and *VO*, 5:37–38.

106. *VO*, 5:80.

107. G. Polvani, *Alessandro Volta* (Pisa: Domus Galilaeana, 1942), pp. 144–45.

108. *VO*, 5:49.

109. Sigaud de la Fond, *Précis historique et expérimental des phénomènes électriques* (Paris, 1781), pp. 81–95, 656–706; Sigaud de la Fond, *Description et usage d'un cabinet de physique expérimentale*, 2nd ed. (Paris, 1784), pp. 423–33; T. Cavallo, *A Complete Treatise on Electricity*, 3d ed. (London, 1786), 1:244–72; G. Adams, *An Essay on Electricity*, 2nd ed. (London, 1785), pp. 171–74, 213–49; G. C. Lichtenberg, in J.C.P. Erxleben, *Anfangsgründe der Naturlehre*, 5th ed., G. C. Lichtenberg ed. (Göttingen, 1791), pp. 482–93, 497; F. A. C. Gren, *Grundriß der Naturlehre* (Halle, 1797), pp. 814–31; J. Cuthbertson, *Practical Electricity, and Galvanism* (London, 1807), pp. 139–48.

110. Venosto Lucati, *Iconografia ed epigrafia di Alessandro Volta* (Como, 1982), p. 13.

## Chapter 5
## The Cosmopolitan Network

1. See under the country names in *VE, Indici*.

2. *VE*, 1:63, *VE*, 2:72 and 128; *VO*, 5:235, *VO*, 7:341.

3. Pasquale Garovaglio, spices and drugs merchant; ASC, Notarile, MS 5221, no. 676, 26 Mar. 1800. Additional examples in *VE*, 1:392–93.

4. For a discussion, with bibliography, see L. Daston, "The Ideal and Reality of the Republic of Letters in the Enlightenment," *Sci. Context*, 4 (1991):367–86. Older studies include: T. J. Schlereth, *The Cosmopolitan Ideal in Enlightenment Thought* (Notre Dame and London: University of Notre Dame Press, 1977), and G. De Beer, *The Sciences Were Never at War* (London: Nelson and Sons, 1960). According to F. Venturi, *Italy and the Enlightenment* (London: Longman, 1972), xx, in 1789 "the cosmopolitan century came to an end and there began the age of nations and the internationals."

5. See ch. 1, above.

6. On Barletti and Volta see ch. 1, above.

7. On Beccaria and Volta see chs. 1 and 3, above.

8. See the sonnet following the dedication to Firmian in C. Barletti, *Nuove sperienze elettriche, secondo la teoria del Sig. Franklin e le produzioni del P. Beccaria* (Milan, 1771), and the dedication to Joseph II in G. B. Beccaria, *Experimenta atque observationes, quibus electricitas vindex late constituitur, atque explicatur* (Turin, 1769).

9. J. Torlais, *Un physicien au siècle des lumières* (Paris: Sipuco, 1954), 53–54.

10. J. A. Nollet, "Journal du Voyage de Piedmont et d'Italie en 1749," Bibliothèque Municipale, Soissons, MS 150.

11. *VE*, 1:42.

12. J. Priestley, *The History and Present State of Electricity* (London, 1767), xiv.

13. "Il bravo inglese . . . ," *VO*, 6:159.

14. *VO*, 3:189.

15. See under Lavoisier and Crawford in *VE, Indici*.

16. Ch. 3, above.

17. Ch. 1, above.

18. The plan is mentioned by Kaunitz Rietberg in a letter to Firmian of 26 June 1777, published in *VE*, 2:178–79.

19. *VE*, 1:178, 180–84, 195.

20. *VE*, 2:178.

21. SAV, Lombardei Korrepondenz, MS 137, fols. 48–50, 174–78, 237–43.

22. See F. Venturi, *Settecento riformatore*, V, *L'Italia dei lumi (1764–1790)*, I (Torino: Einaudi, 1987), pp. 713–25.

23. R. Pasta, *Scienza, politica e rivoluzione. L'opera di Giovanni Fabbroni (1752–1822) intellettuale e funzionario al servizio dei Lorena* (Florence: Olschki, 1989).

24. V. Ferrone, *La Nuova Atlantide e i lumi* (Turin: Meynier, 1988), 59–61.

25. *VE*, 1:159.

26. *VE*, 1:161, 169–70.

27. *VE*, 1:177.

28. *VE*, 1:182–83.

29. On Venini see M. Mamiani, "Francesco Venini: Un *philosophe* a Parma (1764–1772)" *G. Crit. Fil. Ital.*, 68 (70) (1989): 213–24.

30. *VE*, 1:186.

31. *VE*, 1:488, 492.

32. *VE*, 1:493.

33. *VE*, 1:188.

34. *VE*, 1:494.

35. Ibidem.

36. *VE*, 1:229–30.

37. *VE*, 1:495. On the scientific circles Volta became acquainted with in Geneva, see C. Montandon, *Le développement de la science à Genève aux XVIIIe et XIXe siècles* (Vivey: Delta, 1975), and *Les savants genevois dans l'Europe intellectuelle*, ed. J. Trembley (Genève: Journal de Genève, 1987).

38. *VE*, 1:197; *VO*, 6:168.

39. *VE*, 1:496.

40. See ch. 3, above.

41. *VE*, 1:494n.

42. Mentioned in *VO*, 7:175 n and *VE*, 1:326 respectively.

43. *VE*, 1:266.

44. *VE*, 1:229–30.

45. *VE*, 1:195.

46. *VE*. 1:392–93.

47. *VE*, 1:260–61, 420.

48. *VE*, 1:420–21.

49. *VE*, 2:5–10.

50. *VE*, 1:124. On Magellan: S. Pierson, "Magellan," in *DSB*, 9:5–6; J. de Carvalho, "Correspondência ciêntífica dirigida a Joâo de Magalhâes," *Revista da Faculdade de Ciências, Universidade de Coimbra*, 20 (1951): 93–283; S. F. Mason, "Jean Hyacinthe Magellan, F.R.S., and the Chemical Revolution of the Eighteenth Century," *Notes Rec. R. Soc. Lond.*, 45 (1991): 155–64.

51. *VE*, 2:17–19, 31, 35.

52. *VE*, 2:21–22, 25.

53. *VE*, 2:47–48.

54. *VE*, 2:129.

55. *VE*, 2:62, 70, 83, 89, 91, 105, 117.

56. *VE*, 2:79, 84.

57. H. Guerlac, "Sage," *DSB*, 12:63–69.

58. J. B. Gough, "Charles," *DSB*, 3:207–208; C. C. Gillispie, *Science and Polity in France at the End of the Old Regime* (Princeton: Princeton University Press, 1980), 539–43; C. C. Gillispie, *The Montgolfier Brothers and the Invention of Aviation, 1783–1784* (Princeton: Princeton University Press, 1983), 26–29.

59. On Volta and De Luc see ch. 4, above. On Volta and Madame Le Noir de Nateuil, see *VE*, 2:90–91, 100; *VO*, 1:8–12; *VO*, 6:329.

60. See ch. 4, above.

61. *VE*, 2:79.

62. *VE*, 2:94. On the Paris Academy see R. Hahn, *The Anatomy of a Scientific Institution* (Berkeley, Los Angeles, London: University of California Press, 1971), and C. C. Gillispie, *Science and Polity*, cit. For a colorful, contemporary account of the habits within the Academy around the time Volta visited it, see A. Birembaut, "L'Académie royale des Sciences en 1780 vue par l'astronome suédois Lexell," *Rev. Hist. Sci.*, 10 (1957): 148–66.

63. Fascinating accounts of such "salons" in J. Roger, *Buffon: Un philosophe au Jardin du Roi* (Paris: Fayard, 1989).

64. See note 59, above.

65. See ch. 4, above.

66. *VE*, 2:237n.

67. See *VO*, 3:296–98, and *PT*, 72 (1782), Appendix, vii–xxxiii, esp. xxix–xxx.

68. *Oeuvres de Lavoisier* (Paris, 1862–1893), 2:374–76, on 376. An account of the experiments carried out by Volta with Laplace and Lavoisier in H. Guerlac, "Chemistry as a Branch of Physics," *Hist. Stud. Phys. Sci.*, 7 (1976): 193–276, on pp. 234–40.

69. *Oeuvres de Lavoisier*, loc. cit.

70. *VO*, 3:306.

71. Volta's long memoir was read at the meetings of 7, 21, 28 Feb. and 14 Mar. 1782. See "Journal Book," RSA, MS (copy), vol. 30, 497–521. For the news circulating in Paris see *VE* 2:99.

72. *VE*, 2:237.

73. *VE*, 2:224–25, 237.

74. Venturi, *Settecento riformatore*, 5:739.

75. *VE*, 2:129–30.

76. *VE*, 2:112, 123.

77. Fascinating reports of the Society's activities and membership in "Chapter Coffee House Society," Museum of the History of Science, Oxford, MS Gunther 4.

78. *VE*, 2:292.

79. See note 71, above.

80. See ch. 4, above.

81. RSA, Cert. V, 165.

82. RSA, "Journal Book," vol. 35 (1793–96), 284–87. See also M. B. Hall, "The Royal Society and Italy, 1667–1795," *Notes Rec. Roy. Soc. Lond.*, 37 (1982): 63–81. On Galvani and Volta see chapter 6, below.

83. *VE*, 2:472.

84. Ibidem. On the circles Volta came in touch with in Birmingham, see R. E. Schofield, *The Lunar Society of Birmingham* (Oxford: Clarendon, 1963).

85. *VE*, 2:472. On the Manchester Society see A. Thackray, "Natural Knowledge in Cultural Context: The Manchester Model," *Amer. Hist. Rev.*, 79 (1974): 672–709.

86. *VE*, 2:473.

87. *VE*, 2:121.

88. *VE*, 2:472–73.

89. *VE*, 2:119.

90. Ibidem.

91. *VE*, 2:126.

92. *VE*, 2:285.

93. *VE*, 2:73.

94. See *VE*, 2:127.

95. See W. D. Hackmann, "The Electrical Researches of Martinus van Marum (1750–1837)" (Queen's University of Belfast, MA thesis, 1970), 161–89. I am grateful to the author for having lent me his personal copy.

96. Much valuable information on the scientific circles Volta got in touch with in Holland is found in *Martinus van Marum: Life and Work*, 3 vols., ed. R. J. Forbes (Haarlem: Willink, 1969–71), and in *Van Marum's Scientific Instruments in the Teyler's Museum*, ed. G. L'E. Turner and T. H. Levere, published as vol. 4 of the former, ed. E. Lefebvre and J. G. De Bruijn (Leyden: Noordhoof, 1973).

97. *VE*, 2:68.

98. *VE*, 2:72.

99. *VE*, 2:117n.

100. *VE*, 2:217.

101. *VE*, 2:232, 240.

102. *VE*, 2:229, 272.

103. *VE*, 2:215–17, 272.

104. *VE*, 2:275.

105. Ibidem.

106. *Lichtenbergs Briefe* (Leipzig, 1901–1904), 2:153–54.

107. Op. cit., 154.

108. Op. cit., 150.

109. J. L. Heilbron, *Electricity in the Seventeenth and Eighteenth Centuries* (Mineola, New York: Dover Publications, 1999), p. 424. See also C. Volpati, *Scritti voltiani* (Como: Comune di Como, 1974), pp. 55–103.

110. *VE*, 2:496.

111. *VE*, 2:435.

112. *VE*, 2:219, 273, 496–97.

113. *VE*, 2:490–91.

114. *VE*, 2:252.

115. *VE*, 2:232, 240.

116. *VE*, 2:246–47.

117. *VE*, 2:246 and n., 344, 355. See: J. B. Gough, "Born," *DSB*, 2:315–16; [I. von Born], *Specimen monachologiae* (Augsburg, 1783).

118. *VE*, 2:282–84, 291, 295, and *VE*, 3:279. The drawings of the physics theater planned by Volta for the University of Pavia are preserved in Vienna, SAV, Lombardei Collectanea, MS 84, fols. 78, 80–82, 126, 130.

119. The best overall treatment of Volta's political conduct is still C. Volpati, *Alessandro Volta nella gloria e nell'intimità* (Milano: Treves, 1927), pp. 243–323. Among primary sources of the "French" period: G. C. Gattoni, "Giornale Gallo-Cisalpino," BCC, MS 4.6.1.

120. *VE*, 4:9–10, 15–6, 21–23, 32.

121. *VO*, 1:525; *VO*, 2:24–26.

122. *VE*, 4:37.

123. *VE*, 4:44–47.

124. Brugnatelli's diary in *VE*, 4:491, 519–20.

125. *VE*, 4:151.

126. Besides the already mentioned golden medal, the rewards dispensed to Volta by Bonaparte included: a gratuity of 6,000 francs in 1801; the Legion of Honor and the promise of a pension of 3,000 lire yearly in 1805 (the latter never materialized because of administrative and political troubles); the title of Knight of the Iron Crown in 1806; the appointment of Senator of the Kingdom of Italy in 1809; the title of Count of the same Kingdom in 1810; and—through a decree dated from Moscow!—the title of Chairman of the Constituency of Lario (the lake of Como) in 1812. See *VE*, 5:544.

127. Brugnatelli in *VE*, 4:519.

128. *VO*, 2:122–23.

129. See Ken Alder, *Engineering the Revolution: Arms and Enlightenment in France, 1763–1815* (Princeton: Princeton University Press, 1997), esp. ch. 8.

130. Brugnatelli in *VE*, 4:519, 521. On science and weapon research in Paris in those years, see C. C. Gillispie, "Science and Secret Weapons Development in Revolutionary France, 1792–1804: A Documentary History," *Hist. Stud. Phys. Biol. Sci.*, 23 (1992): 35–152.

131. *VE*, 4:519, 521.

132. Crosland, *The Society of Arcueil* (Cambridge, Mass.: Harvard University Press, 1967), pp. 9–12. On the general context in which Volta's second visit to Paris took place: Nicole and Jean Dhombres, *Naissance d'un pouvoir: sciences et savants en France, 1793–1824* (Paris, Payot, 1989).

133. *VE*, 4:85, 92–93.

## Chapter 6
## The Battery

1. On Volta's discovery see: J. L. Heilbron, "Volta's Path to the Battery," in G. Dubpernell and J. H. Westbrook (eds.), *Selected Topics in the History of Electrochemistry* (Princeton, 1978), pp. 39–65, and J. L. Heilbron, *Electricity in the Seventeenth and Eighteenth Centuries: A Study in Early Modern Physics* (Mineola, New York: Dover Publications, 1999; 1st ed., 1979), Preface to the Dover edition. See

also: G. Polvani, *Alessandro Volta* (Pisa: Domus Galilaeana, 1942), ch. 8; John L. Heilbron, "Analogy in Volta's Exact Natural Philosophy," in Fabio Bevilaccqua and Lucio Fregonese (eds.), *Nuova Voltiana: Studies on Volta and His Times*, 1 (2000), pp. 1–23, esp. 15–23; Joost Mertens, "Shocks and Sparks: The Voltaic Pile as a Demonstration Device," *Isis*, 89 (1998): 300–11.

2. Polvani, *Volta*, p. 443, declares he could not consult Nicholson's *Journal*; Heilbron, "Volta's Path," emphasized the torpedo fish, but did not mention Nicholson's paper. See however Heilbron's Preface to the Dover Edition, in J. L. Heilbron, *Electricity*, pp. xi–xxv, esp. xxiv–v, and Heilbron, "Analogy in Volta's Exact Natural Philosophy," p. 20.

3. The best overall account of the controversy over Galvanism is Marcello Pera, *La rana ambigua: La controversia sull'elettricità animale tra Galvani e Volta* (Torino: Einaudi, 1986; Amer. trans., Princeton: Princeton University Press, 1992). See also Walter Bernardi, *I fluidi della vita: Alle origini della controversia sull'elettricità animale* (Florence: Olschki, 1992); and Marco Piccolino and Marco Bresadola, *Rane, torpedini e scintille* (Turin: Bollati Boringhieri, forthcoming).

4. M. Faraday, *Experimental Researches in Electricity*, 3 vols. (London, 1839–1855; reprint Brussels: Culture et Civilisation, 1969), I, 300.

5. *VO*, 2:339.

6. [L. Galvani], *Dell'uso e dell'attività dell'arco conduttore nelle contrazioni dei muscoli* (Bologna, 1794).

7. On this and other instruments mentioned below, see W. D. Hackmann, "Eighteenth Century Electrostatic Measuring Devices," *Annali dell'Istituto e Museo di Storia della Scienza di Firenze*, 3 (1978): 3–58, and W. D. Hackmann, "The Relationship between Concept and Instrument Design in Eighteenth-Century Experimental Science," *Ann. Sci.*, 36 (1979): 205–24.

8. *VE*, 3:294.

9. *VO*, 1:495–96. On the *condensatore* see ch. 4, fig. 4:1, above.

10. *VO*, 1:472.

11. Ibidem.

12. *VO*, 1:493n.

13. Ibid., 469, 473; cfr. ibid. 398.

14. *PT*, 78 (1788): 406–7.

15. *VO*, 1:497.

16. Ibid., 499–500.

17. Ibid., 506.

18. Ibid., 166.

19. Ibid., 134, 137, 139, 238.

20. Ibid., 489.

21. Ibid., 284.

22. Loc. cit.

23. Ibid., 463.

24. Ibid., 464.

25. Ibid., 545.

26. Ibid., 305.

27. Ibid., 304.

28. Ibid., 456–57.

29. Ibid., 475.

30. Ibid., 485–86.

31. T. Cavallo, *A Complete Treatise on Electricity*, 3d. vol. (London, 1795), p. 76ff.

32. T. Cavallo, "Of the Method of Manifesting the Presence, and Ascertaining the Quality, of Small Quantities of Natural or Artificial Electricity," *PT*, 78 (1788): 1–22, on p. 8.

33. A. Bennet, "An Account of a Doubler of Electricity," *PT*, 77 (1787): 288–96.

34. Cavallo, *Complete Treatise*, p. 98 f.

35. Ibid., 105.

36. W. Nicholson, "Description of an Instrument," *Journal of Natural Philosophy, Chemistry, and the Arts*, 1 (1797): 16–18; and W. Nicholson, "Observations" on T. Cavallo, "On the Multiplier of Electricity," ibid., 394–99.

37. *VE*, 3:349.

38. *VO*, 1:443.

39. *VO*, 1:439.

40. (Bologna, 1797).

41. A. von Humboldt, *Expériences sur le galvanisme* (Paris, 1799; original German ed., 1797).

42. *VO*, 1:517–55.

43. See W. C. Walker, "Animal Electricity before Galvani," *Ann. Sci.*, 1 (1937): 84–113. On the prehistory of electric fishes, see Brian P. Copenhaver, "A Tale of Two Fishes: Magical Objects in Natural History from Antiquity through the Scientific Revolution," *J. Hist. Ideas*, 52 (1991), pp. 373–98.

44. *VO*, 1:559.

45. *VO*, 2:205–99.

46. See chapter 1, above.

47. *VO*, 1:19.

48. Isaac Newton, *Opticks*, 4th ed. (1730; London, 1931), pp. 353–54. On the subject discussed in this section, see R. W. Home, "Electricity and the Nervous Fluid," *J. Hist. Biol.*, 3 (1970): 235–71.

49. *VO*, 1:341.

50. See I. B. Cohen, "Introduction" to L. Galvani, *Commentary on the Effects of Electricity on Muscular Motion* (Norwalk: Burndy Library, 1953), p. 10 f, and Marco Bresadola, *Medicina e filosofia naturale in Luigi Galvani* (Ph.D. dissertation, University of Florence, 1999).

51. *VO*, 4:457.

52. *VO*, 1:8–12.

53. *VO*, 4:453–57.

54. G. Mangili, "Autobiografia," Biblioteca Civica di Bergamo, MS 79.R.8.f.11.

55. *VO*, 1:364.

56. Ibid., 125.

57. Loc. cit.

58. A. Scarpa, *Tabulae neurologicae* (Pavia, 1794), p. 6.

59. G. Mangili, *De systemate nerveo hirundinis, lumbrici terrestris, aliorum Vermium* (offprint, 1795), pp. 249–61.

60. *VO,* 1:340.

61. Ibid., 341.

62. For an (unsympathetic) discussion of similar ideas, see Felice Fontana in A. von Haller, *Mémoires sur les parties sensibles et irritables du corps animal* (Lausanne, 1760), pp. 207–8.

63. M. Girardi, "Saggio di osservazioni anatomiche intorno agli organi elettrici della torpedine," *Memorie di matematica e fisica della Società Italiana,* 3 (1786): 553–70, on p. 566.

64. *VO,* 1:341.

65. Ibid., 339.

66. Ibid., 340.

67. *VE,* 3:412–18.

68. See for example J. Walsh, "Of the Electric Property of the Torpedo," *PT,* 63 (1773): 461–80, on p. 476. Galvani himself, in his memoir (Galvani, *Memorie,* p. 65), compared the layers observed in the lateral organs of the torpedo to as many Franklin squares. According to Galvani, in any case, the organs derived their electricity from the brain of the animal.

69. *VO,* 1:340.

70. VMS, J 45, J48.

71. *VO,* 1:540.

72. Ibid., 551–52.

73. VMS, J 53.

74. *VE,* 3:440; and ibid., 436.

75. Ibid., 441 f.

76. Ibid., 376–78.

77. Ibid., 466.

78. W. Nicholson, "Observations on the Electrophore, Tending to Explain the Means by Which the Torpedo and Other Fish Communicate the Electric Shock," *Journal of Natural Philosophy, Chemistry, and the Arts,* 1 (1797): 355–59.

79. Ibid., 358–59.

80. Drafts of Volta's letter of 20 March 1800 to Banks, interspersed with other manuscripts, are in VMS, H 47, J 54–57, J 67–68.

The document known and referred here as Volta's letter of 20 Mar.1800 to Banks *(VO,* 1:565–82) is in fact the combination prepared in London of two documents sent by Volta separately. This was known from Nicholson's report of their arrival in London *(Journal,* iv, 179), and from *VE,* 3:469–70. The portion of Volta's autographed letter I identified in RSA, "Letters and Papers," vol. 11, no. 137, fol. 2, confirms this. It also confirms that Volta in 1800 was working on a longer paper about the battery, of which the MSS in VMS quoted above are drafts. Volta's autographs preserved in RSA also show that Volta let Banks decide whether to publish the part of his description of the battery dealing with Nicholson.

81. On the "imitative experiment" in eighteenth-century science, see Hackmann, "The Relationship," note 7, above.

82. See fig. 6.8.

83. *VO*, 1:285.

84. VMS, J 87ß, and *VO*, 2:369. For the date of the discovery (Dec. 1799), see also *VO*, 2:10.

85. D. J. de Solla Price, "Of Sealing Wax and String," *Natural History*, 93 (1984): 49–56, on p. 49.

86. The example in fig. 68, above, should not be misread: the combinations there represented, dating from 1794–95, are always arcs or circuits, and the "units" are not discs, but conventional representations of the pieces, including frogs (*r*) and people (*p*), forming the circuit. Volta's laboratory notebooks from this period show that he then worked mainly with metal bars or wires, not discs.

87. A. Humboldt, "De l'irritabilité de la fibre nerveuse et musculaire," *Journal de physique*, Prairial an VI (May 1798): 465–74, on p. 470.

88. See G. Tabarroni, "Galvani, Aldini e la corrente elettrica," Istituto Tecnico Industriale Aldini-Valeriani, *Annuario* (1971): 43–54.

89. *VO*, 1:572.

90. *VO*, 2:62–63.

91. VMS, H 48.

92. Ibid., J 82.

93. *VO*, 2:62.

94. See R. J. Haüy, *Traité élémentaire de physique* (Paris, 1803), 2:47–48.

95. J. N. Hallé, "Exposition abrégé des principales expériences répétées par M. Volta," *Bulletin des sciences, par la Société Philomatique*, 3, Nivose an X (Dec. 1801): 74–80 and planche 4.

96. *VO*, 2:205–99.

97. Reporting Volta's discovery, Nicholson apparently did not complain about the treatment Volta had reserved for his model of the torpedo in the letter addressed to Banks ("Sig. Volta makes honourable mention of my conjectural theory of the torpedo . . ."). Yet, he reproached Volta for having ignored the chemical phenomena accompanying the action of the battery: W. Nicholson, "Account of the Electrical or Galvanic Apparatus of Sig. Alex. Volta," *Journal of Natural Philosophy, Chemistry and the Arts*, 4 (1800): 179–87, on p. 181.

98. F. L. Holmes, *Lavoisier and the Chemistry of Life: An Exploration of Scientific Creativity* (Madison: University of Wisconsin Press, 1985), pp. 486–87.

99. Heilbron, *Electricity*, pp. 71–72. See also J. L. Heilbron, "A. Volta," *DSB*, 14:69–82.

100. Heilbron, *Electricity*, 431, and Price, "Of Sealing Wax and String," p. 49.

101. More on unintended consequences in chapter 9, below.

## Chapter 7
## Appropriating Invention

1. Thomas S. Kuhn, "Energy Conservation as an Example of Simultaneous Discovery," in Marshall Clagett (ed.), *Critical Problems in the History of Science* (Madison: University of Wisconsin Press, 1959), pp. 321–56, also in Thomas S. Kuhn, *The Essential Tension* (Chicago: The University of Chicago Press, 1977), pp. 66–104; Thomas S. Kuhn, "The Function of Measurement in Modern Physi-

cal Science," *Isis*, 52 (1961): 161–90, also in Thomas S. Kuhn, *The Essential Tension*, pp. 178–224; Thomas S. Kuhn, "What Are Scientific Revolutions?," in Lorenz Krüger, Lorraine J. Daston, Michael Heidelberger (eds.), *The Probabilistic Revolution*, volume 1: *Ideas in History* (Cambridge, Mass.: MIT Press, 1987), pp. 7–22; John L. Heilbron's *Electricity in the Seventeenth and Eighteenth Centuries: A Study in Early Modern Physics* (Mineola, N.Y.: Dover Publications, 1999; original edition, 1979), pp. 491–94.

2. Ibidem, p. 494.

3. A major exception, focusing on Humphry Davy's use of the battery, is Jan Golinski, *Science as Public Culture: Chemistry and Enlightenment in Britain, 1760–1820* (Cambridge: Cambridge University Press, 1992), chapter 7. Other important exceptions: Helge Kragh, "Confusion and Controversy: Nineteenth-Century Theories of the Voltaic Pile," in Fabio Bevilacqua and Lucio Fregonese (eds.), *Nuova Voltiana: Studies in Volta and His Times*, vol. 1 (2000), pp. 133–57; Nahum Kipnis, "Debating the Nature of Voltaic Electricity," in Fabio Bevilacqua and Lucio Fregonese (eds.), *Nuova Voltiana: Studies in Volta and His Times*, vol. 3 (2001), pp. 121–51; Willem Hackmann, "The Enigma of Volta's 'Contact Tension' and the Development of the 'Dry Pile,'" in Fabio Bevilacqua and Lucio Fregonese (eds.), *Nuova Voltiana: Studies in Volta and His Times*, vol. 1 (2000), pp. 103–19. See also: Otto Sibum, "Die Mechanisierung der Lebensvorgänge—der Weg zum elektrischen Strom," in Jörg Meya, Otto Sibum, *Das fünfte Element: Wirkungen und Deutungen der Elektrizität* (Munich: Deutsches Museum, Rowohlt, 1987), pp. 117–41; Richard H. Schallenberg, *Bottled Energy: Electrical Engineering and the Evolution of Chemical Energy Storage*, American Philosophical Society, Memoirs, vol. 148 (Philadelphia, 1982); Joost, Mertens, "From the Lecture Room to the Workshop: John Frederic Daniell, the Constant Battery and Electrometallurgy around 1840," *Ann. Sci.*, 55 (1998): pp. 241–61.

4. On Napoleon and Europe: Stuart Joseph Woolf, *Napoleon's Integration of Europe* (London and New York: Routledge, 1991); Michael Broers, *Europe under Napoleon, 1799–1815* (London, Arnold, 1996).

5. Royal Society Archives, London, Letters and Papers, XI, no. 137.

6. Royal Society Archives, London, Letters and Papers, XI, no. 133, f. 2, v. For the identification of Pasquale Garovaglio: Archivio di Stato, Como, Notarile, cartella 5221, n. 676.

7. "The latter end of last April," according to W. Nicholson, "Account of the New Electrical or Galvanic Apparatus of Sig. Alex. Volta, and Experiments Performed with the Same," *A Journal of Natural Philosophy, Chemistry, and the Arts* (July 1800), pp. 179–87, 179. "In the beginning of last Month [i.e., April]," according to a public statement agreed to among Garnett, Rumford, and Banks, to be read at the Royal Institution on the evening of 30 May 1800: Rumford to Banks, 30 May 1800, Dartmouth College Library, Hanover, N.H.

8. As evinced from Nicholson's letter to the editor in *The Morning Chronicle*, 3 June 1800, p. 3. Here Nicholson vindicates his own and Carlisle's merits in the discovery of the decomposition of water as if knowing that Volta had *not* treated the issue in the second part of the paper describing the battery announced in the first part, which had been in Nicholson's hands since April.

9. *VE*, 4:1–2.

10. *VE*, 4:6.

11. *VO*, 2:3.

12. On Carlisle: R. J. Cole, "Sir Anthony Carlisle, FRS (1768–1840)," *Ann. Sci.*, 8, no. 3 (1952?): 255–70.

13. Nicholson, "Account of the New Electrical or Galvanic Apparatus," p. 181.

14. Rumford to Banks, 29 May and 30 May 1800, Dartmouth College Library, Hanover, N.H. On the Royal Institution: Morris Berman, *Social Change and Scientific Organization: The Royal Institution, 1799–1844* (London: Heinemann, 1978). On Garnett: S.G.E. Lythe, *Thomas Garnett: 1766–1802* (Glasgow: Polpress, 1984).

15. *The Morning Chronicle*, 30 May 1800, p. 3.

16. William Nicholson, Letter to the Editor, *The Morning Chronicle*, 3 June 1800, p. 3.

17. As reported by Rumford in Rumford to Banks, 29 May 1800, Dartmouth College Library, Hanover, N.H.

18. Rumford to Banks, 30 May 1800, Dartmouth College Library, Hanover, N.H.

19. *Courier de Londres*, 8 August 1800, p. 96.

20. *Gazette nationale ou Le moniteur universel*, 29 thermidor an 8 (17 August 1800), no. 329, p. 1.

21. *VO*, 2:7.

22. E.-G. Robertson, "Expériences nouvelles sur le fluide galvanique, . . . lues à l'Institut National de France, le 11 fructidor an 8," *Annales de chimie*, vol. 37 (December 1800): 132–50. On Robertson's career: E.-G. Robertson, *Mémoires récréatifs, scientifiques et anecdotiques du physicien aéronaute E.-G. Robertson* (Paris, 1831–1834). On the *Annales de chimie*: Maurice Crosland, *In the Shadow of Lavoisier: The* Annales de chimie (British Society for the History of Science, 1994).

23. Jac Remise, Pascale Remise, Regis van de Walle, *Magie lumineuse. Du théatre d'ombres à la lanterne magique* (Balland, 1979), pp. 39–61; Thomas L. Hankins and Robert J. Silverman, *Instruments and the Imagination* (Princeton: Princeton University Press, 1995), pp. 63–64.

24. Years later, when Robertson won new popularity as an aeronaut, a big, friction electrical machine was represented on the illustrations advertising his balloon, called "La Minerve," together with other scientific and technological devices that he supposedly carried on board: Bibliothèque Nationale, Paris, Estampes, Collection Hennin, vol. 164, no. 14356.

25. E.-G. Robertson, "Expériences nouvelles sur le fluide galvanique," p. 134.

26. C. Blondel, "Animal Electricity in Paris: From Initial Support, to Its Discredit and Eventual Rehabilitation," in *Luigi Galvani International Workshop: Proceedings*, ed. Marco Bresadola and Giuliano Pancaldi, *Bologna Studies in History of Science*, vol. 7 (CIS, University of Bologna, 1999), pp. 187–209.

27. Banks to Van Marum, 14 June 1800, British Library, Warren R. Dawson MSs, 68 (I), 28–29, copy.

28. See Willem D. Hackmann, "Electrical Researches," in *Martinus Van Marum: Life and Work*, ed. R. J. Forbes (Haarlem: Willink, 1969–76), 6 vols., vol. 3, pp. 329–78, esp. 359–61.

29. See S. Lilley, "Nicholson's Journal (1797–1813)," *Ann. Sci.*, 6 (1948–50), pp. 78–101.

30. Vol. 14, pp. 398–99.

31. *Bibliothèque britannique*, vol. 15 (1800): 3–4.

32. Ibidem, vol. 14, 102.

33. *Annalen der Physik*, vol. 6 (1800): 340–75; vols. 8–11 (1801–1802); passim.

34. Hans Christian Ørsted, *Selected Scientific Works*, translated and edited by Karen Jelved, Andrew D. Jackson, and Ole Knudsen, with an introduction by Andrew D. Wilson (Princeton: Princeton University Press, 1998), p. 101.

35. H. Davy, "An Account of Some Experiments Made with the Galvanic Apparatus of Signor Volta," *A Journal of Natural Philosophy, Chemistry, and the Arts* (September 1800): 275.

36. Robison to Watt [October 1800], published in *Partners in Science. Letters of James Watt and Joseph Black*, edited with introductions and notes by Eric Robinson and Douglas McKie (Cambridge, Mass.: Harvard University Press, 1970), p. 358.

37. The literature on replication includes: Ludwik Fleck, *Genesis and Development of a Scientific Fact* (Chicago: University of Chicago Press, 1979; original edition: 1935); Harry M. Collins, *Changing Order: Replication and Induction in Scientific Practice* (London: SAGE, 1985); Steven Shapin and Simon Schaffer, *Leviathan and the Air-Pump: Hobbes, Boyle, and the Experimental Life* (Princeton: Princeton University Press, 1985); Andy Pickering, "Forms of Life: Science, Contingency, and Harry Collins," *Brit. J. Hist. Sci.*, 20 (1987): 213–21.

38. *The Morning Chronicle*, 30 May 1800, p. 3.

39. Volta to Banks, 20 March 1800, Royal Society of London, Letters and Papers, no. 137.

40. *A Journal of Natural Philosophy, Chemistry, and the Arts* (July 1800): 179.

41. *VO*, 2:7.

42. A few early voltaic batteries preserved in Lombardy were destroyed by fire during an exhibition in 1899: Cencio Poggi, "Il salone dei cimelii," in Società Storica Comense, *Raccolta Voltiana*, (1899), pp. 12–15. The Musée d'Histoire des Sciences de Genève preserves a battery that is said to have been donated by Volta to Nicholas-Théodore de Saussure in 1801; see "Association pour le Musée d'Histoire des Sciences de Genève, *Les savants genevois dans l'Europe intellectuelle du XVIIe au milieu du XIXe siècle*, ed., Jacques Trembley (Genève: Editions du Journal de Genève, 1987), p. 134. The Royal Institution, London, preserves a battery which is said to have been donated by Volta to Faraday, when the latter visited Volta on the occasion of a trip with Humphry Davy to Italy in 1814.

43. For an in-depth review of the literature on constructivism: Jan Golinski, *Making Natural Knowledge: Constructivism and the History of Science* (Cambridge: Cambridge University Press, 1998); on replication: pp. 28–29, 137–39.

44. Also, in English translation, in *The Philosophical Journal*, vol. 7 (1800), pp. 289–311.

45. Volta's first recorded letters to expert electricians (Beccaria and Nollet) were dated 1763: see chapter 1, above.

46. Chapters 3 and 4, above.

47. Volta, "On the Electricity Excited by the Mere Contact of Conducting Substances of Different Kinds," in *VO*, 1:565–82, esp. 565–67.

48. Ibidem, pp. 565, 566, 576.

49. Ibidem, pp. 573–74, 576–81.

50. Chapter 6, above.

51. *VO*, 1:575.

52. *VO*, 1:572.

53. *VO*, 1:569.

54. Royal Society of London, Archives, Letters and Papers, nos. 133, 137.

55. *A Journal of Natural Philosophy* (July 1800): 180.

56. *VO*, 1:565, 577.

57. See "Volta's Battery, 2," below.

58. Chapter 6, above.

59. *VO*, 1:581–82.

60. *A Journal of Natural Philosophy* (July 1800), 179.

61. Ibidem, 181.

62. Ibidem.

63. Giovanni Fabbroni, "On the Chemical Action of Different Metals on Each Other at the Common Temperature of the Atmosphere," *A Journal of Natural Philosophy, Chemistry and the Arts* (1799): 300–10; and (1800): 120–27. On Fabbroni, Nicholson, and the battery: Ferdinando Abbri, "Il misterioso 'spiritus salis': Le ricerche di elettrochimica nella Toscana napoleonica," *Nuncius*, 2 (1987): 55–88, esp. 67–68.

64. *A Journal of Natural Philosophy* (July 1800): 186.

65. Ibidem, 187.

66. E.-G. Robertson, "Expériences nouvelles sur le fluide galvanique, . . . lues à l'Institut National de France, le 11 fructidor an 8," *Annales de chimie*, 37 (December 1800): 132–50, 132.

67. Ibidem, 136–37.

68. Ibidem, 139.

69. Ibidem, 144; on Volta's frequent contacts with Robertson, see Brugnatelli's diary (note 106, below).

70. On Ritter: Maria Jean Trumpler, *Questioning Nature: Experimental Investigations of Animal Electricity in Germany, 1791–1810* (Ph.D. dissertation, Yale University, 1992), ch. 5; Stuart Walker Strickland, *Circumscribing Science: Johann Wilhelm Ritter and the Physics of Sideral Man* (Ph.D. dissertation, Harvard University, 1992).

71. J. W. Ritter, *Beweis, dass ein beständiger Galvanismus den Lebensprocess in dem thierreich begleite* (Weimar, 1798); *VE*, 3:385–406.

72. Quoted in Wilhelm Ostwald, *Electrochemistry: History and Theory*, American translation (New Delhi: Amerind Publishing, 1980), 1:154, 156.

73. *A Journal of Natural Philosophy* (July 1800): 183.

74. Quoted in Ostwald, *Electrochemistry*, 1:155.

75. *VE* 5:85.

76. Golinski, *Making Natural Knowledge*, ch. 7.

77. Stuart Walker Strickland, "The Ideology of Self-Knowledge and the Practice of Self-Experimentation," *Eighteenth-Century Studies*, 31, no. 4 (1998): 453–71, p. 457.

78. Quoted in Strickland, "The Ideology," 455.

79. *VE*, 3:385ff; *VO*, 4:271.

80. Hans Christian Ørsted, *Selected Scientific Works*, translated and edited by Karen Jelved, Andrew D. Jackson, and Ole Knudsen, with an introduction by Andrew D. Wilson (Princeton: Princeton University Press, 1998), p. 101.

81. L. Pearce Williams, "Oersted, Hans Christian," *DSB*, 10: 182–86.

82. Ørsted, *Selected Scientific Works*, 101.

83. Ibidem.

84. Ørsted, *Selected Scientific Works*, 103.

85. Ibidem, 102–3.

86. Newton's motto was reprinted on the title page of Alessandro Volta's *L'identità del fluido elettrico col così detto fluido galvanico* (Pavia: Capelli, 1814).

87. *VO*, 2:7.

88. *A Journal of Natural Philosophy* (November 1800): 341; Davy's emphasis.

89. *VE*, 4:17.

90. *VE*, 4:24.

91. *VO*, 1:525; *VE*, 2:32n.

92. *VE*, 4:37.

93. *VE*, 4:52–53.

94. *VO*, 2:16.

95. Chapters 4 and 5, above.

96. Chapter 4, above.

97. Alessandro Volta, "De l'électricité dite galvanique," *Annales de chimie*, 30 Frimaire an 10 (Dec. 1801): 225–56. The phrases "expériences fondamentales," or "fait principal," to designate Volta's experiments with a single metallic pair without humid conductors, were first introduced by Biot in his report of Volta's presentation at the Institut (dated 2 Dec. 1801): J. B. Biot, "Rapport sur les expériences du citoyen Volta," *Mémoires de l'Institut National, Sc. mathém. et phys.*, 5 (1804): 195–222, 196. Later, to designate these same experiments, Volta himself occasionally adopted the phrase "sperienza fondamentale" (*VO*, 2:52; from a paper originally published in 1802). I am grateful to Christine Blondel for discussion on this point.

98. Volta, "De l'électricité dite galvanique," p. 235.

99. Ibidem, 236, 238, 241.

100. Ibidem, 241.

101. Ibidem, 232.

102. Ibidem, 241.

103. The most detailed account of the uncertainties involved is Volta's own MS L 31 (likely date: 1808), VMS, published in *VO*, 2:340–41. Another experimental protocol, pointing at a measure of 1/100th rather than 1/60th for each metallic pair is MS I 45, VMS. Here Volta is using Leyden jars to test the tension generated by his battery.

104. See for example J. N. Hallé, "Exposition abregée des principales experiences répétées par M. Volta en présence des Commissaires de l'Institut national," *Bulletin de la Société Philomatique*, Nivose an 10 (Dec.1801): 74–80, esp. 74–75, 76, 78. On Volta and Coulomb's experiments with the torsion balance, see chapter 4, above.

105. M. A. Pictet to a correspondent in London, 15 Oct. 1801, Smithsonian Institution Libraries, Washington, Dibner Collection, MS 1142A; Rumford to Joseph Banks, 11 Nov. 1801, Massachusetts Historical Society, and "Count Rumford's Journal," University Library, Birmingham (transcriptions of Rumford material were kindly supplied by the Dartmouth College Library, Hanover, New Hampshire); H.-R. Wiedemann, "Alexander Volta und Christoph Heinrich Pfaff," *Christiana Albertina*, 23 (1986): 25–34; on Robertson see note 22, above.

106. The information reported in this and the following paragraphs is taken from Brugnatelli's diary (published in *VE*, 4:461–533), and from the newspapers mentioned in the text.

107. Rumford to Banks, 11 November 1801, Massachusetts Historical Society.

108. Quoted in J. Fischer, *Napoleon und die Naturwissenschaften* (Stuttgart: Steiner, 1988), p. 283.

109. Chapter 4, above.

110. *VE*, 4:124, and chapter 5, above.

111. *Le Moniteur*, 2 Nivose an 10 (24 Dec. 1801), pp. 369–70.

112. Bonaparte to Volta, 17 Frimaire an 10 (Dec. 1801), *VE*, 4:123.

113. J. B. Biot, "Rapport sur les expériences du citoyen Volta," *Mémoires de l'Institut National, Sc. mathém. et phys.*, 5 (1804): 195–222. On the report see E. Frankel, "J. B. Biot and the Mathematization of Experimental Physics in Napoleonic France," *Hist. Stud. Phys. Sci.*, 8 (1977): 33–72, on pp. 47–52.

114. G. Sutton, "The Politics of Science in Early Napoleonic France: The Case of the Voltaic Pile," *Hist. Stud. Phys. Sci.*, 11 (1981): 329–66, esp. on p. 363. On the context of Biot's report see R. Fox, "The Rise and Fall of Laplacian Physics," *Hist. Stud. Phys. Sci.*, 4 (1974): 89–130.

115. Chapter 4, above.

116. See Blondel , "Animal Electricity in Paris." On Biot: Eugene Frankel, *Jean Baptiste Biot: The Career of a Physicist in Nineteenth-Century France* (Ph.D. dissertation, Princeton University, 1972); Ivor Grattan-Guinness, *Convolutions in French Mathematics, 1800–1840* (Basel, Boston, Berlin: Birkäuser, 1990), 3 vols., 1:187–9, and *ad indicem*.

117. J.-B. Biot and F. Cuvier, "Sur quelques propriétés de l'appareil galvanique," *Annales de chimie*, 30 Messidor an 9 (July 1801): 242–50. See also *Bulletin des sciences, par la Société Philomatique*, an 9 (1801): 40.

118. J.-B. Biot, "Sur le mouvement du fluide," *Bulletin des Sciences par la Société Philomatique*, Thermidor an 9 (Jul. 1801): 4–48, 48.

119. Ibidem.

120. See J.-B. Biot, "Volta (Alexandre)," in *Biographie universelle ancienne et moderne*, 85 vols. (Paris: Michaud, 1811–1862), vol. 49 (1827), pp. 459–64.

121. See J.-B. Biot, "Sur le mouvement du fluide galvanique," p. 45.

122. J.-B. Biot, "Rapport sur les expériences du citoyen Volta," *Mémoires de l'Institut National des Sciences et Arts. Sciences mathématiques et physiques*, 5, Fructidor an 12 (Aug. 1804): 195–222.

123. Volta, "Lettera terza a Gren" (March 1797), *VO*, 1:437.

124. Biot, "Rapport," pp. 199–200, note.

125. *Measuring* the electricity of the different layers making the column was regarded as almost impossible by Biot himself: see Biot's remarks in his *Traité* (note 128, below), 2:480.

126. Biot, "Rapport," pp. 209–10. Biot's severe judgement of Volta's scientific merits was explicit in the article on Volta that he wrote after his death: J.-B. Biot, "Volta (Alexandre)," in *Biographie universelle ancienne et moderne*, 85 vols. (Paris: Michaud, 1811–1862), vol. 49 (1827), pp. 459–64.

127. J.-B. Biot, "Quelle est l'influence de l'oxidation sur l'électricité développée par la colonne de Volta," *Annales de chimie*, 30 prairial an 11 (June 1803): 2–45, 16. On some implications following the adoption of Coulomb's balance in Biot's treatment of the battery, see Eugene Frankel, "J. B. Biot and the Mathematization of Experimental Physics in Napoleonic France," *Hist. Stud. Phys. Sci.*, 8 (1977): 33–72, 51.

128. J.-B. Biot, *Traité de physique experimentale et mathematique*, 4 vols. (Paris: Deterville, 1816), vol. 2, pp. 478–504. On mathematics and physics in France over this period: Elizabeth Garber, "Siméon-Denis Poisson: Mathematics versus Physics in Early-Nineteenth-Century France," in Elizabeth Garber (ed.), *Beyond History of Science: Essays in Honor of Robert E. Schofield* (Bethlehem: Lehigh University Press, 1990), pp. 156–76; Elizabeth Garber, *The Language of Physics: The Calculus and the Development of Theoretical Physics in Europe, 1750–1914* (Basel, Boston, Berlin: Birkhäuser, 1998).

129. An offer Volta declined, perhaps because of the frustrations he had suffered on the occasion of an earlier, similar offer by De Luc; ch. 4, above.

130. VMS, J 89, published in *VO*, 2:351–54, esp. 353.

131. VMS, J 89, published in *VO*, 2:353–54.

132. See Thomas Archibald, "Tension and Potential from Ohm to Kirchhoff," *Centaurus*, 31 (1988): 141–63; Sungook Hong, "Controversy over Voltaic Contact Phenomena, 1862–1900," *Arch. Hist. Exact Sci.*, 47 (1994): 233–89; Olivier Darrigol, *Electrodynamics from Ampère to Einstein* (Oxford: Oxford University Press, 2000), pp. 272–73.

133. *VO*, 1:576.

134. *VO*, 1:566.

135. Ibidem.

136. Nicholson, "Account of the New Electrical or Galvanic Apparatus of Sig. Alex. Volta," p. 179 (emphasis added), pp. 182–84.

137. Landriani to Volta, *VO*, 2:3.

138. *VO*, 2:169.

139. *VO*, 1:566.

140. *OED*.

141. H. Davy, Royal Institution, London, MSs. Notebook 22B, fol. 9ff.

142. T. Cavallo, *The Elements of Natural or Experimental Philosophy*, 4 vols. (London, 1803).

143. From 1806, according to the *OED*.

144. From 1828, according to the *OED*.

145. On the early history of the patent system: Christine MacLeod, *Inventing the Industrial Revolution: The English Patent System, 1660–1800* (Cambridge: Cambridge University Press, 1988).

146. On Edward Nairne: Paola Bertucci, *Sparks of Life: Medical Electricity and Natural Philosophy in England, c. 1746–1792* (D. Phil. dissertation, University of Oxford, 2001), pp. 236–49.

147. *Catalogue des specifications, moyens et procédés* (Paris: Boucher, 1826), p. 103.

148. On Nicholson and his patents: Samuel P. Oliver, "Nicholson, William," *DNB*, 14:473–75.

149. Specification of Frederick De Moleyns, *Production or Development of Electricity, and Its Application to Illuminating and Motive Purposes*, London Patent Office, 1841, no. 9053 (London, Spottiswoode, 1855). On the history of electricity and lighting: Brian Bowers, *Lengthening the Day: A History of Lighting Technology* (Oxford: Oxford University Press, 1998).

150. Ibidem, pp. 2–3, emphasis added.

151. Ibidem, p. 5, emphasis added.

152. Chapter 6, above.

153. Chapter 3, above.


## Chapter 8
## The Scientist as Hero

1. Alessandro Volta, *Collezione dell'opere del cavaliere Alessandro Volta*, edited by Vincenzio Antinori, 2 vols. (Florence: Piatti, 1816). For a glimpse on Antinori's personality, and his (limited) scientific work in the field of electricity and magnetism, see Vincenzio Antinori, *Scritti editi e inediti*, edited by Marco Tabarrini (Florence: Barbera, 1868).

2. On the "Tribuna di Galileo" see Giuseppe Boffito, *Gli strumenti della scienza e la scienza degli strumenti: con l'illustrazione della Tribuna di Galileo* (Rome, Multigrafica, 1982; reprint of the Florentine edition, 1929), and Alessandro Gambuti, *La Tribuna di Galileo* (Florence: Alinea, 1990).

3. On the annual meetings of Italian scientists before national unification, see Giuliano Pancaldi (ed.), *I congressi degli scienziati italiani nell'età del positivismo* (Bologna: CLUEB, 1983).

4. On Cianfanelli: P. Spadini, "Cianfanelli, Niccola," *DBI*, 25: 172–74.

5. Auguste Comte, *Cours de philosophie positive*, 6 vols. (1830–1842; reprint, Paris: Anthropos, 1968–69), vol. 2, pp. 538–39. On the science of electricity compared with other branches of physics, ibidem, p. 529.

6. Auguste Comte, *Calendrier positiviste ou Système générale de commémoration publique* (Paris, 1849), quoted in Mary Pickering, *Auguste Comte: An Intellectual Biography*, vol. 1 (Cambridge: Cambridge University Press, 1993), p. 157.

7. Auguste Comte, *Calendrier positiviste* (originally published in 1852), as reprinted in Auguste Comte, *Catéchisme positiviste*, edited by Pierre Arnaud (Paris: Garnier-Flammarion, 1966).

8. See *Dizionario del Risorgimento nazionale* (Milan: Vallardi, 1933), vol. 3, p. 348.

9. On Barabino: G. Di Genova, "Barabino, Niccolò," *DBI*, 5: 767–69; Davide Roscelli, *Nicolò Barabino: Maestro dei maestri* (Sampierdarena, Associazione Operaia "G: Mazzini," 1982); *Nicolò Barabino: "Il segno in trappola"* (Genoa: Marietti, 1990).

10. When not otherwise stated, the information that follows is taken from the journal *Voltiana*, published in Como in 33 issues from 1926 through 1927.

11. See Comitato Voltiano, "Relazioni e Rendiconti Esposizioni Voltiane," dated 29 May 1930, preserved in Archivio Municipale, Como, Cart. 11, Agricoltura, Industria, Commercio, Fasc. 2.

12. For these comparisons see the data published in Silvana Galdabini and Giuseppe Giuliani, "Physics in Italy between 1900 and 1940: The Universities, Physicists, Funds, and Research," *Hist. Stud. Phys. Sci.*, 19 (1988): part 1, pp. 115–36, esp. 124–25.

13. *Atti del congresso internazionale dei fisici*, edited by the Comitato [per le onoranze ad Alessandro Volta] (Bologna: Zanichelli, 1928), 2 vols.

14. See note 12, above.

15. On Carlo Baragiola: Edoardo Savino, *La nazione operante: Profili e figure* (Milan, 1934), p. 342.

16. On Enrico Musa: *Chi è?* (Rome, Formiggini, 1936), p. 630; Ernesto Codignola (ed.), *Pedagogisti ed educatori*, in *Enciclopedia bio-bibliografica italiana* (Milano: IEI, 1939), p. 353.

17. The following reconstruction is based on the rich documentation preserved in Archivio Centrale dello Stato (ACS), Rome, Presidenza del Consiglio dei Ministri, Anno 1927, fasc. 14/5, no. 1–1140, "Como, Onoranze ad Alessandro Volta."

18. "Il Prefetto di Como alla Presidenza del Consiglio dei Ministri," 13 September 1927, ACS, Rome, 14/5, 198.

19. "Pro Memoria" by Carlo Baragiola (n.d., but before September 1927), in ACS, Rome.

20. Giacinto Motta to Gian Giacomo Ponti, 16 November 1926, copy in ACS, Rome.

21. Giacinto Motta to Gian Giacomo Ponti, 16 November 1926, copy in ACS, Rome. On Motta: Piero Bolchini, "Giacinto Motta, la Società Edison e il fascismo," in *Storia in Lombardia*, n. 1–2 (1989), pp. 349–76.

22. Adrian Lyttelton, *La conquista del potere. Il fascismo dal 1919 al 1929*, Bari, Laterza, 1974, p. 517.

23. Several drafts of the message, including the one with Mussolini's note, are preserved in ACS, Rome, 14/5, 198 ("Messaggio del Capo del Governo agli italiani"). The request for money was submitted to Mussolini (through his own cabinet's head) on 8 September 1927, and again by Carlo Baragiola on 13 September 1927 (ACS, Rome, 14/5, 198).

24. On Quirino Majorana: Giorgio Dragoni and Giulio Maltese, "Quirino Majorana's Research on Gravitational Absorption: A Case Study in the Misinterpreted Experiments Tradition," *Centaurus*, 39 (1997): 141–87.

25. On Bohr's papers submitted to the Como and the Solvay (Bruxelles) conferences of 1927, and the discussions that followed, see Sandro Petruccioli, *Atomi, metafore, paradossi: Niels Bohr e la costruzione di una nuova fisica* (Roma-Napoli: Theoria, 1988), pp. 19–26 (English translation by Ian McGilvray, Cambridge, Cambridge University Press, 1993).

26. *Atti del Congresso Internazionale dei Fisici*, 2 vols., (Bologna: Zanichelli, 1928), vol. 2, p. 590.

27. Archivio Municipale, Como.

28. Petruccioli, *Atomi, metafore, paradossi*, p. 20.

29. Niels Bohr, "The Quantum Postulate and the Recent Development of Atomic Theory," *Nature* (Supplement, 14 April 1928): 121; (1928), 580–90.

30. Niels Bohr, "The Quantum Postulate and the Recent Development of Atomic Theory," in *Atti del Congresso Internazionale dei Fisici*, 2 vols. (Bologna: Zanichelli, 1928), vol. 2, pp. 565–88. These proceedings, printed in December 1928, circulated in 1929.

31. Comitato Voltiano, "Relazioni e Rendiconti Esposizioni Voltiane," dated 29 May 1930, preserved in Archivio Municipale, Como, cart. 11, Agricoltura, Industria, Commercio, fasc. 2.

## Chapter Nine
## Conclusion: Science, Technology, and Contingency

1. Boscovich Archives, Bancroft Library, Berkeley, MS 72/238, B43; my translation.

2. Quoted in B. Dibner, *Alessandro Volta and the Electric Battery* (New York: Franklin Watts, 1964), p. 103. A comparison between the battery, the telescope, and the steam engine was drawn as early as 1831 by François Arago: see François Arago, "Alexandre Volta," *Oeuvres*, ed. J. A. Barral, 13 vols. (2nd ed., Paris: Gide, 1865), 1:187–240.

3. M. Faraday, *Experimental Researches in Electricity*, 3 vols. (London, 1839–1855; reprint, Brussels: Culture et Civilisation, 1969), 1:300.

4. Cavallo to Lind, 13 June 1800, BL, Add. MS. 22,898.

5. For some penetrating remarks on the interplay between individual and collective contributions to scientific discovery, Ludwik Fleck, *Genesis and Development of a Scientific Fact*, edited by Thaddeus J. Trenn and Robert K. Merton (Chicago: The University of Chicago Press, 1979), pp. 78–79 (originally published in 1935).

6. I shall assume throughout that Volta's innovation and its aftermath belong to an epoch characterized by an increasing application of science to economic production, as suggested by economic historians and historians of technology since Simon Kuznets, *Modern Economic Growth* (New Haven: Yale University Press, 1966), esp. pp. 8–12. See also Nathan Rosenberg, *Perspectives on Technology* (Cambridge: Cambridge University Press, 1976), esp. pp. 260–89, and Peter Mathias and John A. Davis (eds.), *Innovation and Technology in Europe* (Oxford, Blackwell, 1991).

7. For a recent discussion of the many usages of "utility" in enlightened Europe, see Lissa Roberts, "Going Dutch: Situating Science in the Dutch Enlight-

enment," in William Clark, Jan Golinski, and Simon Schaffer (eds.), *The Sciences in Enlightened Europe* (Chicago: The University of Chicago Press, 1999), pp. 350–88. On utility and progress in eighteenth-century Britain, Roy Porter, *Enlightenment: Britain and the Creation of the Modern World* (London: Allen Lane, The Penguin Press, 2000), pp. 424–45. On invention in eighteenth-century France and Britain, Liliane Hilaire-Pérez, *L'invention technique au siècle des Lumières* (Paris: Albin Michel, 2000).

8. Frängsmyr, Tore, J. L. Heilbron, and R. E. Rider (eds.), *The Quantifying Spirit in the Eighteenth Century* (Berkeley, Los Angeles, London: University of California Press, 1990), and John L. Heilbron, "A Mathematicians' Mutiny, with Morals," in *World Changes: Thomas Kuhn and the Nature of Science*, ed. Paul Horwich (Cambridge, Mass.: MIT Press, 1993), pp. 81–129.

9. Heilbron, *The Quantifying Spirit*, p. 2. See also John L. Heilbron, "A Mathematicians' Mutiny, with Morals," p. 82: ". . . we should not look upon the momentous turn in the history of physics that brought about its quantification as the result of the accretion of subjects mathematized seriatim by mathematicians. Rather, we should consider it the invention of people who did not regard themselves as mathematicians but who embodied that esprit geometrique, that quantitative spirit, that can be followed through many aspects of the life and thought of the late Enlightenment."

10. *VE, Indici*, p. 790.

11. *VE*, 4:305–6, and *VE, Indici*, p. 576.

12. *VE*, 4:7.

13. *VE*, 4:15–16.

14. *VE*, 4:127.

15. For a general discussion of contingency in the historiography of science: Ian Hacking, *The Social Construction of What?* (Cambridge, Mass.: Harvard University Press, 1999). See also Mario Biagioli, "From Relativism to Contingentism," in *The Disunity of Science: Boundaries, Contexts, and Power*, ed. Peter Galison and David J. Stump (Stanford: Stanford University Press, 1996), pp. 189–206. For discussions on determinism and indeterminacy in the history of technology: Merritt Roe Smith and Leo Marx (eds.), *Does Technology Drive History? The Dilemma of Technological Determinism* (Cambridge, Mass.: MIT Press, 1994), especially Philip Scranton, "Determinacy and Indeterminacy in the History of Technology," pp. 143–68; Robert McC. Adams, *Paths of Fire: An Anthropologist's Inquiry into Western Technology* (Princeton: Princeton University Press, 1996) and Sungook Hong, "Unfaithful Offspring? Technologies and Their Trajectories," *Perspect. Sci.*, 6 (1998): 259–87. On unintended consequences in sociology: Raymond Boudon, *The Unintended Consequences of Social Action* (New York: St. Martin's Press, 1982).

16. See Richard Rorty, *Contingency, Irony, and Solidarity* (Cambridge: Cambridge University Press, 1989), chapter 3: "The Contingency of a Liberal Community," pp. 44–69.

17. *VE*, 3:377.

18. For recent assessments of Enlightenment historiography and legacies, focusing on the history of science and technology, see: William Clark, Jan Golinski, and Simon Schaffer, "Introduction," in William Clark, Jan Golinski,

and Simon Schaffer (eds.), *The Sciences in Enlightened Europe* (Chicago: The University of Chicago Press, 1999), pp. 4–31; Dorinda Outram, "The Enlightenment Our Contemporary," ibidem, pp. 32–40; Lorraine Daston, "Afterword: The Ethos of Enlightenment," ibidem, pp. 495–504.

19. See for example David Hume's careful discussion of the roles of "chance" and "causes" in "Of The Rise and Progress of the Arts and Sciences" (1742) in David Hume, *Essays: Moral, Political, and Literary*, ed. T. H. Green and T. H. Grose, 2 vols. (reprint of the London 1882 edition, Scientia Verlag Aalen, 1964).

20. The situation was best captured perhaps by Ralf Dahrendorf when he wrote:

"Science and technology are part of the prevailing structure of socio-economic relations in an industrial society, capitalist or otherwise, but they are at the same time a motive force in the development of a new potential for satisfying human life chances. . . . Possibly, the charismatic elements of science as an activity have something to do with this dual position. Certainly, the practical ambiguity of scientific theory has a story to tell: for quite often, the uses of scientific research are unknown when it is begun; apparently highly theoretical and irrelevant findings can turn out to be directly applicable to topical problems; which means that the impact of research findings can be unexpected in its direction and unwanted by its sponsors. The very fact that science thrives on a medium of uncertainty makes it a doubtful ally of any status quo" (Ralf Dahrendorf, "Observations on Science and Technology in a Changing Socio-Economic Climate," in *Scientific-Technological Revolution: Social Aspects*, SAGE Studies in International Sociology, 8 [London and Beverly Hills: SAGE, 1977], pp. 73–82, 75–86).

21. *VE*, 4:52–53.

## Manuscripts

*Bergamo*
Biblioteca Civica

G. Mangili. "Autobiografia." MS 79.R.8.f.11.

*Berkeley*
Bancroft Library

Boscovich Archives, MS. 72/238.

*Cambridge, Massachusetts*
Burndy Library, Dibner Institute

Books from Volta's personal library; Marginalia.

*Como*
Archivio di Stato (Notarile)

Gio. Batt. Stampa, Gravedona, 28 June 1756.
Gravedona, 13 July 1756.
Giacomo Filippo Clerici, Como, 12 Aug. 1794.
Como, 26 Mar. 1800.

Archivio Municipale

Comitato Voltiano. "Relazioni e Rendiconti Esposizioni Voltiane." Dated 29 May 1930, cart. 11, Agricoltura, Industria, Commercio, fasc. 2.

Archivio Storico della Diocesi

"Catalogo de' regolari di Città, & Diocesi" (Nov. 1767).
"Stato del Clero e delle anime 1768."

Biblioteca Comunale

Antonio Maria Odescalco. "Diverse notizie spettanti al Rv.mo Capitolo della Cattedrale di Como." MS 2.3.19.
Giulio Cesare Gattoni. "Giornale Gallo-Cisalpino." MS 4.6.1.

Seminario Arcivescovile

Seminario Santa Caterina. "Inventario di crediti, debiti e mobili." 27 July 1760. Folder 2, B, 2.
Seminario Santa Caterina. Folder 1, I-D.
Seminario Santa Caterina. Folder 2, E, 10.

*Downe, Kent*
Down House

E. Darwin. "Commonplace Book."

*Hanover, New Hampshire*
Dartmouth College Library

Count Rumford to Joseph Banks. 29 and 30 May 1800.

*London*
British Library

Cavallo Correspondence. Add. MSs 22,897/8.

Royal Society Archives

"Certificates."
"Journal Book."
"Letters and Papers."

*Milan*
Archivio di Stato

Letters exchanged between Alessandro Volta and Girolamo Bonesi. MS PADD, 9.
"Indice della Libreria nel Coll.o degl'Exgesuiti di Como." 1774. MS Studi, Parte antica, folder 230.

Istituto Lombardo, Accademia di Scienze e Lettere

Manoscritti del Volta: The Volta Papers, amounting to about 9,000 pages, arranged in nineteen classes under the letters A–T—including the correspondence, laboratory notes, drafts of published and unpublished memoirs, etc.—plus the Volta Family Papers.

Politecnico, Biblioteca

Paolo Frisi. "Piano del regolamento di studj dell'Università di Pavia e delle Scuole Palatine di Milano." Frisi Coll., 34, no. 10.
———. "Stato odierno della letteratura." Frisi Coll., 35, no. 7.

*New Haven, Connecticut*
Yale University Library

Jean André De Luc Papers. Series 2, box 13.

*Oxford*
Bodleian Library

William Nicholson Jr. "Life of William Nicholson." MS Don.d.175.

Museum of the History of Science

"Chapter Coffee House Society." MS Gunther 4.

*Pavia*
Archivio di Stato

Fondo Università, Rettorato. MS 180.

*Rome*
Archivio Centrale dello Stato

Presidenza del Consiglio dei Ministri. Anno 1927. Fasc. 14/5, no. 1–1140. "Como, Onoranze ad Alessandro Volta."

Archivio Storico della Società di Gesù

"Catalogus personarum, et officiorum provinciae mediolanensis Societatis Jesu, Anno 1761."
"Mediol. Cat. Triennal, 1761–1764." MS Med.69.

*Soissons*
Bibliothèque Municipale

J. A. Nollet. "Journal du Voyage de Piedmont et d'Italie en 1749." MS 150.

*Turin*
Archivio di Stato

Minutes of letters from Cigna to Priestley and Lagrange. Mar. 1781. Archivio Saluzzo di Monesiglio. Folder no. 6.

*Uppsala*
University Library

Waller Collection, Volta to De Luc. 30 March 1784.
Waller Collection, Volta to Bertola. 24 July 1788.

*Vienna*
Staatsarchiv

Lombardei Collectanea. MS 840.
Lombardei Korrepondenz. MS 137.

*Washington*
Smithsonian Institution Libraries

M. A. Pictet to a correspondent in London. 15 Oct. 1801. Dibner Collection. MS 1142A.


**Printed Primary Sources**

Achard, Franz Karl. "Expériences sur l'électrophore avec une théorie de cet instrument." In *Nouveaux mémoires de l'Académie Royale des sciences et belles-lettres*. Berlin, 1776, but 1779.
Adams, George. *An Essay on Electricity.* 2nd ed. London, 1785.
*Aepinus's Essay on the Theory of Electricity and Magnetism.* Introductory Monograph and Notes by R. W. Home. Trans. by P. J. Connor. Princeton, 1979.
Aepinus, Franz Ulrich Theodor. *De similitudine vis electricae atque magneticae.* St. Petersburg, 1758.
———. *Recueil de differents mémoires sur la tourmaline.* St. Petersburg, 1762.
———. *Tentamen theoriae electricitatis et magnetismi.* St. Petersburg, 1759.

Algarotti, Francesco. *Dialoghi sopra l'ottica neutoniana.* Ed. E. Bonora. Torino, 1977.

———. *Opere.* 17 vols. Venice, 1792.

Antinori, Vincenzio. *Scritti editi e inediti.* Ed. Marco Tabarrini. Florence, 1868.

Arago, Dominique François Jean. "Alexandre Volta." *Oeuvres*, ed. J. A. Barral. 13 vols. 2nd ed., Paris, 1865.

Barletti, Carlo. *Nuove sperienze elettriche, secondo la teoria del Sig. Franklin e le produzioni del P. Beccaria.* Milan, 1771.

———. *Physica specimina.* Milan, 1772.

Beccaria, Giambatista. "De athmosphera electrica." *PT*, 60 (1770): 277–301.

———. *Dell'elettricismo artificiale e naturale libri due.* Turin, 1753.

———. *Elettricismo artificiale.* Turin, 1772.

———. *L'elettricismo atmosferico.* 2nd ed. Bologna, 1758.

———. *Experimenta, atque observationes, quibus electricitas vindex late constituitur, atque explicatur.* Turin, 1769.

———. "Novorum quorundam in re electrica experimentorum specimen." *PT*, 55 (1765): 105–18.

———. "Novorum quorundam in re electrica experimentorum specimen." *PT*, 57 (1767): 297–311.

———. *A Treatise upon Artificial Electricity.* London, 1776.

Beckmann, Johann. *A History of Inventions and Discoveries.* 3 vols. London, 1797.

Bennet, Abraham. "An Account of a Doubler of Electricity." *PT*, 77 (1787): 288–96.

———. "Description of a New Electrometer." *PT*, 76 (1786): 26–34.

Bettinelli, Saverio. *Opere complete.* Venice, 1780.

Berzelius, Jöns Jacob. *De electricitatis galvanicae apparatu cel. Volta excitae in corpora organica effectu.* Uppsala: Edman, 1802.

*Bibliotheca Firmiana.* 6 vols. Milan, 1783.

Biot, Jean-Baptiste. "Quelle est l'influence de l'oxidation sur l'électricité développée par la colonne de Volta." *Annales de chimie*, 30 prairial an 11 (June 1803): 2–45.

———. "Rapport sur les expériences du citoyen Volta." *Mémoires de l'Institut National, Sc. mathém. et phys.*, 5 (1804): 195–222.

———. "Volta (Alexandre)." in *Biographie universelle ancienne et moderne.* 85 vols. Paris: Michaud, 1811–1862. Vol. 49 (1827), pp. 459–64.

Biot, Jean-Baptiste, and F. Cuvier. "Sur quelques propriétés de l'appareil galvanique." *Annales de chimie*, 30 Messidor an 9 (July 1801): 242–50.

Born, Ignaz von. *Specimen monachologiae.* Augsburg, 1783.

Boscovich, Roger Joseph. *Philosophiae naturalis theoria / A Theory of Natural Philosophy.* Latin-English ed. Trans. J. M. Child in the 2nd ed. Venice, 1763; Chicago and London, 1922.

Cantù, Cesare. *Storia della città e della diocesi di Como.* Firenze, 1856.

Cavallo, Tiberius. *A Complete Treatise on Electricity.* 4th ed. 3 vols. London, 1793–95.

———. "Of the Method of Manifesting the Presence, and Ascertaining the Quality, of Small Quantities of Natural or Artificial Electricity." *PT*, 78 (1788): 1–22.

———. *Trattato d'elettricità*, Italian trans. (Florence, 1779).

Cavendish, Henry. "An Account of Some Attempts to Imitate the Effects of the *Torpedo* by Electricity." *PT*, 66 (1776): 196–225.

———. "An Attempt to Explain Some of the Principal Phenomena of Electricity by Means of an Elastic Fluid." *PT*, 61 (1771): 584–677.

———. *The Electrical Researches of Henry Cavendish*. Ed. J. C. Maxwell. Cambridge, 1879.

Cigna, Giovanni Francesco. "De novibus quibusdam experimentis electricis." *Miscellanea taurinensia*, vol. 3. 1763–65.

———. "Esperienze elettriche." *Scelte di opuscoli interessanti* (Turin), 9 (1775): 68–85.

Comte, Auguste. *Calendrier positiviste ou Système génèrale de commémoration publique*. Paris, 1849.

———. *Catéchisme positiviste*. Ed. Pierre Arnaud. Paris: Garnier-Flammarion, 1996.

———. *Cours de philosophie positive*. 6 vols. 1830–42; reprint, Paris: Anthropos, 1968–69.

Coulomb, Charles-Augustin. *Mémoires*. Paris, 1884.

Cuthbertson, John. *Practical Electricity, and Galvanism*. London, 1807.

De Luc, Jean André. *Idées sur la météorologie*. 2 vols. London, 1786–87.

De Tipaldo, Emilio. *Biografia degl Italiani illustri* 10 vols. Venice, 1834–45.

Erxleben, Johan Christian Polycarp. *Anfangsgründe der Naturlehre*. 5th ed. Ed. G. C. Lichtenberg. Göttingen, 1791.

Fabbroni, Giovanni. "On the Chemical Action of Different Metals on Each Other at the Common Temperature of the Atmosphere." *A Journal of Natural Philosophy, Chemistry and the Arts* (1799): 300–10; and (1800): 120–27.

Faraday, Michael. *Experimental Researches in Electricity*. 3 vols. London, 1839–1855; reprint, Brussels, 1969.

———. *Faraday's Diary*. Ed. T. Martin. 8 vols. London, 1932–36.

———. *The Selected Correspondence of Michael Faraday*. Ed. Pearce L. Williams. 2 vols. Cambridge, 1971.

Franklin, Benjamin. *Benjamin Franklin's Experiments*. Ed. I. B. Cohen. Cambridge, Mass., 1941.

———. *Experiences et observations sur l'électricité*. French translation edited by Thomas François Dalibard. 2d edition. Paris, 1756.

———. *Papers*. Ed. L. W. Labaree et al. New Haven, 1959–.

———. *Scelta di lettere e di opuscoli del Signor B. F.*, Ital. trans., abridged, by C. G. Campi from the 3d French ed. Milan, 1774.

Frisi, Paolo. *De existentia et motu aetheris seu de theoria electricitatis ignis et luci dissertatio*. In *Dissertationes* by J. A. Euler et al. St. Petersburg, 1755.

Galvani, Luigi. *Dell'uso e dell'attià dell'arco conduttore nelle contrazioni dei muscoli*. Bologna, 1794.

———. *De viribus electricitatis in motu musculari*. Latin (1791) ed. with Eng. trans. and Introduction by I. B. Cohen. Norwalk, Conn. 1953.

———. *Il taccuino di Luigi Galvani*. Bologna, 1937.

———. *Memorie ed esperimenti inediti*. Bologna, 1937.

———. *Opere edite e inedite*. Bologna, 1841.

Geoffroy Saint-Hilaire, Étienne. "Sur l'anatomie comparée des organes électriques de la raie torpille, du gymnote engourdissant, et du silure trembleur." *Annales du Muséum d'Histoire Naturelle*, 1 (1802): 392–407.

Gherardi, Silvestro. "Relazione ragionata su i fatti e le cognizioni più vere, o interessanti che si possedavano intorno alla singolare virtù dei pesci elettrici." Offprint, Bologna, 1838.

Giovio, Giovanni Battista. *Gli uomini della comasca diocesi*. Modena, 1784.

Girardi, Michele. "Saggio di osservazioni anatomiche intorno agli organi elettrici della Torpedine." *Memorie di matematica e fisica della società italiana*, 3 (1786): 553–70.

Gravesande, Willelm Jacob van's. *Philosophiae newtonianae institutiones*. Bassano, 1749.

Gren, Friedric Albrecht Carl. *Grundriß der Naturlehre*. Halle, 1797.

Griselini, Francesco. *Dizionario delle arti e de' mestieri*. 18 vols. Venice, 1768.

Hallé, Jean-Noël. "Exposition abrégé des principales expériences répétées par M. Volta." *Bulletin des sciences, par la Société Philomatique*. 3, Nivose an X (Dec. 1801): 74–80 and planche 4.

Haller, Albrecht von. *Elementa physiologiae corporis humani*. Vol. 8: "Cerebrum, Nervi, Musculi." Lausanne, 1762.

———. *Mémoires sur les parties sensibles et irritables du corps animal*. Lausanne, 1760.

Harrington, Robert. *Some Experiments and Observations on Sig. Volta's Electrical Pile*. London, 1801.

Hauksbee, Francis. *Esperienze fisico-meccaniche sopra vari soggetti*. Ital. trans. of *Physico-Mechanical Experiments on Various Subjects*. Florence, 1716.

Haüy, René Just. *Exposition raisonné de la théorie de l'électricité et du magnétisme*. Paris, 1787.

———. *Traité élémentaire de physique*. Paris, 1803.

Henley, William. "Experiments and Observations on a new Apparatus, Called, a Machine for Exhibiting Perpetual Electricity." *PT*, 66 (1776): 513–22.

Humboldt, Alexander. "De l'irritabilité de la fibre nerveuse et musculaire." *Journal de physique*, Prairial an VI (May 1798): 465–74.

———. *Expériences sur le galvanisme*. French trans. by J.F.N. Jadelot. Paris, 1799.

Hume, David. "Of The Rise and Progress of the Arts and Sciences" (1742). In his *Essays: Moral, Political, and Literary*. Ed. T. H. Green and T. H. Grose. 2 vols. Reprint of the London 1882 edition, Scientia Verlag Aalen, 1964.

Hunter, John. "An Account of the Gymnotus Electricus." *PT*, 65 (1775): 395–407.

———. "Anatomical Observations on the Torpedo." *PT*, 63 (1773): 481–89.

Ingenhousz, John. "Electrical Experiments, to Explain How Far the Phaenomena of the Electrophorus May Be Accounted for by Dr. Franklin's Theory." *PT*, 68 (1778): 1027–55.

Jacopi, Giuseppe. *Elementi di fisiologia e notomia comparativa*. 3 vols. Milan, 1808–09.

Kinnersley, Ebenezer. "New Experiments in Electricity." *PT*, 53 (1763): 84–97.

Lalande, Joseph-Jérôme Le Français de. *Voyage en Italie*. 9 vols. Yverdon, 1787.

Lavoisier, Antoine-Laurent. *Correspondence*. 6 vols. Paris, 1955–97.

————. *Oeuvres de Lavoisier*. Paris, 1862–93.

Lichtenberg, Georg Christoph. *Briefe*. 3 vols. Leipzig, 1901–4.

————. *Vermischte Schriften*. 9 vols. Göttingen, 1800–1806.

Mangili, Giuseppe. *De systemate nerveo hirundinis, lumbrici terrestris, aliorumque Vermium*. Offprint, 1795.

Marum, Martinus van. "Description d'un très grande machine électrique placé dans le Muséum de Teyler." *Journal de Physiques*, 27 (1785): 145–55.

Moleyns, Frederick De, *Production or Development of Electricity, and Its Application to Illuminating and Motive Purposes* (London Patent Office, 1841, no. 9053). London: Spottiswoode, 1855.

Musschenbroek, Pieter van. *Elementa physicae*. 2 vols. Naples, 1745, $1771^2$, and Bassano, 1774.

————. *Introductio ad philosophiam naturalem*. Padua 1768.

Newton, Isaac. *Optices libri tres: accedunt Lectiones opticae*. 3 vols. Padua, 1749, $1773^2$.

————. *Opticks*, 4th ed. 1730; London, 1931.

————. *Philosophiae Naturalis Principia Mathematica*. 3d ed., 1726, with variant readings, Eds. Alexandre Koyré and I. B. Cohen. Cambridge, 1972.

Nicholson, William. "Account of the Electrical or Galvanic Apparatus of Sig. Alex. Volta." *Journal of Natural Philosophy, Chemistry and the Arts*, 4 (1800): 179–87.

————. "Description of an instrument." *Journal of Natural Philosophy, Chemistry, and the Arts*, 1 (1797): 16–18.

————. *Introduzione alla filosofia naturale*. Ital. trans. of *An Introduction to Natural Philosophy*, 4th ed., London, 1797. 3 vols. Florence, 1800.

————. "Observations on the Electrophore, Tending to Explain the Means by Which the Torpedo and Other Fish Communicate the Electric Shock." *Journal of Natural Philosophy, Chemistry, and the Arts*, 1 (1797): 355–59.

————. "Observations on T. Cavallo, 'On the Multiplier of Electricity.'" *Journal of Natural Philosophy, Chemistry, and the Arts*, 1 (1797): 394–99.

Nollet, Jean-Antoine. *L'arte dell'esperienza*. Ital. trans. 4 vols. Cesena, 1780.

————. *L'arte dell'esperienza*. Ital. trans. 4 vols. Venice, 1783.

————. *Lettere intorno all'elettricità*. Ital. trans. 2 vols. Naples, 1761.

————. *Lezioni di fisica sperimentale*. Ital. trans. 5 vols. Venice, 1746.

————. *Lezioni di fisica sperimentale*. Ital. trans. 5 vols. Venice, 1762.

————. *Ricerche sopra le cause particolari de' fenomeni elettrici*. Ital. trans. Venice, 1750.

————. *Saggio intorno all'elettricità de' corpi*. Ital. trans. Venice, 1747.

Ohm, Georg Simon. *The Galvanic Circuit Investigated Mathematically* (Berlin, 1827). Trans. William Francis. New York: D. van Nostrand Company, 1891.

Ørsted, Hans Christian. *Selected Scientific Works*. Translated and edited by Karen Jelved. Andrew D. Jackson, and Ole Knudsen, with an introduction by Andrew D. Wilson. Princeton University Press, 1998.

Paulian, Aimé-Henri. *Dictionnaire de physique*. 3 vols. Avignon, 1761.

Priestley, Joseph. *Histoire de l'électricité*. French trans. 3 vols. Paris, 1771.

————. *The History and Present State of Electricity*. London 1967.

Priestley, Joseph. *Osservazioni sopra differenti specie d'aria*. Ital. trans. by G. F. Fromond. Milan, 1774.

———. *A Scientific Autobiography of Joseph Priestley, 1733–1804*. Ed. R. E. Schofield. Cambridge, Mass., 1966.

———. *Sperienze e osservazioni sopra varj rami della fisica*. Ital. trans. 2 vols. Naples, 1785.

Pujoulx, Jean Baptiste. *Paris à la fin du 18e siècle*. Paris, 1801.

Réaumur, René Antoine Ferchault de. "Des effets que produit le poisson appellé en François Torpille." In Académie Royale des Sciences, *Histoire*. Paris, 1714, but 1717.

Ritter, Johann Wilhelm. *Beweis, dass ein beständiger Galvanismus den Lebensprocess in dem thierreich begleite*. Weimar, 1798.

Robertson, Étienne-Gasparde. "Expériences nouvelles sur le fluide galvanique, . . . lues à l'Institut National de France, le 11 fructidor, an 8," *Annales de chimie*, vol. 37 (Decemeber 1800): 132–50.

———. *Mémoires récréatifs, scientifiques et anecdotiques du physicien aéronaute E.-G. Robertson*. Paris, 1831–34.

Rutter, J.O.N. *Gas-Lighting: Its Progress and Its Prospects . . . With a Note on the Electric-Light*. London, Parker, 1849.

Scarpa, Antonio. *Tabulae neurologicae*. Pavia, 1794.

Sigaud de la Fond. *Description et usage d'un cabinet de physique expérimentale*. 2nd ed., Paris, 1784.

———. *Précis historique et expérimental des phénomènes électriques*. Paris, 1781.

Symmer, Robert. "New Experiments and Observations Concerning Electricity," *PT*, 51 (1759): 340–89.

Villa, Angelo Teodora. *Caroli Comitis Firmiani Vita*. Milan, 1783.

Volta Alessandro. *Aggiunte alle opere e all'epistolario di Alessandro Volta*. Ed. F. Massardi and A. Ferretti Torricelli. Bologna, 1966.

———. *Collezione dell' opere del cavaliere Alessandro Volta*. Ed. V. Antinori. 2 vols. Florence, 1816.

———. *Edizione Nazionale delle Opere e dell'Epistolario in 15 volumi (1918–1976)*. CD-Rom ed. by Fabio Bevilacqua, Gianni Bonera, and Lidia Falomo, Dipartimento di Fisica "A. Volta." Università di Pavia, Milan: Hoepli, 2000–2002.

———. *Elettricità: Scritti scelti*. Ed. G. Pancaldi. Biblioteca della Scienza Italiana, Florence: Giunti, 1999.

———. *Epistolario*. Ed. F. Massardi. 5 vols. Bologna, 1949–55.

———. *Indici delle opere e dell'epistolario di Alessandro Volta*. Ed. A. Ferretti Torricelli. 2 vols. Milan, 1974–76.

———. *L'opera di Alessandro Volta*. Ed. F. Massardi. Milan, 1927.

———. *Le opere*. 7 vols. Milan, 1918; reprint: The Sources of Science, no. 70, New York and London, 1968.

———. *Opere scelte*. Ed. M. Gliozzi. Turin, 1967.

Walker, Adam. *Ideas Suggested on the Spot in a Late Excursion through Flanders, Germany, France, and Italy*. London, 1790.

Walsh, John. "Of the Electric Property of the Torpedo." *PT*, 63 (1773): 461–80.

Wilcke, Johan Carl. *Disputatio physica experimentalis de electricitatibus contrariis*. Rostock, 1757.

————. "Osservazioni sulla forma della neve." *Scelta di opuscoli interessanti,* 2 (1775): 117–18.

————. "Untersuchung der bey Herrn Voltas *neuen* Electrophoro perpetuo vorkommenden elektrischen Erscheinungen." Kungl. Svenska Vetenskapsakademien der Wissenschaften, *Abhandlungen,* (1777) 54–78, 116–30, 200–16.

## Secondary Sources

Abbri, Ferdinando. "Il misterioso 'spiritus salis': Le ricerche di elettrochimica nella Toscana napoleonica," *Nuncius,* 2 (1987): 55–88.

————. "Volta's Chemical Theories: The First Two Phases." In *Nuova Voltiana: Studies on Volta and His Times,* ed. Fabio Bevilacqua and Lucio Fregonese. Pavia and Milan: Hoepli, 2000. 2: 1–14.

Adams, Robert McC. *Paths of Fire: An Anthropologist's Inquiry into Western Technology.* Princeton: Princeton University Press, 1996.

Altieri Biagi, Maria Luisa. "Il linguaggio scientifico italiano in Europa." In *Lingua e cultura italiana in Europa,* ed. Vincenzo Lo Cascio. Florence: Le Monnier, 1990.

Anderson, R.G.W., J. A. Bennett, and W. F. Ryan, eds. *Making Instruments Count: Essays on Historical Scientific Instruments Presented to Gerard L'Estrange Turner.* Aldershot: Variorum, 1993.

Annoni, Ada. "Cultura e scuola nell'epoca del Volta." In *Conferenze Voltiane.* Como: Società Storica Comense, 1978.

Archibald, Thomas. "Tension and Potential from Ohm to Kirchhoff." *Centaurus,* 31 (1988): 141–63.

Baldini, Ugo. "L'attività scientifica nel primo Settecento." In *Storia d'Italia, Annali 3, Scienza e tecnica nella cultura e nella società dal Rinascimento a oggi.* Turin: Einaudi, 1980.

Baldini, Ugo, and Pietro Nastasi, eds. *R. G. Boscovich: Lettere a M. Lorgna, 1765–1785.* Rome: Accademia Nazionale delle Scienze detta dei XL, 1988.

Barbarisi, Gennaro, ed. *Ideologia e scienza nell'opera di Paolo Frisi.* 2 vols. Milan: Angeli, 1987.

Barsanti, D. and L. Rombai. *Leonardo Ximenes.* Florence: Osservatorio Ximeniano, 1987.

Basile, Bruno, and Maria Luisa Altieri-Biagi. *Scienziati italiani del Settecento.* Milan and Naples: Ricciardi, 1983.

Bedi, Joyce E., et al., compilers. *Sources in Electrical History: Archives and Manuscript Collections in U.S. Repositories.* New York: Center for the History of Electrical Engineering, 1989.

Bedini, Silvio A. *Science and Instruments in Seventeenth-Century Italy.* Aldershot: Variorum, 1994.

Bektas, M. Yakup, and Maurice Crosland. "The Copley Medal, 1731–1839." *Notes Rec. Roy. Soc. Lond.,* 46 (1992): 43–76.

Bellodi, Giuliano, and Paolo Brenni. "The 'Arms of the Physicist': Volta and Scientific Instruments." In *Nuova Voltiana: Studies on Volta and His Times,* ed. Fabio Bevilacqua and Lucio Fregonese. Pavia and Milan: Hoepli, 2000. 3: 1–40.

Bellone, Enrico. "Alessandro Volta." In *Economia, istituzioni, cultura in Lombardia nell'età di Maria Teresa*, ed. Aldo De Maddalena et al. 3: 451–59.

Bellone, Enrico. "Newtonianesimo e leggi di Coulomb." In *Teorie e filosofie della materia nel Settecento*, ed. Antonio Di Meo and Silvano Tagliagambe. Rome: Editori Riuniti, 1993.

Ben-Chaim, Michael. "Social Mobility and Scientific Change: Stephen Gray's Contribution to Electrical Research." *Brit. J. Hist. Sci.*, 22 (1990): 3–24.

Benguigui, Isaac. *Théories électriques du XVIIIe siècle*. Geneva: Georg, 1984.

Benjamin, Andrew E., Geoffrey N. Cantor, and John R. R. Christie, eds. *The Figural and the Literal: Problems of Language in the History of Science and Philosophy*. Manchester: Manchester University Press, 1987.

Bensaude-Vincent, Bernadette. "Pneumatic Chemistry Viewed from Pavia." *Nuova Voltiana: Studies on Volta and His Times*, ed. Fabio Bevilacqua and Lucio Fregonese. Pavia and Milan: Hoepli, 2000–. 2: 15–31.

Beretta, Marco. *A History of Non-Printed Science: A Select Catalogue of the Waller Collection*. Uppsala: Acta Universitatis Upsaliensis, 1993.

———. "From Nollet to Volta: Lavoisier and Electricity." *Rev. Hist. Sci.*, 54 (2001): 29–52.

———. "Pneumatics vs. 'Aerial Medicine': Salubrity and Respirability of Air at the End of the Eighteenth Century." In *Nuova Voltiana: Studies on Volta and His Times*, ed. Fabio Bevilacqua and Lucio Fregonese. Pavia and Milan: Hoepli, 2000–. 2: 49–71.

Berlin, Isaiah. "The Philosophers of the Enlightenment." In *The Power of Ideas*. Ed. Henry Hardy. London, Chatto & Windus, 2000.

Berman, Morris. *Social Change and Scientific Organization: The Royal Institution, 1799–1844*. London: Heinemann, 1978.

Bernardi, Walter. *I fluidi della vita: Alle origini della controversia sull'elettricità animale*. Florence: Olschki, 1992.

———. *Le metafisiche dell'embrione: Scienze della vita e filosofia da Malpighi a Spallanzani (1672–1793)*. Florence: Olschki, 1986.

Bertucci, Paola. *Sparks of Life: Medical Electricity and Natural Philosophy in England, c. 1746–1792*. D. Phil. dissertation, University of Oxford, 2001.

Bertucci, Paola, and Giuliano, Pancaldi eds. *Electric Bodies: Episodes in the History of Medical Electricity. Bologna Studies in History of Science*, vol. 9. CIS, University of Bologna, 2001.

Bevilacqua, Fabio. *The Principle of Conservation of Energy and the History of Classical Electromagnetic Theory*. Pavia: La goliardica pavese, 1983.

Bevilacqua, Fabio, and Lucio Fregonese, eds. *Nuova Voltiana: Studies on Volta and His Times*. Università degli Studi di Pavia. Pavia and Milan: Hoepli, 2000–.

Biagioli, Mario. "From Relativism to Contingentism." In Peter Galison and David J. Stump (eds.), *The Disunity of Science: Boundaries, Contexts, and Power*. Stanford University Press, 1996.

Blondel, Christine. *A.-M. Ampère et la création de l'électrodynamique: 1820–1827*. Paris: Bibliothèque nationale, 1982.

———. "Animal Electricity in Paris: From Initial Support, to Its Discredit and Eventual Rehabilitation," in *Luigi Galvani International Workshop: Proceedings*, ed. Marco Bresadola and Giuliano Pancaldi. *Bologna Studies in History of Science*, vol. 7. CIS, University of Bologna, 1999.

———. *Histoire de l'electricité*. Paris, Cité des Sciences, 1994.

Blondel, Christine and Matthias Dörries, eds. *Restaging Coulomb: Usages, controverses et réplications autour de la balance de torsion*. Florence: Olschki, 1994.

Boden, Margaret A., ed. *Dimensions of Creativity.* Cambridge, Mass.: MIT Press, 1994.

Boffito, Giuseppe. *Gli strumenti della scienza e la scienza degli strumenti: con l'illustrazione della Tribuna di Galileo*. Florentine edition, 1929. Reprint, Rome: Multigrafica, 1982.

Bolchini, Piero. "Giacinto Motta, la Società Edison e il fascismo." In *Storia in Lombardia*, n. 1–2 (1989): 349–76.

Bossi Michele and Pasquale Tucci, eds. *Bicentennial Commemoration of R. G. Boscovich*. Milan: UNICOPLI, 1988.

Boudon, Raymond. *The Unintended Consequences of Social Action*. New York: St. Martin's Press, 1982.

Bowers, Brian. *Lengthening the Day: A History of Lighting Technology.* Oxford: Oxford University Press, 1998.

Brambilla, Elena. "Scientific and Professional Education in Lombardy, 1760–1803: Physics between Medicine and Engineering." In *Nuova Voltiana: Studies on Volta and His Times*, ed. Fabio Bevilacqua and Lucio Fregonese. Pavia and Milan: Hoepli, 2000–. 1: 51–100.

Brazier, Mary A. B. *A History of Neurophysiology in the Seventeenth and Eighteenth Centuries*. New York: Raven Press, 1984.

Brenna, G. G. "Pranzi di carnevale in casa del conte Filippo Maria Volta." *Omaggio a Volta nel centocinquantenario della morte*. Como: Nani, 1978.

Bresadola, Marco. *Medicina e filosofia naturale in Luigi Galvani*. Ph.D. dissertation, University of Florence, 1999.

Bresadola, Marco, and Giuliano Pancaldi, eds. *Luigi Galvani International Workshop: Proceedings*. Bologna Studies in History of Science, vol. 7 CIS, University of Bologna, 1999.

Brock, Helen. "The Many Facets of Dr. William Hunter (1718–83)." *Hist. Sci.*, 32 (1994): 387–408.

Brockliss, Laurence W. B, "Curricula." In *A History of the University in Europe*, general editor Walter Rüegg. Vol. 2. *Universities in Early Modern Europe (1500–1800)*, editor Hilde de Ridder-Symoens. Cambridge: Cambridge University Press, 1996.

———. *French Higher Education in the Seventeenth and Eighteenth Centuries*. Oxford: Clarendon, 1987.

Broers, Michael. *Europe under Napoleon, 1799–1815*. London: Arnold, 1996

Brown, Sanborn C. *Benjamin Thompson, Count Rumford*. Cambridge, Mass.: MIT Press, 1979.

Brown, Theodore M. "Galvani." *DSB*, 5: 267–69.

Brown, Theodore M. "The Electric Current in Early Nineteenth-Century French Physics." *Hist. Stud. Phys. Sci.*, 1 (1969): 61–103.

Buchwald, Jed Z., ed. *Scientific Practice: Theories and Stories of Doing Physics*. Chicago: The University of Chicago Press, 1995.

Caizzi, Bruno. *Industria, commercio e banca in Lombardia nel XVIII secolo*. Milan: Banca Commerciale Italiana, 1968.

Caneva, Kenneth L. "From Galvanism to Electrodynamics: The Transformation of German Physics and Its Social Context." *Hist. Stud. Phys. Sci.*, 9 (1978): 63–159.

Cantor, Geoffrey N. "The Eighteenth Century Problem." *Hist. Sci.*, 20 (1982): 44–63.

———. *Michael Faraday: Sandemanian and Scientist*. Basingstoke: Macmillan, 1991.

Cantor, Geoffrey N., and M.J.S. Hodge, eds. *Conceptions of Ether: Studies in the History of Ether Theories, 1740–1900*. Cambridge: Cambridge University Press, 1981.

Cappelletti, Vincenzo. "Barletti." *DBI*, 6: 401–405.

Cardwell, Donald S. *James Joule: A Biography*. Manchester: Manchester University Press, 1989.

Carozzi, Albert V., "Saussure, H.B. de." *DSB*, 12: 119–23.

Carpanetto, Dino, and Giuseppe Ricuperati. *Italy in the Age of Reason, 1685–1789*. London: Longman, 1987.

Carvalho, Joaquim de. "Correspondência ciêntífica dirigida a João de Magalhães." *Revista da Faculdade de Ciências, Universidade de Coimbra*, 20 (1951): 93–283.

Casini, Paolo. "Boscovich." *DBI*, 13: 221–30.

———. *Newton e la coscienza europea*. Bologna: 1983.

———. *Scienza, utopia e progresso: profilo dell'Illuminismo*. Bari: Laterza, 1994.

Cavazza, Marta. *Settecento inquieto: Alle origini dell'Istituto delle Scienze di Bologna* Bologna: il Mulino, 1990.

Chinea, Eleuterio. *L'istruzione pubblica e privata nello Stato di Milano dal Concilio tridentino alla riforma teresiana (1563–1773)*. Florence: La Nuova Italia, 1953.

Ciancio, Luca. *Autopsie della terra: illuminismo e geologia in Alberto Fortis (1741–1803)*. Florence: Olschki, 1995.

Ciardi, Marco. *L'atomo fantasma: Genesi storica dell'ipotesi di Avogadro*. Florence: Olschki, 1995.

Clark, William. "German Physics Textbooks in the *Goethezeit*." Parts 1 and 2. *Hist. Sci.*, 35 (1997): 219–39, 295–363.

Clark, William, Jan Golinski, and Simon Schaffer, eds. *The Sciences in Enlightened Europe*. Chicago: The University of Chicago Press, 1999.

Codignola, Ernesto, ed. *Pedagogisti ed educatori*, in *Enciclopedia bio-bibliografica italiana*. Milano: IEI, 1939.

Cohen, I. Bernard. *Benjamin Franklin's Science*. Cambridge, Mass.: Harvard University Press, 1990.

———. *Franklin and Newton*. Philadelphia: American Philosophical Society, 1956.

———. Introduction to L. Galvani. *Commentary on the Effects of Electricity on Muscular Motion*. Norwalk: Burndy Library, 1953.

———. *Science and the Founding Fathers: Science in the Political Thought of Jefferson, Franklin, Adams, and Madison*. New York: W. W. Norton & Company, 1995.

Collins, Harry M. *Changing Order: Replication and Induction in Scientific Practice.* London: SAGE, 1985.

*Contributi alla storia dell'Università di Pavia.* Pavia: Tipografia Cooperativa, 1925.

Copenhaver, Brian P. "A Tale of Two Fishes: Magical Objects in Natural History from Antiquity through the Scientific Revolution." *J. Hist. Ideas,* 52. 1991: 373–98.

Costa, Gustavo. "I rapporti Frisi-Boscovich alla luce di lettere inedite di Frisi, Boscovich, Mozzi, Lalande e Pietro Verri." *Rivista storica italiana,* 79 (1967): 819–76.

Crosland, Maurice. *In the Shadow of Lavoisier: The* Annales de chimie. British Society for the History of Science, 1994.

———. *The Society of Arcueil: A View of French Science at the Time of Napoleon I.* London: Heinemann, 1967.

Cunningham, Andrew, and Nicholas Jardine, eds. *Romanticism and the Sciences.* Cambridge: Cambridge University Press, 1990.

Dahrendorf, Ralf. "Observations on Science and Technology in a Changing Socio-Economic Climate." In *Scientific-Technological Revolution: Social Aspects.* SAGE Studies in International Sociology, vol. 8. London and Beverly Hills: SAGE, 1977.

Dainville, François de. *L'éducation des Jésuites (XVIe–XVIIIe siècles).* Paris: Éd. de Minuit, 1978.

Darnton, Robert. *The Business of the Enlightenment: A Publishing History of the Encyclopédie, 1775–1800.* Cambridge: Belknap Press, 1979.

———. *The Kiss of Lamourette: Reflections in Cultural History.* New York, Norton, 1990.

———. *Mesmerism and the End of the Enlightenment in France.* Cambridge, Mass.: Harvard University Press, 1968.

Darrigol, Olivier. *Electrodynamics from Ampère to Einstein.* Oxford: Oxford University Press, 2000.

Daston, Lorraine. "Afterword: The Ethos of Enlightenment." In *The Sciences in Enlightened Europe,* ed. William Clark, Jan Golinski, and Simon Schaffer. Chicago: The University of Chicago Press, 1999.

———. "The Ideal and Reality of the Republic of Letters in the Enlightenment." *Sci. Context,* 4 (1991): 367–86.

———, ed. *Biographies of Scientific Objects.* Chicago: The University of Chicago Press, 2000.

Daumas, Maurice. *Scientific Instruments of the Seventeenth and Eighteenth Centuries and Their Makers.* London: Batsford, 1972.

Davidson, Nicholas. "Toleration in Enlightenment Italy." In *Toleration in Enlightenment Europe,* ed. Ole Peter Grell and Roy Porter. Cambridge: Cambridge University Press, 2000.

De Beer, Gavin. *The Sciences Were Never at War.* London: Nelson and Sons, 1960.

della Bella, R. "Il seminario della diocesi di Como e la sua biblioteca." Università Cattolica di Milano, "Laurea" thesis, 1969–1970.

della Torre, S. "Lo sviluppo di Como tra 700 e 800." In *Omaggio a Volta nel centocinquantanario della morte* Como: Nani, 1978.

De Maddalena, Aldo, et al., eds. *Economia, istituzioni, cultura in Lombardia nell'età di Maria Teresa*. 3 vols. Bologna: il Mulino, 1982.

De Santillana, Giorgio. "Volta." *Scientific American*, 212 (1965): 82–91.

Dhombres, Nicole, and Jean Dhombres. *Naissance d'un pouvoir: sciences et savants en France, 1793–1824*. Paris: Payot, 1989.

Di Bella, Saverio. *Chiesa e società civile nel Settecento italiano*. Milano: Giuffrè, 1982.

Dibner, Bern. *Alessandro Volta and the Electric Battery*. New York: F. Watts: 1964.

Dooley, Brendan. "The Communication Revolution in Italian Science." *Hist. Sci.*, 33 (1995): 469–96.

———. *Science, Politics, and Society in Eighteenth-Century Italy: The* Giornale de'Letterati d'Italia *and its World*. New York: Garland, 1991.

Dragoni, Giorgio, and Giulio Maltese. "Quirino Majorana's Research on Gravitational Absorption: A Case Study in the Misinterpreted Experiments Tradition." *Centaurus*, 39 (1997): 141–87.

Elias, Norbert. *Mozart: Portrait of a Genius*. Ed. Michael Schröter. Trans. Edmund Jephcott. Berkeley: University of California Press, 1993. [*Mozart: Zur Soziologie eines Genies* (Frankfurt, 1991)].

Feingold, Mordechai, ed. *Jesuit Science and the Republic of Letters*. Cambridge, Mass.: MIT Press, 2002.

Ferguson, Allan, ed. *Natural Philosophy through the Eighteenth Century and Allied Topics*. London: Taylor & Francis, 1972[2].

Ferrone, Vincenzo. *La Nuova Atlantide e i lumi: Scienza e politica nel Piemonte di Vittorio Amedeo III*. Turin: Meynier, 1988.

———. "Riflessioni sulla cultura illuministica napoletana e l'eredità di Galilei." In *Galileo e Napoli*. Eds. F. Lomonacao and M. Torrini. Naples: Guida, 1987.

———. *Scienza, natura, religione: mondo newtoniano e cultura italiana nel primo Settecento*. Naples: Jovene, 1982.

Findlen, Paula. *Possessing Nature: Museums, Collecting, and Scientific Culture in Early Modern Italy*. Berkeley: University of California Press, 1994.

Fiocca, Alessandra, and Luigi Pepe. "L'Università e le scuole per gli Ingegneri a Ferrara." *Annuario dell'Università di Ferrara, Sc. Mat.* 32 (1986): 125–66.

Fischer, Joachim. *Napoleon und die Naturwissenschaften*. Stuttgart: Steiner, 1988.

Fleck, Ludwik. *Genesis and Development of a Scientific Fact*. Ed. Thaddeus J. Trenn and Robert K. Merton. Chicago: The University of Chicago Press, 1979.

Forbes, Robert J., ed. *Martinus van Marum: Life and Work*. 3 vols. Haarlem: Willink, 1969–71.

Fox, Robert. *The Caloric Theory of Gases from Lavoisier to Regnault*. Oxford: Clarendon, 1971.

———. *The Culture of Science in France, 1700–1900*. Aldershot: Variorum, 1992.

———, ed. *Technological Change: Methods and Themes in the History of Technology*. Amsterdam: Harwood, 1996.

———. "Learning, Politics and Polite Culture in Provincial France: The *Sociétés Savantes* in the Nineteenth Century," *Réflexions historiques/Historical Reflections*, 7 (1980): 543–64.

————. "The Rise and Fall of Laplacian Physics." *Hist. Stud. Phys. Sci.*, 4 (1974): 89–130.

Fox, Robert, and Anna Guagnini. *Laboratories, Workshops, and Sites: Concepts and Practices of Research in Industrial Europe, 1800–1914.* Berkeley: Office for History of Science and Technology, University of California, 1999.

Frängsmyr, Tore, J. L. Heilbron, and R. E. Rider, eds. *The Quantifying Spirit in the Eighteenth Century.* Berkeley, Los Angeles, London: University of California Press, 1990.

Frankel, Eugene. "J. B. Biot and the Mathematization of Experimental Physics in Napoleonic France." *Hist. Stud. Phys. Sci.*, 8 (1977): 33–72.

————. *Jean Baptiste Biot: The Career of a Physicist in Nineteenth-Century France.* Ph.D. dissertation, Princeton University, 1972.

Fregonese, Lucio. *Volta's Electrical Programme.* Ph.D. dissertation, University of Cambridge, 1999.

Freudenthal, Gad. "Early Electricity between Chemistry and Physics." *Hist. Stud. Phys. Sci.*, 11 (1981): 203–29.

Frost, Alfred J. *Catalogue of Books and Papers Relating to Electricity, Magnetism, and the Electric Telegraph.* London: Spon, 1880.

Fuchs, James L. "Nationality and Knowledge in Eighteenth-Century Italy." *Stud. 18th-Cent. Cult.*, 21 (1991): 207–18.

Fullmer, June Z. *Young Humphry Davy.* Philadelphia: American Philosophical Society, 2000.

Fulton, John Farquhar. *Muscular Contraction and the Reflex Control of Movement.* Baltimore: Williams and Wilkins, 1926.

Galdabini, Silvana and Giuseppe Giuliani. "Physics in Italy between 1900 and 1940: The Universities, Physicists, Funds, and Research." *Hist. Stud. Phys. Sci.* 19 (1988): part 1, pp. 115–36.

Galison, Peter. *Image and Logic: A Material Culture of Microphysics.* Chicago: University of Chicago Press, 1997.

Galison, Peter, and David J. Stump, eds. *The Disunity of Science: Boundaries, Contexts, and Power.* Stanford: Stanford University Press, 1996.

Garber, Elizabeth. *The Language of Physics: The Calculus and the Development of Theoretical Physics in Europe, 1750–1914.* Boston, Basel, Berlin: Birkhäuser, 1998.

————. "Siméon-Denis Poisson: Mathematics versus Physics in Early Nineteenth-Century France." In *Beyond History of Science: Essays in Honor of Robert E. Schofield*, ed. Elizabeth Garber. Bethlehem: Lehigh University Press, 1990.

Garin, Eugenio. "Antonio Genovesi e la sua introduzione storica agli *Elementa physicae* di Pietro van Musschenbroek." *Physis*, 11 (1969): 211–22.

Gartrell, Ellen G. *Electricity, Magnetism and Animal Electricity: A Checklist of Printed Sources, 1600–1850.* Wilmington, Del.: Scholarly Resources, 1975.

Gascoigne, John. *Joseph Banks and the English Enlightenment.* Cambridge: Cambridge University Press, 1994.

Gavroglu, Kostas, ed. *The Sciences in the European Periphery during the Enlightenment.* Monographic issue, *Archimedes*, 2 (1999).

Giarrizzo, Giuseppe. *Massoneria e Illuminismo nell'Europa del Settecento.* Venezia: Marsilio, 1994.

Gigerenzer, Gerd, and Reinhard Selten, eds. *Bounded Rationality: The Adaptive Toolbox.* Cambridge, Mass.: MIT Press, 2001.

Gigli Berzolari, A. *Alessandro Volta e la cultura scientifica e tecnologica tra '700 e '800.* Milano: Cisalpino, 1993.

Gill, Sydney. "A Voltaic Enigma and a Possible Solution to It." *Ann. Sci.*, 33 (1976): 351–70.

Gillispie, Charles Coulston. *The Montgolfier Brothers and the Invention of Aviation, 1783–1784.* Princeton: Princeton University Press, 1983.

———. *Science and Polity in France at the End of the Old Regime.* Princeton: Princeton University Press, 1980.

———. "Science and Secret Weapons Development in Revolutionary France, 1792–1804: As Documentary History." *Hist. Stud. Phys. Biol. Sci.*, 23 (1992): 35–152.

Gillmor, C. Steward. *Coulomb and the Evolution of Physics and Engineering in Eighteenth-Century France.* Princeton: Princeton University Press, 1971.

Gini, P. "L'età posttridentina, secc. XVII–XVIII." In *Storia religiosa della Lombardia, Diocesi di Como*, ed. A. Caprioli, A. Rimoldi, and L. Vaccaro. Brescia: La Scuola, n.d.

Gliozzi, Mario. "Consonanze e dissonanze tra l'elettrologia di Cavendish e quella di Volta." *Physis*, 2 (1969): 231–48.

———. *L'elettrologia fino al Volta.* 2 vols. Naples: Loffredo, 1937.

———. "Giambatista Beccaria nella storia dell'elettricità." *Archeion*, 17 (1935): 15–47.

———. "I fisici piemontesi del Settecento nel movimento filosofico del tempo." *Filosofia*, Quaderni (1962): 558–71.

———. "Il Volta della seconda maniera." *Cultura e scuola*, 5 (1966): 235–39.

Golinski, Jan. *Making Natural Knowledge: Constructivism and the History of Science.* Cambridge: Cambridge University Press, 1998.

———. "Science in the Enlightenment." *Hist. Sci.*, 34 (1986): 410–24.

———. *Science as Public Culture: Chemistry and Enlightenment in Britain, 1760–1820.* Cambridge: Cambridge University Press, 1992.

Gooding, David, Trevor Pinch, and Simon Schaffer, eds. *The Uses of Experiment.* Cambridge: Cambridge University Press, 1989.

Graf, Arturo. *L'anglomania e l'influsso inglese in Italia nel XVIII secolo.* Turin: Loescher, 1911.

Grattan-Guinness, Ivor. *Convolutions in French Mathematics, 1800–1840.* 3 vols. Basel, Boston, Berlin: Birkäuser, 1990.

Guerlac, Henry. "Chemistry as a Branch of Physics." *Hist. Stud. Phys. Sci.*, 7 (1976): 193–276.

———. *Essays and Papers in the History of Modern Science.* Baltimore and London: Johns Hopkins University Press, 1977.

Habermas, Jürgen. *The Structural Transformation of the Public Sphere.* Trans. Thomas Burger. London: Polity Press, 1989.

Hacking, Ian. *Representing and Intervening.* Cambridge: Cambridge University Press, 1983.

————. *The Social Construction of What?* Cambridge, Mass.: Harvard University Press, 1999.

Hackmann, Willem Dirk. *Catalogue of Pneumatical, Magnetical, and Electrical Instruments.* Florence: Istituto e Museo di Storia della Scienza, 1995.

————. "Eighteenth-Century Electrostatic Measuring Devices." *Annali dell'Istituto e Museo di Storia della Scienza di Firenze,* 3 (1978): 3–58.

————. "Electrical Researches." In *Martinus van Marum: Life and Work.* ed. R. G. Forbes. Haarlem: Willink, 1969–71.

————. "The Electrical Researches of Martinus van Marum (1750–1837)." Queen's University of Belfast, MA thesis, 1970.

————. *Electricity from Glass: The History of the Frictional Electrical Machine, 1600–1850.* Alphen aan den Rijin: Sijthoff and Noordshoff, 1978.

————. "The Enigma of Volta's 'Contact Tension' and the Development of the 'Dry Pile,' " *Nuova Voltiana: Studies on Volta and His Times,* ed. Fabio Bevilacqua and Lucio Fregonese. Pavia and Milan: Hoepli, 2000–. 3: 103–19.

————. "The Relationship between Concept and Instrument Design in Eighteenth-Century Experimental Science," *Ann. Sci.,* 36 (1979): 205–24.

Hahn, Roger. *The Anatomy of a Scientific Institution: The Paris Academy of Sciences, 1666–1803.* Berkeley, Los Angeles, London: University of California Press, 1971.

————. *Calendar of the Correspondence of Pierre Simon Laplace.* Berkeley: Office for History of Science and Technology, University of California, 1982.

————. "Science and the Arts in France: The Limitations of an Encyclopedic Ideology." *Studies 18th-Cent. Cult.,* 10 (1981): 77–93.

Hall, Mary Boas. "The Royal Society and Italy, 1667–1795." *Notes Rec. Roy. Soc. Lond.,* 37 (1982): 63–81.

Hankins, Thomas L. "In Defence of Biography: The Use of Biography in the History of Science." *Hist. Sci.,* 17 (1979): 1–16.

————. *Science and the Enlightenment.* Cambridge: Cambridge University Press, 1985.

Hankins, Thomas L., and Robert J. Silverman. *Instruments and the Imagination.* Princeton: Princeton University Press, 1995.

Harris, Dianne Suzette. *"Lombardia Illuminata": The Formation of an Enlightenment Landscape in Eighteenth-Century Lombardy.* Ph.D. dissertation, University of California at Berkeley, 1996.

Heilbron, John L. "Analogy in Volta's Exact Natural Philosophy." In *Nuova Voltiana: Studies on Volta and His Times,* ed. Fabio Bevilacqua and Lucio Fregonese. Pavia and Milan: Hoepli, 2000–. 1: 1–23.

————. "Beccaria." *DSB,* 1: 546–48.

————. "Cavallo." *DBS,* 3: 155–59.

————. "The Contributions of Bologna to Galvanism." *Hist. Stud. Phys. Biol. Sci.,* 22 (1991): 57–85.

————. *Electricity in the Seventeenth and Eighteenth Centuries.* Berkeley: University of California Press, 1979; Mineola, N.Y.: Dover Publications (with a new preface), 1999.

Heilbron, John L. "Experimental Natural Philosophy." In *The Ferment of Knowledge: Studies in the Historiography of Eighteenth-Century Science*, ed. George S. Rousseau and Roy Porter. Cambridge: Cambridge University Press, 1980.

———. "Galvani, Volta, and the Uses of Centennials." In *Luigi Galvani International Workshop: Proceedings*, ed. Marco Bresadola and Giuliano Pancaldi. *Bologna Studies in History of Science*, vol. 7. CIS, University of Bologna, 1999.

———. "A Mathematicians' Mutiny, with Morals." In *World Changes: Thomas Kuhn and the Nature of Science*, ed. Paul Horwich. Cambridge, Mass.: MIT Press, 1993.

———. "The Measure of Enlightenment." In *The Quantifying Spirit in the Eighteenth Century*, ed. Tore Frängsmyr, John L. Heilbron, and R. E. Rider.

———. "Robert Symmer and the Two Electricities." *Isis*, 67 (1976): 7–20.

———. "Some Connections Among the Heroes." *Rev. Hist. Sci.*, 54 (2001), 5–28.

———. "Symmer." *DSB*, 13: 224–25.

———. "Volta." *DSB*, 14: 69–82.

———. "Volta's Path to the Battery," in *Selected Topics in the History of Electrochemistry*, ed G. Dubpernell and J. H. Westbrook. Princeton: Electrochemical Society, 1978.

———. *Weighing Imponderables and Other Quantitative Science around 1800*. Supplement to vol. 24 of *Hist. Stud. Phys. Biol. Sci.* Part 1, 1993.

Hilaire-Pérez, Liliane. *L'invention technique au siècle des Lumières*. Paris: Albin Michel, 2000.

Hofmann, James R. *André-Marie Ampère*. Oxford: Blackwell, 1995.

Holmes, Frederic L. "Do We Understand Historically How Experimental Knowledge Is Acquired?" *Hist. Sci.*, 30 (1992): 119–36.

——— "The Fine Structure of Scientific Creativity." *Hist. Sci.*, 19 (1981): 60–70.

——— *Lavoisier and the Chemistry of Life: An Exploration of Scientific Creativity*. Madison: University of Wisconsin Press, 1985.

——— "Phlogiston in the Air." In *Nuova Voltiana: Studies on Volta and His Times*, ed. Fabio Bevilacqua and Lucio Fregonese. Pavia and Milan: Hoepli, 2000–. 2: 73–113.

Holmes, Frederic L., and Trevor H. Levere, eds. *Instruments and Experimentation in the History of Chemistry*. Cambridge, Mass.: MIT Press, 2000.

Home, Roderick W. "Aepinus and the English Electricians: The Dissemination of a Scientific Theory." *Isis*, 63 (1972): 190–204.

———. "Aepinus, the Tourmaline Crystal, and the Theory of Electricity and Magnetism." *Isis*, 67 (1976): 21–30.

———. *The Effluvial Theory of Electricity*. New York: Arno Press, 1981.

———. "Electricity and the Nervous Fluid." *J. Hist. Biol.*, 3 (1970): 235–51.

———. Essay Review of Heilbron, *Electricity*. In *Ann. Sci.*, 38 (1981): 477–82.

———. "Franklin's Electrical Atmospheres." *Brit. J. Hist. Sci.*, 6 (1972): 131–51.

———. " 'Newtonianism' and the Theory of the Magnet." *Hist. Sci.*, 15 (1977): 252–66.

———. "Nollet and Boerhaave: A Note on Eighteenth-Century Ideas about Electricity and Fire." *Ann. Sci.*, 36 (1979): 171–76.

————. *Electricity and Experimental Physics in Eighteenth-Century Europe*. Aldershot: Variorum, 1992.

————. "Volta's English Connections." In *Nuova Voltiana: Studies on Volta and His Times*, ed. Fabio Bevilacqua and Lucio Fregonese. Pavia and Milan: Hoepli, 2000–. 1: 115–32.

Hong, Sungook. "Controversy over Voltaic Contact Phenomena." *Arch. Hist. Exact Sci.*, 47 (1994): 233–89.

————. "Unfaithful Offspring? Technologies and Their Trajectories." *Perspect. Sci.*, 6 (1998): 259–87.

Horn, Jeff, and Margaret C. Jacob. "Jean-Antoine Chaptal and the Cultural Roots of French Industrialization." *Technol. Cult.*, 39 (1998): 671–98.

Hughes, Thomas P. *Networks of Power: Electrification in Western Society, 1880–1930*. Baltimore: Johns Hopkins University Press, 1983.

Hulme, Peter, and Ludmilla Jordanova. *The Enlightenment and Its Shadows*. London: Routledge, 1990.

Hunt, Bruce J. "The Ohm Is Where the Art Is: British Telegraph Engineers and the Development of Electrical Standards." *Osiris*, 9 (1993): 48–63.

*Immagini del Settecento in Italia*. Bari: Laterza, 1980.

Israel, Jonathan I. *Radical Enlightenment: Philosophy and the Making of Modernity*. Oxford: Oxford University Press, 2001.

James, Frank A.J.L., ed. *The Development of the Laboratory*. Basingstoke: Macmillan, 1989.

Jemolo, Arturo Carlo. *Il giansenismo in Italia prima della rivoluzione*. Bari: Laterza, 1928.

Julia, Dominique, and Jaques Revel, eds. *Histoire sociale des populations étudiantes*. Tome 1 de la série "Les universitées européennes du XVIe au XVIIIe siècle." Paris: École des hautes études en sciences sociales, 1986.

Jungnickel, Christa, and Russell McCormmach. *Cavendish: The Experimental Life*. Cranbury: Bucknell University Press, 1999.

————. *Intellectual Mastery of Nature: Theoretical Physics from Ohm to Einstein*. 2 vols. University of Chicago Press, 1986.

Ken, Alder. *Engineering the Revolution: Arms and Enlightenment in France: 1763–1815*. Princeton: Princeton University Press, 1997.

King, W. J. "The Quantification of the Concepts of Electric Charge and Electric Current." *Natural Philosopher*, 2 (1963): 107–27.

Kipnis, Nahum. "Debating the Nature of Voltaic Electricity." In *Nuova Voltiana: Studies in Volta and His Times*, ed. Fabio Bevilacqua and Lucio Fregonese. Pavia and Milan: Hoepli, 2000–. 3: 121–51.

————. "Luigi Galvani and the Debate on Animal Electricity, 1791–1800." *Ann. Sci.*, 64 (1987): 107–42.

Kragh, Helge. "Confusion and Controversy: Nineteenth-Century Theories of the Voltaic Pile." In *Nuova Voltiana: Studies in Volta and His Times*, ed. Fabio Bevilacqua and Lucio Fregonese. 3 vols. Pavia and Milan: Hoepli, 2000–2001. 1: 133–57.

Kuhn, Thomas S. "Energy Conservation as an Example of Simultaneous Discovery." In Marshall Clagett (ed.), *Critical Problems in the History of Science*.

Madison: University of Wisconsin Press, 1959. Also in Thomas H. Kuhn, *The Essential Tension*. Chicago, 1977.

Kuhn, Thomas S. "The Function of Measurement in Modern Physical Science." *Isis*, 52 (1961): 161–93. Now in T. H. Kuhn, *The Essential Tension*. Chicago: University of Chicago Press, 1977.

———. "What Are Scientific Revolutions?" In *The Probabilistic Revolution*, ed. Lorenz Krüger, Lorraine J. Daston, Michael Heidelberger. Volume 1: *Ideas in History*. Cambridge, Mass.: MIT Press, 1987.

Kuznets, Simon. *Modern Economic Growth*. New Haven: Yale University Press, 1966.

Landes, David S. *The Unbound Prometheus: Technology Change and Industrial Development in Western Europe from 1750 to the Present*. Cambridge: Cambridge University Press, 1969.

———. *The Wealth and Poverty of Nations*. London: Little, Brown and C., 1998.

Langford, Paul. *A Polite and Commercial People: England, 1727–1783*. Oxford: Oxford University Press, 1990.

Langley, Pat, et al., eds. *Scientific Discovery: Computational Explorations of the Creative Processes*. Cambridge, Mass.: MIT Press, 1987.

Latour, Bruno. *Pandora's Hope: Essays on the Reality of Science Studies*. Cambridge, Mass.: Harvard University Press, 1999.

———. *Science in Action: How to Follow Scientists and Engineers through Society*. Cambridge, Mass.: Harvard University Press, 1987.

Lemay, J. A. Leo, ed. *Reappraising Benjamin Franklin: A Bicentennial Perspective*. Cranbury, N.J.: Associated University Presses for University of Delaware Press, 1993.

Lenoir, Timothy. "Practice, Reason, Context: The Dialogue between Theory and Experiment." *Sci. Context*, 2 (1988): 3–22.

Liley, S. "Nicholson's Journal, 1797–1813." *Ann. Sci.*, 6 (1948): 78–101.

Longatti, Alberto. "Il Volta poeta." *Omaggio a Volta nel centocinquantenario della morte*. Como: Nani, 1978.

Lowood, Henry Ernest. "Patriotism, Profit, and the Promotion of Science in the German Enlightenment, 1760–1815." University of California, Ph.D. thesis 1987.

Lucati, Venosto. *Iconografia ed epigrafia di Alessandro Volta*. Como: Nani, 1982.

Lythe, S.G.E. *Thomas Garnett: 1766–1802*. Glasgow: Polpress, 1984.

MacLeod, Christine. *Inventing the Industrial Revolution: The English Patent System, 1660–1800*. Cambridge: Cambridge University Press, 1988.

Maffioli, Cesare, S. "Italian Hydraulics and Experimental Physics in Eighteenth-Century Holland: From Poleni to Volta." In *Italian Scientists in the Low Countries in the Seventeenth and Eighteenth Centuries*, ed. C. S. Maffioli, L. C. Palm. Amsterdam: Rodopi, 1989.

Malaquias, Isabel Maria, and Manuel Fernandes Thomaz. "Scientific Communication in the Eighteenth Century: The Case of John Hyacinth De Magellan." *Physis*, 31 NS (1994): 817–834.

Maluf, Ramez Bajige. "Jean Antoine Nollet and Experimental Natural Philosophy." University of Oklahoma, Ph.D. thesis, 1985.

Mamiani, Maurizio. "Francesco Venini: Un *philosophe* a Parma." *Giornale critico della filosofia italiana*, 9 (1989): 213–24.

Manzoni, Cesare. *Biografia Italica*. Osnarbrück: Biblio Verlag, 1981.

Mason, Stephen F. "Jean Hyacinthe Magellan, F.R.S., and the Chemical Revolution of the Eighteenth Century." *Notes Rec. R. Soc. Lond.* 45 (1991): 155–64.

Massardi, Francesco. "Concordanze di risultati e formule emergenti da manoscritti inediti del Volta con quelli ricavata dalla fisico-matematica nella risoluzione del problema generale dell'elettrostatica." *Rendiconti dell'Istituto Lombardo*, 2nd. series, 56 (1923): 293–308.

———. "Sull'importanza dei concetti fondamentali esposti dal Volta nel 1769 nella sua prima memoria scientifica 'De vi attractiva ignis electrici.' " *Rendiconti dell'Istituto Lombardo*, 2nd series, 59 (1926): 373–81.

Mathias, Peter and John A. Davis, eds. *Innovation and Technology in Europe: From the Eighteenth Century to the Present Day.* Oxford: Blackwell, 1991.

Mauro, Alexander. "The Role of the Voltaic Pile in the Galvani-Volta Controversy Concerning Animal vs. Metallic Electricity." *J. Hist. Med.* 24 (1969): 140–50.

Mazzolini, Renato G. *The Iris in Eighteenth-Century Physiology*. (Bern: H. Huber, 1980).

Mazzotti, Massimo. "Maria Gaetana Agnesi: Mathematics and the Making of the Catholic Enlightenment." *Isis*, 92 (2001): 657–83.

McClellan III, James E. *Science Reorganized: Scientific Societies in the Eighteenth Century.* New York: Columbia University Press, 1985.

McCormmach, Russell. "Cavendish." *DSB*, 3: 155–59.

McEvoy, John G. "Electricity, Knowledge and the Nature of Progress in Priestley's Thought." *Brit. J. Hist. Sci.*, 12 (1979), 1–30.

Mertens, Joost. "From the Lecture Room to the Workshop: John Frederic Daniell, the Constant Battery and Electrometallurgy around 1840." *Ann. Sci.*, 55 (1998): 241–61.

———. "Shocks and Sparks: The Voltaic Pile as a Demonstration Device." *Isis*, 89 (1998): 300–11.

Mesini, Candido. *Luigi Galvani*. Bologna: S. Francesco, 1958.

Mieli, Aldo. *Alessandro Volta*. Rome: Formiggini, 1927.

Miller, David Philip. " 'Puffing Jamie': The Commercial and Ideological Importance of Being a 'Philosopher' in the Case of the Reputation of James Watt (1736–1819)." *Hist. Sci.*, 2000, 1–24.

———. "The Usefulness of Natural Philosophy: The Royal Society and the Culture of Practical Utility in the Later Eighteenth Century." *Brit. J. Hist. Sci.*, 32 (1999): 185–201.

Mitchell, Trent A. "The Politics of Experiment in the Eighteenth Century: The Pursuit of Audience and the Manipulation of Consensus in the Debate over Lightning Rods." *Eighteenth-Cent. Stud.*, 31 (1998): 307–31.

Moiso, Francesco. "Magnetismus, Elektrizität, Galvanismus." In *Wissenschaftshistorischer Bericht zu Schellings Naturphilosophischen Schriften, 1797–1800, Ergänzungsband zu Werke Band 5 bis 9*. Stuttgart: Frommann-Holzboog, 1994.

Money, John. "From Leviathan's Air Pump to Britannia's Voltaic Pile." *Canadian Journal of History*, 28 (1993): 521–44.

Money, John. "Joseph Priestley in Cultural Context: Philosophic Spectacle, Popular Belief and Popular Politics in Eighteenth-Century Birmingham." *Enlightenment and Dissent*, 7 (1988): 57–82, and 8 (1989): 69–89.

Montandon, Cléopâtre. *Le développement de la science à Genève aux XVIIIe et XIXe siècles*. Vevey: Delta, 1975.

Mor, Carlo Guido. *Storia della Università di Modena*. Florence: Olschki, 1975.

Moravia, Sergio. *Il tramonto dell'illuminismo*. Bari: Laterza, 1968.

Morselli, Mario. *Amedeo Avogadro: A Scientific Biography*. Dordrecht: Reidel, 1984.

Morus, Iwan Rhys. *Frankenstein's Children: Electricity, Exhibition, and Experiment in Early-Nineteenth-Century London*. Princeton: Princeton University Press, 1998.

Mottelay, Paul Fleury. *Bibliographical History of Electricity and Magnetism Chronologically Arranged*. London: Griffith, 1922.

Munck, Thomas. *The Enlightenment: A Comparative Social History, 1721–1794*. London: Arnold, 2000.

*Newtonianesimo [Il] nel Settecento*. Rome: Istituto della Enciclopedia Italiana, 1983.

Nicolò Barabino: "Il segno in trappola." Genoa: Marietti, 1990.

Nielsen, Keld. "Another Kind of Light: The Work of T. J. Seebeck and His Collaboration with Goethe. Part I." *Hist. Stud. Phys. Sci*, 20 (1989): 107–78.

Oliver, Samuel P. "Nicholson, William." *DNB*, 14: 473–75.

Olmi, Giuseppe. *L'inventario del mondo: catalogazione della natura e luoghi del sapere nella prima età moderna*. Bologna: il Mulino, 1992.

*Omaggio a Volta nel centocinquantenario della morte*. Como: Nani, 1978.

Ostwald, Wilhelm. *Electrochemistry: History and Theory*. American translation. 2 vols. New Delhi: Amerind Publishing, 1980.

Outram, Dorinda. *The Enlightenment*. Cambridge: Cambridge University Press, 1995.

———. "The Enlightenment Our Contemporary." In *The Sciences in Enlightened Europe*, ed. William Clark, Jan Golinski, and Simon Schaffer. Chicago: The University of Chicago Press, 1999.

———. *Georges Cuvier. Vocation, Science and Authority in Post-Revolutionary France*. Manchester: Manchester University Press, 1984.

———. "The Ordeal of Vocation: The Paris Academy of Sciences and the Terror, 1793–95." *Hist. Sci.*, 21 (1983): 251–73.

———. "Scientific Biography and the Case of Georges Cuvier." *Hist. Sci.*, 14 (1976): 101–37.

Owen, Dawn. "The Constant Battery and the Daniell-Becquerel-Grove Controversy." *Ambix*, 48 (2001): 25–90.

Pace, Antonio. "Beccaria." *DBI*, 7: 469–71.

———. *Benjamin Franklin and Italy*. Philadelphia: American Philosophical Society, 1958.

———. "The Manuscripts of Giambatista Beccaria." *Proc. Amer. Phil. Soc.*, 156 (1952): 406–16.

Pachter, Marc. ed. *Telling Lives: The Biographer's Art*. Washington, D.C.: New Republic Books, 1979.

Palter, Robert. "Some Impressions of Recent Work on Eighteenth-Century Science." *Hist. Stud. Phys. Biol. Sci.*, 19 (1989): 349–401.

Pancaldi, Giuliano. "Electricity and Life: Volta's Path to the Battery." *Hist. Stud. Phys. Biol. Sci.*, 21 (1990): 123–60.

———, ed. *I congressi degli scienziati italiani nell'età del positivismo*. Bologna: CLUEB, 1983.

Pasta, Renato. *Scienza politica e rivoluzione: L'opera di Giovanni Fabbroni (1752–1822) intelletuale e funzionario al servizio dei Lorena*. Florence: Olschki, 1989.

Pepe, Luigi. "Scienziati e stabilimenti scientifici a Ferrara." *Museologia scientifica*, 3 (1986): 113–19.

Pera, Marcello. *La rana ambigua: La controversia sull'elettricità animale tra Galvani e Volta*. Torino: Einaudi, 1986; Amer. trans. Princeton: Princeton University Press, 1992.

Petronio, Ugo. *Il Senato di Milano: Istituzioni giuridiche ed esercizio del potere nel Ducato di Milano da Carlo V a Giuseppe II*. Rome: Giuffrè, 1972.

Petruccioli, Sandro. *Atomi, metafore, paradossi*. Rome-Napoli: Theoria, 1988. Trans. Ian McGilvray. Cambridge: Cambridge University Press, 1993.

Piacenza, Mario. "Note biografiche e bibliografiche e nuovi documenti su G. B. Beccaria." *Bollettino storico bibliografico subalpino*, 9 (1904): 209–28, 340–54.

Picccolino, Marco, and Marco Bresadola. *Rane, torpedini e scintille*. Turin: Bollati Boringhieri, forthcoming.

Pickering, Andrew. "Forms of Life: Science, Contingency, and Harry Collins." *Brit. J. Hist. Sci.* 20 (1987): 213–21.

Pickering, Andrew. *The Mangle of Practice: Time, Agency, and Science*. Chicago: The University of Chicago Press, 1995.

Pickering, Mary. *Auguste Comte: An Intellectual Biography*. Vol. 1. Cambridge: Cambridge University Press, 1993.

Pino, Francesca. "Patriziato e decurionato a Milano nel secolo XVIII." *Società e storia*, 5 (1979): 339–78.

Polanyi, Michael. *Personal Knowledge: Towards a Post-Critical Philosophy*. London: Routledge, 1998 (first published 1958).

Polvani, Giovanni. *Alessandro Volta*. Pisa: Domus Galilaeana, 1942.

Poni, Carlo. "All'origine del sistema di fabbrica: tecnologia e organizzazione produttiva dei mulini da seta nell'Italia settentrionale (Sec. XVII–XVIII)." *Rivista storica italiana*, 88 (1976): 444–96.

———. "Standard, fiducia e conversazione civile: misurare lo spessore e la qualità del filo di seta." *Quaderni storici*, 32 (1997): 717–34.

Poni, Carlo, and Giorgio Mori. "Italy in the *longue durée*: The Return of an Old First-Comer." In *The Industrial Revolution in National Context: Europe and the USA*, eds. Mikuláš Teich and Roy Porter. Cambridge: Cambridge University Press, 1996.

Porter, Roy. *English Society in the Eighteenth Century*. Rev. ed., London: Allen Lane, 1990.

———. *Enlightenment: Britain and the Creation of the Modern World*. London: Allen Lane, The Penguin Press, 2000.

Porter, Roy, and Mikuláš Teich, eds. *The Enlightenment in National Context.* Cambridge: Cambridge University Press, 1981.

Porter, Roy, et al. *Science and Profit in Eighteenth-Century London.* Cambridge: Cambridge University Press, 1985.

Price, D. J. de Solla. "Of Sealing Wax and String." *Natural History,* 93 (1984): 49–56.

Ramazzotti, S. and L. Briatore, "Didattica e ricerca fisica nell'ateneo Torinese nel XVIII secolo: G. Cigna scienziato illuminista." *Giornale di fisica,* 17 (1976): 232–38.

Ramazzotti, S. and L. Briatore, "Le ultime sperimentazioni elettriche di Giovanni Cigna." *Giornale di fisica,* 18 (1977): 149–57.

Redondi, Pietro. "Cultura e scienza dall'illuminismo al positivismo." In *Storia d'Italia: Annali 3, Scienza e tecnica nella cultura e nella società dal Rinascimento a oggi.* Turin: Einaudi, 1980.

Remise, Jac, Pascale Remise, and Régis Van de Walle. *Magie lumineuse du théâtre d'ombres à la lanterne magique.* Paris: Balland, 1979.

Rettaroli, R., and F. Tassinari. "Studenti e docenti dell'Ateneo tra VIII e IX Centenario." In *Lo studio e la città: Bologna, 1888–1988,* ed. W. Tega. Bologna: Nuova Alfa, 1987.

Rheinberger, Hans-Jörg. "Experimental Systems: Historiality, Narration, and Deconstruction." *Sci. Context,* 7 (1994): 65–81.

Ricuperati, Giuseppe. "Giornali e società nell'Italia dell'Ancien Régime" (1668–1789)." In *La stampa italiana dal '500 all'800.* Bari: Laterza, 1976.

———, ed. *La reinvenzione dei lumi: Percorsi storiografici del Novecento.* Florence: Olschki, 2000.

———. *L'organizzazione della cultura nell'Italia del '700: Istruzione e accademie.* Turin: Tirrenia, 1976.

Rider, Robin. "Bibliographical Afterward." In *The Quantifying Spirit of the 18th Century,* ed. Tore Frängsmyr, John L. Heilbron, Rider, and Robin R. Rider.

Riskin, Jessica. "Poor Richard's Leyden Jar: Electricity and Economy in Franklinist France." *Hist. Stud. Phys. Biol. Sci.,* 28 (1998): 301–36.

Roberts, John Morris. "Lombardy." In *The European Nobility in the Eighteenth Century,* ed. Albert Goodwin. 2d ed. London, 1967.

Roberts, Lissa. "Going Dutch: Situating Science in the Dutch Enlightenment." In *The Sciences in Enlightened Europe,* ed. William Clark, Jan Golinski, and Simon Schaffer. Chicago: The University of Chicago Press, 1999.

———. "Science Becomes Electric: Dutch Interaction with the Electrical Machine during the Eighteenth Century." *Isis,* 90 (1999): 680–714.

Rocchi, G. *Profilo storico della chiesa di S. Anaznzio e del Gesù in Como.* Como: Noseda, 1968.

Roche, Daniel. *Le siècle des lumières en province: Académies et académiciens provinciaux, 1680–1789.* 2 vols. Paris: Mouton: 1978.

Roche, John. *Physicists Look Back: Studies in the History of Physics.* Bristol and New York: Hilger, 1990.

Roger, Jacques. *Buffon: Un philosophe au Jardin du Roi.* Paris: Fayard, 1989; Amer. Trans., Ithaca: Cornell University Press, 1997.

Roggero, Marina. "Professori e studenti nelle università tra crisi e riforme." *Storia d'Italia: Annali 4, Intellettuali e potere.* Ed. C. Vivanti. Torino: Einaudi, 1981.

———. *Scuola e riforme nello stato sabaudo.* Turin: Deputazione Subalpina di Storia Patria, 1981.

Roller, Duane, and Duane H. D. Roller. "The Development of the Concept of Electric Charge." In *Harvard Case Histories in Experimental Science*, ed. J. B. Conant and L. K. Nash. 2 vols. Cambridge, Mass.: Harvard University Press, 1957.

Romani, Mario. "La ripressa dell'agricoltura nel fervore del riforme (1760–1768)." In *Storia di Milano*, 16 vols. and Index. Rome: Enciclopedia Italiana, 1953–66. 17: 525–43.

Rorty, Richard. *Contingency, Irony, and Solidarity.* Cambridge: Cambridge University Press, 1989.

Rosa, Mario, ed. *Cattolicesimo e lumi nel Settecento italiano.* Rome: Herder, 1981.

Roscelli, Davide. *Nicolò Barabino: Maestro dei masetri.* Sampierdarena, Associazione Operaia "G: Mazzini," 1982.

Rosenberg, Nathan. *Perspectives on Technology.* Cambridge: Cambridge University Press, 1976.

Rossetti, Francesco, and G. Cantoni. *Bibliografia italiana di elettricità e magnetismo.* Padua: 1881.

Rossi, Paolo. *Philosophy, Technology, and the Arts in the Early Modern Era.* Translated by Salvator Attanasio. Edited by Benjamin Nelson. New York: Harper & Row, 1970.

Rota, Ettore. *Il giansenismo in Lombardia.* Pavia: Fusi, 1907.

Rotelli, Ettore. "Gli ordinamenti locali della Lombardia preunitaria 1755–1859." *Archivio storico lombardo*, 100 (1974): 171–234.

Rousseau, George S. and R. Porter, eds. *The Ferment of Knowledge: Studies in the Historiography of Eighteenth Century Science.* Cambridge: Cambridge University Press, 1980.

Russo, Arturo. "Science and Industry in Italy between the Two World Wars." *Hist. Stud. Phys. Biol. Sci.*, 16 (1986): 281–320.

Savino, Edoardo. *La nazione operante: Profili e figure.* Milan, 1934.

Savorgnan di Brazzà, Francesco. "Alessandro Volta." In *Gli scienziati italiani in Francia*, a volume in the 11th Series ("Gli scienziati") of *L'opera del genio italiano all'estero.* Rome: Libreria dello Stato, 1941.

Schaffer, Simon. "Accurate Measurement is an English Science." In *The Values of Precision*, ed. M. Norton Wise. Princeton: Princeton University Press, 1995.

———. "The Consuming Flame: Electrical Showmen and Tory Mystics in the World of Goods." In *Consumption and the World of Goods*, ed. John Brewer and Roy Porter. London: Routledge, 1993. 488–526.

———. "Natural Philosophy." In *The Ferment of Knowledge. Studies in the Historiography of Eighteenth Century Science*, ed. George S. Rousseau and Roy Porter.

———. "Self Evidence." *Critical Inquiry*, 18 (1992): 327–62.

Schallenberg, Richard H. *Bottled Energy.* Philadelphia: American Philosophical Society, 1982.

———. *The Electric Battery, 1800–1930.* Ph.D. dissertation, Yale University, 1973.

Schlereth, Thomas John. *The Cosmopolitan Ideal in Enlightenment Thought*. Notre Dame and London: University of Notre Dame Press, 1977.

Schofield, Robert E. "Electrical Researches of Joseph Priestley." *Arch. Int. Hist. Sci.*, 16 (1963): 277–86.

———. *The Lunar Society of Birmingham*. Oxford: Clarendon, 1963.

Scolari, Bice. "Como nell'età di Volta." *Periodico della Società Storica Comense*, n.s. 7 (1951).

Scolari, Felice. *Alessandro Volta*. Fondazione Leonardo, Guide Bibliografiche. Rome, 1927.

Scranton, Philip. "Determinism and Indeterminacy in the History of Technology." In Merritt Roe Smith and Leo Marx (eds.), *Does Technology Drive History? The Dilemma of Technological Determinism*. Cambridge, Mass.: MIT Press, 1994.

Seligardi, Raffaella. "Volta and the Synthesis of Water: Some Reasons of a Missed Discovery." In *Nuova Voltiana: Studies on Volta and His Times*, ed. Fabio Bevilacqua and Lucio Fregonese. Pavia and Milan: Hoepli, 2000–. 2: 33–46.

Sella, Domenico and C. Capra. *Il Ducato di Milano dal 1533 al 1796*. Turin: UTET, 1984.

Shapin, Steven, and Simon Schaffer. *Leviathan and the Air-Pump: Hobbes, Boyle, and the Experimental Life*. Princeton: Princeton University Press, 1985.

Sheets-Pyenson, Susan. "New Directions for Scientific Biography: The Case of Sir William Dawson." *Hist. Sci.*, 28 (1990): 399–410.

Shinn, Terry. *L'école Polytechnique, 1794–1914*. Paris: Presses de la Fondation Nationale des Sciences Politiques, 1980.

Sibum, Otto. "Die Mechanisierung der Lebensvorgänge—der Weg zum elektrischen Strom." In Jörg Meya and Otto Sibum, *Das fünfte Element: Wirkungen und Deutungen der Elektrizität*. Deutsches Museum, Rowohlt, 1987.

———. "Reworking the Mechanical Value of Heat: Instruments of Precision and Gestures of Accuracy in Early Victorian England." *Stud. Hist. Phil. Sci.*, 26 (1995): 73–106.

Sleigh, Charlotte. "Life, Death and Galvanism." *Stud. Hist. Phil. Biol. Biomed. Sci.*, 29 (1998): 219–48.

Smith, Merritt Roe, and Leo Marx, eds. *Does Technology Drive History? The Dilemma of Technological Determinism*. Cambridge, Mass.: MIT Press, 1994.

*Società [Tra] e Scienza. 200 anni di storia dell'Accademia delle Scienze di Torino*. Turin: Allemandi, 1988.

Söderqvist, Thomas. "The Architecture of a Biographical Pathway." *Hist. Stud. Phys. Biol. Sci.*, 25 part 1 (1994): 165–75.

Sokolow, Jayme. "Count Rumford and Late Enlightenment Science, Technology, and Reform." *Eighteenth Century*, 21 (1980): 67–86.

Sorrenson, Richard John. *Scientific Instrument Makers at the Royal Society of London, 1720–1780*. Ph.D. dissertation, Princeton University, 1993.

Stansfield, Dorothy A. *Thomas Beddoes, M.D., 1760–1808: Chemist, Physician, Democrat*. Dordrecht: Reidel, 1984.

Stewart, Larry. *The Rise of Public Science: Rhetoric, Technology, and Natural Philosophy in Newtonian Britain, 1660–1750*. Cambridge: Cambridge University Press, 1992.

Stone, Irving. *The Science, and the Art, of Biography.* Los Angeles: Naumburg Memorial Lecture, 1986.

Strickland, Stuart Walker. *Circumscribing Science: Johann Wilhelm Ritter and the Physics of Sideral Man,* Ph.D. dissertation, Cambridge, Mass.: Harvard University, 1992.

———. "Galvanic Disciplines." *Hist. Sci.* 33 (1995): 450–68.

———. "The Ideology of Self-Knowledge and the Practice of Self-Experimentation." *Eighteenth-Cent. Stud.* 31, no. 4 (1998): 453–71, p. 457.

Sudduth, William M. "Eighteenth-Century Identifications of Electricity with Phlogiston." *Ambix,* 25 (1978): 131–47.

———. "The Voltaic Pile and Electro-Chemical Theory in 1800." *Ambix,* 27 (1980): 26–35.

Sutton, Geoffroy. "The Politics of Science in Early Napoleonic France: The Case of the Voltaic Pile." *Hist. Stud. Phys. Sci.,* 11–12 (1981): 329–66.

Szabo, Franz A. J. *Kaunitz and Enlightened Absolutism, 1753–1780.* Cambridge: Cambridge University Press 1997.

Tabarroni, Giorgio. "Galvani, Aldini e la corrente elettrica." Istituto Tecnico Industriale Aldini-Valeriani, *Annuario* (1971): 43–54.

Teichmann, Jürgen. "Volta and the Quantitative Conceptualisation of Electricity: From Electrical Capacity to the Preconception of Ohm's Law." In *Nuova Voltiana: Studies on Volta and His Times,* ed. Fabio Bevilacqua and Lucio Fregonese. Pavia and Milan: Hoepli, 2000–. 3: 53–80.

*Tempio [Il] voltiano in Como.* Como: Cavalleri, 1939.

Thackray, Arnold. " 'Matter in a Nutshell': Newton's *Opticks* and Eighteenth-Century Chemistry." *Ambix,* 15 (1968): 29–53.

———. "Natural Knowledge in Cultural Context: The Manchester Model." *Amer. Hist. Rev.,* 79 (1974): 672–709.

Thomas, J.-André, ed. *La Société Philomatique de Paris et deux siècles d'histoire de la science en France.* Paris: PUF, 1990.

Thomas, Keith. *Man and the Natural World.* London: Allen Lane, 1983.

Tomani, Silvana. *I manoscritti filosofici di Paolo Frisi.* Florence: La Nuova Italia, 1968.

Torlais, Jean. *Un physicien au siècle des lumières, l'abbé Nollet.* Paris: Sipuco, 1954.

Tortarolo, Edoardo. *L'illuminismo: Ragioni e dubbi della modernità.* Rome: Carocci, 1999.

Trembley, J., ed. *Les savants genevois dans l'Europe intellectuelle.* Genève: Journal de Genève, 1987.

Trumpler, Maria Jean. *Questioning Nature: Experimental Investigations of Animal Electricity in Germany, 1791–1810.* Ph.D. dissertation, Yale University, 1992.

Tunbridge, Paul. *Lord Kelvin.* London: Peter Peregrinus on behalf of IEE, 1992.

Turner, Gerard L'Etranger. "The Portuguese Agent: J. H. de Magellan." *Antiquarian Horology,* 9 (1974): 74–77.

Turner, Gerard L'Etranger and T. H. Levere, eds. *Van Marum's Scientific Instruments in the Teyler's Museum.* Haarlem, 1973. Published as vol. 4 of *Martinus van Marum: Life and Work,* ed. E. Lefebvre and J. G. De Bruijn.

Vaccari, Pietro. *Storia della Università di Pavia.* Pavia: Il Portale, 1948.

Valsecchi, Franco. *L'Assolutismo illuminato in Austria e in Lombardia*. Bologna: Zanichelli, 1931–34.

Venturi, Franco. *Italy and the Enlightenment*. London: Longman, 1972.

———. *Settecento riformatore*, V, *L'Italia dei lumi (1764–1790)*, I. Turin: Einaudi, 1987.

Verrecchia, Anacleto. "Lichtenberg und Volta." *Sudhoffs Archiv*, 51 (1967): 349–60.

Visconti, Alessandra. *La storia dell'Università di Ferrara*. Bologna: Zanichelli, 1950.

Volpati, Carlo. *Alessandro Volta nella gloria e nell'intimità*. Milan: Treves, 1927.

———. *Scritti voltiani*. Ed. Venosto Lucati. Como: Comune di Como, 1974.

Volta, Zanino. *Alessandro Volta*. Milano: Civelli, 1875.

———. *Alessandro Volta a Parigi*. Milano: Vallardi, 1879.

———. "Alessandro Volta e l'Università di Pavia dal 1778 al 1799." *Archivio storico lombardo*, 11 (1899): 393–447.

*Voltiana*. Como, 1926–27.

Walker, W. Cameron. "Animal Electricity before Galvani." *Ann. Sci*, 2 (1937): 84–113.

———. "The Detection and Estimation of Electric Charges in the Eighteenth Century." *Ann. Sci.*, 1 (1936): 66–100.

Warner, Deborah Jean. "What Is a Scientific Instrument, When Did It Become One, and Why?" *Brit. J. Hist. Sci.* 28 (1990): 83–93.

Weaver, William D. *Catalogue of the Wheeler Gift of Books, Pamphlets and Periodicals in the Library of the American Institute of Electrical Engineers*. 2 vols. New York: American Institute of Electrical Engineers, 1909.

Whittaker, Edmund. *A History of the Theories of Aether and Electricity*. New York: Humanities Press, 1973.

Whyte, Lancelot Law. *Roger Joseph Boscovich*. London: Allen & Unwin, 1961.

Wiedemann, Hans Rudolf. "Alexander Volta und Christoph Heinrich Pfaff." *Christiana Albertina*, 23 (1986): 25–34.

Wiener, Norbert. *Invention*. Cambridge, Mass.: MIT Press, 1993.

Williams, L. Pearce. "The Life of Science and Scientific Lives." *Physis*, 28 (1991): 199–213.

———. *Michael Faraday: A Biography*. London: Basic Books, 1965.

———. "Oersted, Hans Christian." *DSB*, 10: 182–86.

Winsor, Mary P. "The Practitioner of Science: Everyone Her Own Historian." *J. Hist. Biol.*, 34 (2001): 229–45.

Wise, M. Norton. "Mediating Machines." *Sci. Context*, 2, 1 (1988): 77–113.

———, ed. *The Values of Precision*. Princeton: Princeton University Press, 1995.

Wolf, Abraham. *A History of Science, Technology, and Philosophy in the Eighteenth Century*. London: Allen & Unwin, 1938.

Woolf, Stuart. *A History of Italy, 1700–1860*. London: Methuen, 1979.

———. *Napoleon's Integration of Europe*. London and New York: Routledge, 1991.

Wurzbach, Constantin von. *Biographisches Lexikon des Kaiserthums Österreich*. 60 vols. Vienna: Hof-und Staatsdruckerie, 1856–91.

Yamazaki, Eizo. "L'Abbé Nollet et Benjamin Franklin." *Jap. Stud. Hist. Sci.*, 15 (1976): 37–64.

Yeo, Richard. *Encyclopedic Visions: Scientific Dictionaries and Enlightenment Culture*. Cambridge: Cambridge University Press, 2001.

Yeo, Richard and Michael Shortland, eds. *Telling Lives: Studies of Scientific Biographies*. Cambridge: Cambridge University Press, 1995.

Ziman, John, ed. *Technological Innovation as an Evolutionary Process*. Cambridge: Cambridge University Press, 2000.

Bassano, 64

Bassora, 49

Bath, 161

Battery, voltaic, 1, 111, 142–43, 169, 171,
178–210; big, 207, 221, 228–29; as an almost useless machine, 276; circulation
of, 212–24; compared to the steam engine, the telescope, and the microscope,
273, 333n.2; and contingency, 285–86;
costs and value of, 280–81; crown of
cups, 203–4, 214, 226, 246; and diversity, 285; a family of instruments, 226;
impact of, 211–57, 287; mathematical
treatment of, 242–43; names of, 246–48;
non-metal, 205–6; patents of, 248–50;
philosophical implications of, 254–55;
pile, 203–4, 226, 246–47; pile of cups,
205; pocket, 206, 214, 235; and politics,
239–40, 279, 281; popular reception of,
254–55, 259–60; theories of, 215–45, 251;
"trough," 215, 247–48; uncertain status
of, 274–75, 282, 287; unusual, 233–34

Beccaria, Giambatista, 17, 18, 20, 27, 28,
29, 31, 40, 42, 46, 49, 54, 55, 60, 63, 64,
65, 76, 80, 88, 90–91, 95, 104, 119–20,
149, 209, 273, 288; on actuation, 120; on
attraction, 90; on "vindicating electricity", 83–86; on Volta, 90

Beckmann, Johann, 286

Becquerel, Antoine César, 253

Beddoes, Thomas, 221

Bellisomi, Pio, Marquis, 114

Bellodi, Giuliano, 305n.83

Ben-Chaim, Michael, 302n.7

Bennet, Abraham, 122, 125, 186–87, 200

Bensaude-Vincent, Bernadette, 308n.2

Beretta, Marco, 309n.8, 311n.65

Berlin, 140, 166, 168; Academy of
Sciences, 48, 64, 107, 167

Berman, Morris, 325n.14

Bern, 154

Bernardi, Walter, 298n.3, 320n.3

Bernoulli, Daniel, 153

Bernoulli, Jacques, 261

Berthollet, Claude-Louis, 170, 235, 238

Bertucci, Paola, 312n.73, 331n.146

Bevilacqua, Fabio, 305n.83

Biagioli, Mario, 291n.5, 334n.15

Bible, 23

Billaum, or Billaux, 124

Biography, 7–9, 41–43, 273–74; and prosopography, 56–62

Biot, Jean-Baptiste, 224, 238, 240–43, 278–
79, 285

Birmingham, 151, 161

Biron, physician in Paris, 239

Blondel, Christine, 325n.26, 328n.97,
329n.116

Body, human, 9, 23–24, 33, 40–41, 43, 103,
194, 219; and the battery, 230–33. *See
also* animal electricity; self-experimentation; Volta, manipulative abilities of

Boerhaave, Herman, 79

Bohr, Niels, 264, 269

Bologna, 64, 162; Academy of Sciences at,
19, 49, 50, 52, 67, 68, 69; University of,
50, 51, 70, 268

Bonaparte, Napoléon, 5, 25, 36, 39, 48, 60,
71, 144, 168, 248, 255, 262–63, 276, 281,
283; and political use of Volta's achievements, 170–72, 238–40; views of on electricity and galvanism, 171–72, 206, 230,
234–35, 238–40

Bonati, Teodoro Massimo, 55

Bonesi, Girolamo, 16–17, 294n.34

Bonioli, Camillo, 55

Bonnet, Charles, 23, 153, 154

Books, and libraries, 17, 19, 31, 47, 50,
62–65, 94, 104, 144, 163, 167–68, 199,
286

Borelli, Giovanni Alfonso, 17

Born, Ignaz Edler, 168

Born, Max, 264, 269

Boscovich, Ruggero Giuseppe, 17, 30, 49,
53, 66, 86, 132, 137, 273; on electricity,
88–89; on force 87; on saturation, 88

Bossi, Luigi, 55

Botanic gardens, 49, 70

Boudon, Raymond, 334n.15

Boulton, Matthew, 161

Bourbons, 48

Boyle, Robert, 64

Bradley, James, 261

Bragg, William L., 264

Brain, 191, 239; and contact electricity,
194; an electrostatic machine, 194; and
will, 198

Brenni, Paolo, 305n.83

Bresadola, Marco, 320n.3, 321n.50

Bridgnorth, 161

Brisson, Mathurin-Jacques, 240